Non Coding RNA Biology

Non Coding RNA Biology

Editor: Boston Carter

R CALLISTO REFERENCE

www.callistoreference.com

Callisto Reference,
118-35 Queens Blvd., Suite 400,
Forest Hills, NY 11375, USA

Visit us on the World Wide Web at:
www.callistoreference.com

ISBN: 978-1-64116-761-1 (Hardback)

Cataloging-in-Publication Data

Non coding RNA biology / edited by Boston Carter.
    p. cm.
Includes bibliographical references and index.
ISBN 978-1-64116-761-1
1. Non-coding RNA. 2. RNA. 3. Molecular biology. I. Carter, Boston.
QP623 .N66 2023
572.88--dc23

Table of Contents

Preface

Every book is a source of knowledge and this one is no exception. The idea that led to the conceptualization of this book was the fact that the world is advancing rapidly; which makes it crucial to document the progress in every field. I am aware that a lot of data is already available, yet, there is a lot more to learn. Hence, I accepted the responsibility of editing this book and contributing my knowledge to the community.

Non-coding RNAs (ncRNAs) are RNA molecules that are not translated into proteins. They are divided into two categories, namely, housekeeping ncRNAs and regulatory ncRNAs. The main function of the ncRNAs is to maintain the normal cell functionalities. Small RNAs, ribosomal RNAs (rRNAs), long ncRNAs and transfer RNAs are some types of non-coding RNAs, which are abundant and functionally significant. Numerous types of small and long non-coding RNAs are considered important regulators of gene expression in a variety of processes, which range from embryonic development to innate immunity. This promotion of gene expression is done by a special ncRNA named enhancer RNA, which is transcribed from the enhancer region of the gene. The human genome is encoded with thousands of small and long RNA transcripts that do not code for proteins. This book explores all the important aspects of regulatory non-coding RNAs. It will prove to be immensely beneficial to students and researchers studying this topic.

While editing this book, I had multiple visions for it. Then I finally narrowed down to make every chapter a sole standing text explaining a particular topic, so that they can be used independently. However, the umbrella subject sinews them into a common theme. This makes the book a unique platform of knowledge.

I would like to give the major credit of this book to the experts from every corner of the world, who took the time to share their expertise with us. Also, I owe the completion of this book to the never-ending support of my family, who supported me throughout the project.

Editor

Conserved miRNAs and their Response to Salt Stress in Wild Eggplant *Solanum linnaeanum* Roots

Yong Zhuang *, Xiao-Hui Zhou and Jun Liu

Institute of Vegetable Crops, Jiangsu Academy of Agricultural Sciences, Nanjing 210014, China; E-Mails: xhzhou1984@sina.com (X.-H.Z.); Kehl_lau@foxmail.com (J.L.)

* Author to whom correspondence should be addressed; E-Mail: jaaszy@163.com

Abstract: The Solanaceae family includes some important vegetable crops, and they often suffer from salinity stress. Some miRNAs have been identified to regulate gene expression in plant response to salt stress; however, little is known about the involvement of miRNAs in Solanaceae species. To identify salt-responsive miRNAs, high-throughput sequencing was used to sequence libraries constructed from roots of the salt tolerant species, *Solanum linnaeanum*, treated with and without NaCl. The sequencing identified 98 conserved miRNAs corresponding to 37 families, and some of these miRNAs and their expression were verified by quantitative real-time PCR. Under the salt stress, 11 of the miRNAs were down-regulated, and 3 of the miRNAs were up-regulated. Potential targets of the salt-responsive miRNAs were predicted to be involved in diverse cellular processes in plants. This investigation provides valuable information for functional characterization of miRNAs in *S. linnaeanum*, and would be useful for developing strategies for the genetic improvement of the Solanaceae crops.

Keywords: salt stress; miRNA; *Solanum linnaeanum*; high-throughput sequencing

1. Introduction

Salt stress is one of the most common abiotic stresses of crops. It was estimated that salt stress may affect half of all arable lands and will be a major factor of agriculture production for the coming decades [1]. Unlike other abiotic stresses, salt stress brings both osmotic stress and ion toxicity to crops. Under salt stress, crops can respond via cascades of molecular networks to change gene expression profile and posttranslational modifications involved in a broad spectrum of biochemical, cellular and physiological processes [2,3]. Therefore, an understanding of the basis of the salt stress response is important for strategies aimed at improving crop tolerance to salt stress.

miRNAs are endogenous non-coding small RNAs that are regulators of gene expression in organisms. They are known to play negative regulatory functions at the post-transcription level by inhibiting gene translation or cleaving target mRNAs via base-pairing their target mRNAs [4–6]. Many investigations indicated that plant miRNAs are involved in various important physiological processes, such as seed germination and root development [7–9]. In addition, increasing evidence has shown that miRNAs play important roles in the response of plants to biotic and abiotic stresses [10]; the expression levels of miRNAs were changed in plants infected with virus and fungus [11–13], and miRNAs were identified to be involved in plant response to abiotic stresses such as temperature [14,15], drought [16,17], metals [18,19], and salt [20–22].

The Solanaceae family includes some agriculturally important crops such as potato (*Solanum tuberosum*), eggplant (*S. melongena*), tomato (*S. lycopersicum*), and pepper (*Capsicum annuum*), and they often suffer from salt stress that can cause reduction of production, especially in greenhouse production. *S. linnaeanum*, which was used to construct a comparative genetic linkage map of eggplant, has tolerance to salt stress [23,24], however, little is known about the mechanism in response to salt stress. Comparative genomic studies revealed that relatively few genome rearrangements and duplications occurred in the evolutionary history of the Solanaceae species [25–28]. Although little information is known about the genomes of *S. linnaeanum* and *S. melongena*, the published data of other plants, especially those from Solanaceae family, may provide sufficient reference.

In the present study, using high-throughput sequencing, a large number of miRNAs and their response to salt stress in *S. linnaeanum* roots are identified and characterized. The results lay the foundation for further investigation and better understanding of the regulatory mechanisms for the plant response to salt stress. In addition, it also provides important information for genetic improvement of Solanaceae crops to salt stress.

2. Results and Discussion

2.1. Deep Sequencing Results of Small RNAs from S. linnaeanum Roots

To identify the miRNAs and their response to salt in *S. linnaeanum*, two small RNA libraries were generated from roots of NaCl-free (CK) and NaCl-treated (TR). Deep sequencing generates 21,284,496 and 13,989,100 raw reads in two libraries. After removal of low-quality and corrupted adapter sequences, 8,462,890 and 8,999,145 mappable reads remain in two libraries. The size distribution of mappable reads is assessed (Figure 1, Table S1). The data show that 24 nt small RNA is the major size class, followed by 21, 23, 30 and 22 nt small RNA. Similar results were reported in some other plant species, such as *Arabidopsis thaliana* [29,30], *Medicago truncatula* [31], *Oryza sativa* [32], *Arachis hypogaea* [33], *Cucumis Sativus* [34], *Nicotiana tabacum* [35], and *Citrus trifoliate* [36].

Because details of *S. linnaeanum* genome are limited, these mappable reads are analyzed with genome information of tomato and other plants. The results show that 5.51% reads of CK and 4.86% reads of TR are mapped to known plant pre-miRNAs in miRbase. Reads from CK (24.17%) and TR

(24.29%) are mapped to plant repeats, mRNA, and other RNAs including tRNA, rRNA, snRNA and snoRNA. In addition, some reads that cannot be mapped to pre-miRNAs in miRbase and other RNAs are mapped to tomato genome sequences, and a fraction of them potentially form hairpins. Also, nearly half of these reads have no mapping information (Table 1). To eliminate possible sequencing errors, only those sequences with more than five reads in either of the two libraries are further analyzed.

Figure 1. Length distribution of mappable small RNAs in two databases of *S. linnaeanum* roots. TR represents library of NaCl treatment, and CK represents library of control. The number in vertical axis is the total reads of all small RNAs in a certain length.

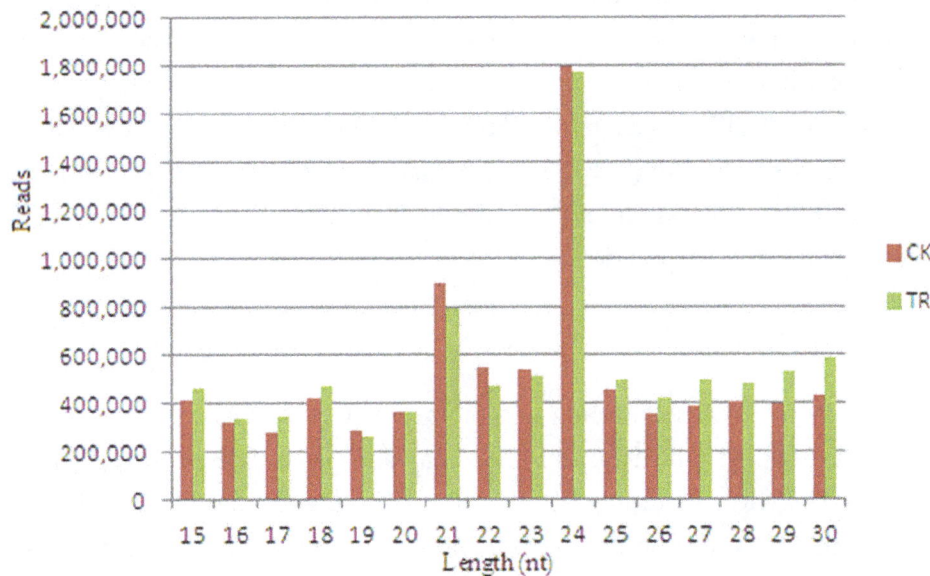

Table 1. Statistical analysis of sequencing reads in the two libraries.

Category	CK	Percent (%)	TR	Percent (%)
Raw reads	13,989,100		21,284,496	
Mappable reads	8,462,890	100.00	8,999,145	100.00
Mapped to miRNA	466,136	5.51	437,144	4.86
Mapped to mRNA	723,297	8.55	793,475	8.82
Mapped to RFam	1,315,886	15.55	1,387,942	15.42
Mapped to Repbase	5,674	0.07	4,745	0.05
Mapped to genome	1,763,934	20.84	2,063,801	22.93
No hit	4,187,963	49.49	4,312,038	47.92

2.2. Conserved miRNAs in S. linnaeanum Roots

To identity the conserved miRNAs in *S. linnaeanum* roots, small RNA sequences are mapped to tomato and other plant miRNAs in miRBase. Based on sequence homology (number of mismatch < 3), 98 known miRNAs and 7 miRNAs* are found (Table S2). The majority of these miRNAs are 20–22 nt long, and 56 of them are 21 nt long. These identified conserved miRNAs correspond to 37 families. The number of miRNA members in each known family shows significant divergence. The miR166 family is the largest one with 11 members, and for the family of miR171, miR396, and miR156, each of them has 7, 6, and 5 members respectively. Six families including miR159, miR162, miR167, miR168, miR319, and miR390 contain four members, and the remaining 27 miRNA families contain one to three members.

The read counts of miRNAs in sequencing libraries can be used as an index to estimate their relative abundance. In this study, the read counts differ among the miRNAs, which indicate that their expressions varied. Counting redundant miRNA reads reveals that 18 out of 98 known miRNAs and 2 miRNAs* are represented by more than 1000 reads in both libraries, and 5 of them, sli-miR166e (201,378 reads), sli-miR2911c (48,948 reads), sli-miR396d (29,823 reads), sli-miR166f (29,594 reads), and sli-miR403a (28,676 reads) are the most frequent. In addition, sequence analysis shows that the relative abundance of certain member within the miRNA families varies greatly, suggesting functional divergence within the family. For instance, reads of the sli-miR166 family vary from 10 reads (sli-miR166k) to 201,378 reads (sli-miR166e). Similar results are observed in some other miRNA families, such as sli-miR396 (7-29,823 reads) and sli-miR2911 (256-48,948 reads). The above results indicate the different expression levels of different miRNAs in roots, and may be the result of tissue specific or developmental expression.

2.3. Validation of miRNAs in S. linnaeanum Roots

To verify the results of RNA sequencing and bioinformatics analysis, six miRNAs (sli-miR156c, sli-miR166i, sli-miR167a, sli-miR397a, sli-miR403a and sli-miR5300) are selected randomly for validation by qRT-PCR. According to the Illumina sequencing results, these miRNAs are four down-regulated miRNAs, one up-regulated miRNA and one no responsive miRNA. As shown in the Figure 2 and Table S3, the expression changes detected by qRT-PCR for 4 miRNAs (sli-miR156c, sli-miR166i, sli-miR397a and sli-miR403a) are similar to the results of Illumina sequencing. For sli-miR167a and sli-miR5300, the results have small differences, but they all show down regulation. This may be induced by sequencing error or sampling difference. Above results suggest that miRNAs and their expression changes under NaCl stress have been successfully discovered from *S. linnaeanum* roots by Illumina sequencing.

2.4. NaCl-Responsive miRNAs in S. linnaeanum Roots

A deep sequencing approach can be used as a powerful tool for profiling miRNA expression [15,31]. The changes in the frequency of miRNAs between the NaCl-treated and control libraries might indicate that their expression is regulated in response to NaCl stress. To minimize noise and improve accuracy, only the 18–24 nt miRNAs with normalized sequence reads over 10 in at least one library are selected for comparison. miRNAs with $\log_2(TR/CK) > 1$ and $p < 0.05$ are designated as up-regulated. Similarly, miRNAs with $\log_2(TR/CK) < -1$ and $p < 0.05$ are designated as down-regulated. As showed in Table 2, under the stress of NaCl treatment, 11 miRNAs belonging to eight families are down-regulated, and three miRNAs belonging to three families are up-regulated. The above results indicate that the number of NaCl-induced down-regulated miRNAs is more than that of up-regulated miRNAs.

Figure 2. Validation of selected miRNAs in roots by qRT-PCR. The data are the average of three qRT-PCR replicates for each sample from three biological repeats. Small nuclear RNA U6 is used as an internal reference. Error bars indicate one standard deviation of three different biological replicates. The expression changes of six miRNAs detected by qRT-PCR are consistent with the Illumina sequencing results.

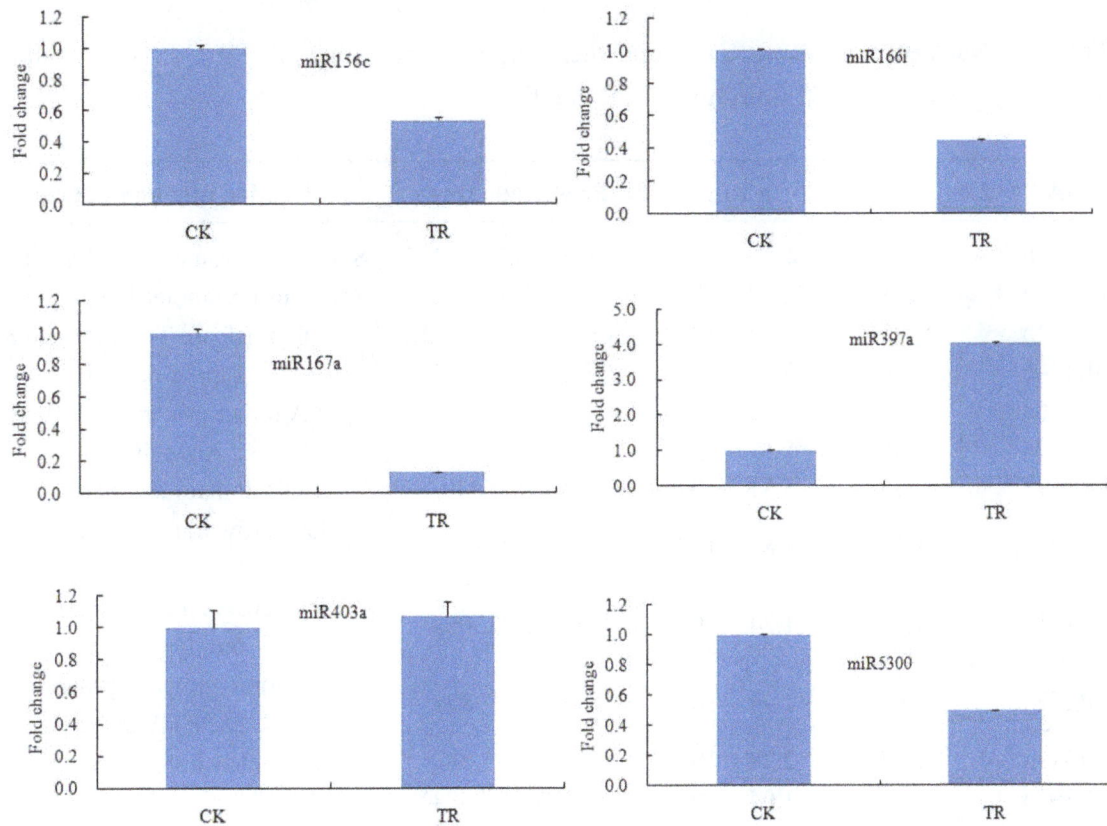

To understand the potential functions of NaCl-responsive miRNAs, 31 target genes (Table S4) for these miRNAs are predicted, and the representative results are listed in Table 2. These genes were reported to be involved in many plant physiological processes, such as plant development, metabolism, and defense. Interestingly, different members in a miRNA family may target the same or different genes. For example, both sli-miRNA156b and sli-miRNA156c can target genes encoding squamosa promoter-binding protein-like, which indicates that they are functionally conservative. The same results are observed for sli-miR171b and sli-miR171e, and they all target the gene encoding scarecrow transcription factor family protein. However, for sli-miR167a and sli-miR167b, which can target different genes, their functions may be differentiated by sequence variation.

As *S. linnaeanum* is salt-tolerant, these salt responsive miRNAs may play an important role for salt tolerance. Some miRNAs, such as *Zea mays* miR166, miR159, miR156 and miR319, and Arabidopsis miR393, miR397b, and miR402, have been reported to show altered expression profile under salt stress [21,37]. In the present study, one of the up-regulated miRNA, sli-miR397a, is predicted to target a laccase gene, which was reported to reduce root growth under dehydration [38]. Similarly, another up-regulated miRNA, sli-miR166d, is predicted to target a DNA repair protein RAD4 family gene which was previously found as a key repair factor that directly recognizes DNA damage and initiates

DNA repair, and recently it was found to regulate protein turnover at a postubiquitylation step [39]. Because of the salt tolerance of *S. linnaeanum*, it is possible that the roots do not suffer with serious injuries such as reduced growth and DNA damage. Therefore, the two potential targets do not need to show high expression to alleviate the injuries. However, further investigations are needed to confirm the above hypothesis.

Table 2. NaCl-responsive miRNAs and their targets. The value of TR/CK is the ratio between normalized count from library TR and CK.

miRNA	Log$_2$(TR/CK)	*p* value	Predicted target	Putative function of target
sli-miR156b	−1.01	2.46×10^{-127}	SGN-U325281	Squamosa promoter-binding protein
sli-miR156c	−1.18	2.72×10^{-52}	SGN-U317176	Squamosa promoter-binding protein
sli-miR162b	−1.25	8.35×10^{-99}	Solyc10g005130.2.1	Ribonuclease 3-like protein 3
sli-miR164c	1.03	3.57×10^{-16}	SGN-U327571	Lipase-related
sli-miR166d	1.97	1.02×10^{-45}	Solyc11g011150.1.1	DNA repair protein Rad4 family
sli-miR167a	−1.45	4.26×10^{-166}	SGN-U313907	Annexin 1
sli-miR167b	−1.25	5.57×10^{-15}	Solyc03g095940.1.1	LOB domain family protein
sli-miR171b	−1.18	1.89×10^{-47}	SGN-U333058	Scarecrow transcription factor family protein
sli-miR171e	−1.08	1.24×10^{-12}	SGN-U333058	Scarecrow transcription factor family protein
sli-miR172a	−1.66	1.94×10^{-44}	SGN-U563871	Floral homeotic protein APETALA2
sli-miR319a	−1.07	2.28×10^{-28}	SGN-U31990	TCP family transcription factor
sli-miR397a	1.91	1.04×10^{-43}	SGN-U327694	Laccase
sli-miR399b	−1.14	9.82×10^{-14}	Solyc03g031410.1.1	Unknown Protein
sli-miR5300	−1.92	1.55×10^{-155}	SGN-U336733	CC-NBS-LRR protein

Unlike the targets of up-regulated miRNAs *in S. linnaeanum*, some of the down-regulated miRNAs target mRNAs of transcription factors, indicating an upstream regulation of miRNAs during the response to salt stress. sli-miR171b and sli-miR171e are predicted to target a scarecrow transcription factor gene which was reported to be involved in ground tissue formation in Arabidopsis root [40]. sli-miR172a is predicted to target an Floral homeotic protein APETALA2 gene. SlAP2a, the true ortholog of AP2 in tomato has been found to control fruit ripening via regulation of ethylene biosynthesis and signaling [41]. However, the role of AP2 in response to salt stress has not been described in detail. sli-miR319a is predicted to target a TCP family transcription factor gene which was reported to play a pivotal role in the control of morphogenesis of shoot organs by negatively regulating the expression of boundary-specific genes in Arabidopsis [42]. The above results indicate that the function involved in the response to salt stress of these potential targets needs to be explored in depth. The identification of salt-responsive miRNAs that target these genes may suggest additional roles for the defense against salt stress.

3. Experimental Section

3.1. Plant Materials and NaCl Treatment

A wild eggplant species, *S. linnaeanum* (PI388846) is used in this study. The seeds are surface-sterilized with 70% ethanol, and allowed to germinatein 30 °C. The uniform germinated seeds are sown in pots containing commercial nursery substrate. The seedlings are grown in an incubator with a 16 h photoperiod at a temperature regime of 25 °C. When the seedlings develop five true leaves, uniform seedlings are picked out and irrigated with 150 mM NaCl for salt treatment or the distilled water as a control. The roots of the NaCl treated and control plants are harvested after 24 h. The collected roots are pooled with ten plants and immediately frozen in liquid nitrogen for RNA extraction.

3.2. Small RNA Library Construction and Sequencing

Total RNA is extracted with the Total RNA Purification Kit (Norgen Biotek, Thorold, Canada and treated with *DNase I* according to the manufacturer's instructions. The small RNA libraries are constructed using the Truseq™ Small RNA Preparation kit (Illumina, San Diego, CA, USA). The purified cDNA library from 15 to 32 nt small RNAs is used for cluster generation on Illumina's Cluster Station and then sequenced on Illumina GAIIx (San Diego, CA, USA). Raw sequencing reads are obtained using Illumina's Sequencing Control Studio software version 2.8 (SCS v2.8, San Diego, CA, USA) following real-time sequencing image analysis and base-calling by Illumina's Real-Time Analysis version 1.8.70 (RTA v1.8.70).2.1.1 (Illumina, San Diego, CA, USA).

3.3. Analysis of Small RNA Sequencing Data

A proprietary pipeline script, ACGT101-miR v4.2 (LC Sciences, Houston, TX, USA), is used for sequencing data analysis. The "impurity" reads due to sample preparation, sequencing chemistry and processes, and the optical digital resolution of the sequencer detector are removed. Those remaining sequences are grouped by families (unique sequences). Thereafter, families that match known plant repeats, mRNA, rRNAs, tRNAs, snRNAs, and snoRNAs were removed. The remaining unique sequences are mapped to known plant miRNAs from miRBase and Pre-miRBase (Version 17.0, ftp://mirbase.org/pub/mirbase/CURRENT, University of Manchester, Manchester, UK) and *S. lycopersicum* genome database (PlantGDB, ftp://ftp.plantgdb.org/download/Genomes/SlGDB/ITAG2_genomic.fasta).

The number of read copies from each sample is tracked during mapping and normalized for comparison. The normalization of sequence counts in each sample is achieved by dividing the counts by a library size parameter of the corresponding sample. The library size parameter is a median value of the ratio between the counts a specific sample and a pseudo-reference sample. A count number in the pseudo-reference sample is the count geometric mean across two samples. For miRNA expression analysis, p value calculation is performed with the method introduced by Audic and Claverie [43].

3.4. miRNA Validation by Quantitative Real-Time PCR

The identified *S. linnaeanum* miRNAs are validated by using quantitative real time PCR (qRT-PCR). In this study, six conserved miRNAs (sli-miR156c, sli-miR166i, sli-miR167a, sli-miR397a, sli-miR403a and sli-miR5300) are validated. Total RNA is isolated from roots of CK and TR, which are samples of parallel experiments for RNA sequencing. For determination of miRNA expression, RNAs are reverse-transcribed by miScript II Reverse Transcription Kit (Qiagen, Germantown, MD, USA), which adds a poly (A) tail to the 3'-end of miRNA and with transcription led by a known oligo-dT ligate. SuperReal PreMix (SYBR Green, TIANGEN, Beijing, China) is used for qRT-PCR. Small nuclear RNA U6 is used as an internal reference. The primers for the 6 miRNAs are universal primers (QIAGEN, Germantown, MD, USA) and corresponding miRNA sequences. qRT-PCR experiments are performed on Roche LightCycler 480 II. PCR program is set as: (1) 95 °C, 15 min; (2) 95 °C, 10 s, thereafter 60 °C, 30 s, 40 cycles. All reactions are run in three replicates for each sample from three biological repeats.

3.5. Prediction of miRNA Target Genes

The putative target sites of miRNA are identified using the psRNATarget program (http://plantgrn.noble.org/psRNATarget/) with default parameters [44]. Because there is not enough genome information for *S. melongena* and *S. linnaeanum*, the database of tomato *S. lycopersicum* is used as the sequence library for target search.

4. Conclusions

In this study, by using high-throughput sequencing and taking advantage of the genome information of other plants, 98 known miRNAs were discovered in *S. linnaeanum* roots, and 14 of them show response to salt stress. The potential targets of the identified salt responsive miRNAs are also predicted based on sequence homology search. However, the further investigation for the function of potential target genes still needs to be performed. As more salt tolerance related miRNAs are confirmed, artificial miRNA will be a powerful tool to create elite plant germplasm with salt tolerance.

Acknowledgments

This research was partially supported by the National Natural Science Foundation (31101542); Natural Science Foundation (BK2011675); and Independent Innovation Foundation of Agricultural Sciences (CX (13)2003) of Jiangsu Province, China.

References

1. Wang, W.; Vinocur, B.; Altman, A. Plant responses to drought, salinity and extreme temperatures: Towards genetic engineering for stress tolerance. *Planta* **2003**, *218*, 1–14.
2. Borsani, O.; Zhu, J.; Verslues, P.E.; Sunkar, R.; Zhu, J.K. Endogenous siRNAs derived from a pair of natural cis-antisense transcripts regulate salt tolerance in Arabidopsis. *Cell* **2005**, *123*, 1279–1291.

3. Vinocur, B.; Altman, A. Recent advances in engineering plant tolerance to abiotic stress: Achievements and limitations. *Curr. Opin. Biotechnol.* **2005**, *16*, 123–132.

4. Bartel, D.P. MicroRNAs: Genomics, biogenesis, mechanism, and function. *Cell* **2004**, *116*, 281–297.

5. Jones-Rhoades, M.W.; Bartel, D.P.; Bartel, B. MicroRNAs and their regulatory roles in plants. *Annu. Rev. Plant Biol.* **2006**, *57*, 19–53.

6. Voinnet, O. Origin, biogenesis, and activity of plant microRNAs. *Cell* **2009**, *136*, 669–687.

7. Liu, P.P.; Montgomery, T.A.; Fahlgren, N.; Kasschau, K.D.; Nonogaki, H.; Carrington, J.C. Repression of AUXIN RESPONSE FACTOR10 by microRNA160 is critical for seed germination and post-germination stages. *Plant J.* **2007**, *52*, 133–146.

8. Reyes, J.L.; Chua, N.H. ABA induction of miR159 controls transcript levels of two MYB factors during Arabidopsis seed germination. *Plant J.* **2007**, *49*, 592–606.

9. Gutierrez, L.; Bussell, J.D.; Pacurar, D.I.; Schwambach, J.; Pacurar, M.; Bellini, C. Phenotypic plasticity of adventitious rooting in Arabidopsis is controlled by complex regulation of AUXIN RESPONSE FACTOR transcripts and microRNA abundance. *Plant Cell* **2009**, *21*, 3119–3132.

10. Sunkar, R.; Chinnusamy, V.; Zhu, J.; Zhu, J.K. Small RNAs as big players in plant abiotic stress responses and nutrient deprivation. *Trends Plant Sci.* **2007**, *12*, 301–309.

11. Bazzini, A.A.; Hopp, H.E.; Beachy, R.N.; Asurmendi, S. Infection and coaccumulation of tobacco mosaic virus proteins alter microRNA levels, correlating with symptom and plant development. *Proc. Natl. Acad. Sci. USA* **2007**, *104*, 12157–12162.

12. Lu, S.; Sun, Y.H.; Amerson, H.; Chiang, V.L. MicroRNAs in loblolly pine (*Pinus taeda* L.) and their association with fusiform rust gall development. *Plant J.* **2007**, *51*, 1077–1098.

13. Yang, L.; Jue, D.; Li, W.; Zhang, R.; Chen, M.; Yang, Q. Identification of MiRNA from eggplant (*Solanum melongena* L.) by small RNA deep sequencing and their response to *Verticillium dahliae* infection. *PLoS One* **2013**, *8*, e72840.

14. Tang, Z.; Zhang, L.; Xu, C.; Yuan, S.; Zhang, F.; Zheng, Y.; Zhao, C. Uncovering small RNA-mediated responses to cold stress in a wheat thermosensitive genic male-sterile line by deep sequencing. *Plant Physiol.* **2012**, *159*, 721–738.

15. Yu, X.; Wang, H.; Lu, Y.; de Ruiter, M.; Cariaso, M.; Prins, M.; van Tunen, A.; He, Y. Identification of conserved and novel microRNAs that are responsive to heat stress in *Brassica rapa*. *J. Exp. Bot.* **2012**, *63*, 1025–1038.

16. Barrera-Figueroa, B.E.; Gao, L.; Diop, N.N.; Wu, Z.; Ehlers, J.D.; Roberts, P.A.; Close, T.J.; Zhu, J.K.; Liu, R. Identification and comparative analysis of drought-associated microRNAs in two cowpea genotypes. *BMC Plant Biol.* **2011**, *11*, 127.

17. Li, B.; Qin, Y.; Duan, H.; Yin, W.; Xia, X. Genome-wide characterization of new and drought stress responsive microRNAs in *Populus euphratica*. *J. Exp. Bot.* **2011**, *62*, 3765–3779.

18. Ding, Y.; Chen, Z.; Zhu, C. Microarray-based analysis of cadmium-responsive microRNAs in rice (*Oryza sativa*). *J. Exp. Bot.* **2011**, *62*, 3563–3573.

19. Chen, L.; Wang, T.; Zhao, M.; Tian, Q.; Zhang, W.H. Identification of aluminum-responsive microRNAs in *Medicago truncatula* by genome-wide high-throughput sequencing. *Planta* **2012**, *235*, 375–386.

20. Liu, H.H.; Tian, X.; Li, Y.J.; Wu, C.A.; Zheng, C.C. Microarray-based analysis of stress-regulated microRNAs in *Arabidopsis Thaliana*. *RNA* **2008**, *14*, 836–843.

21. Ding, D.; Zhang, L.; Wang, H.; Liu, Z.; Zhang, Z.; Zheng, Y. Differential expression of miRNAs in response to salt stress in maize roots. *Ann. Bot.* **2009**, *103*, 29–38.

22. Covarrubias, A.A.; Reyes, J.L. Post-transcriptional gene regulation of salinity and drought responses by plant microRNAs. *Plant Cell Environ.* **2010**, *33*, 481–489.

23. Collonnier, C.; Fock, I.; Kashyap, V.; Rotino, G.L.; Daunay, M.C.; Lian, Y.; Mariska, I.K.; Rajam, M.V.; Servaes, A.; Ducreux, G.; *et al.* Applications of biotechnology in eggplant. *Plant Cell Tissue Organ* **2001**, *65*, 91–107.

24. Doganlar, S.; Frary, A.; Daunay, M.C.; Lester, R.N.; Tanksley, S.D. A comparative genetic linkage map of eggplant (*Solanum melongena*) and its implications for genome evolution in the solanaceae. *Genetics* **2002**, *161*, 1697–1711.

25. Mueller, L.A.; Solow, T.H.; Taylor, N.; Skwarecki, B.; Buels, R.; Binns, J.; Lin, C.; Wright, M.H.; Ahrens, R.; Wang, Y.; *et al.* The SOL Genomics Network: A comparative resource for solanaceae biology and beyond. *Plant Physiol.* **2005**, *138*, 1310–1317.

26. Daniell, H.; Lee, S.B.; Grevich, J.; Saski, C.; Quesada-Vargas, T.; Guda, C.; Tomkins, J.; Jansen, R.K. Complete chloroplast genome sequences of *Solanum bulbocastanum*, *Solanum lycopersicum* and comparative analyses with other Solanaceae genomes. *Theor. Appl. Genet.* **2006**, *112*, 1503–1518.

27. Wu, F.; Eannetta, N.T.; Xu, Y.; Durrett, R.; Mazourek, M.; Jahn, M.M.; Tanksley, S.D. A COSII genetic map of the pepper genome provides a detailed picture of synteny with tomato and new insights into recent chromosome evolution in the genus Capsicum. *Theor. Appl. Genet.* **2009**, *118*, 1279–1293.

28. Wu, F.; Eannetta, N.T.; Xu, Y.; Tanksley, S.D. A detailed synteny map of the eggplant genome based on conserved ortholog set II (COSII) markers. *Theor. Appl. Genet.* **2009**, *118*, 927–935.

29. Rajagopalan, R.; Vaucheret, H.; Trejo, J.; Bartel, D.P. A diverse and evolutionarily fluid set of microRNAs in *Arabidopsis thaliana*. *Genes Dev.* **2006**, *20*, 3407–3425.

30. Fahlgren, N.; Howell, M.D.; Kasschau, K.D.; Chapman, E.J.; Sullivan, C.M.; Cumbie, J.S.; Givan, S.A.; Law, T.F.; Grant, S.R.; Dangl, J.L.; *et al.* High-throughput sequencing of Arabidopsis microRNAs: Evidence for frequent birth and death of MIRNA genes. *PLoS One* **2007**, *2*, e219.

31. Szittya, G.; Moxon, S.; Santos, D.M.; Jing, R.; Fevereiro, M.P.; Moulton, V.; Dalmay, T. High-throughput sequencing of Medicago truncatula short RNAs identifies eight new miRNA families. *BMC Genomics* **2008**, *9*, 593.

32. Morin, R.D.; Aksay, G.; Dolgosheina, E.; Ebhardt, H.A.; Magrini, V.; Mardis, E.R.; Sahinalp, S.C.; Unrau, P.J. Comparative analysis of the small RNA transcriptomes of *Pinus contorta* and *Oryza sativa*. *Genome Res.* **2008**, *18*, 571–584.

33. Chi, X.; Yang, Q.; Chen, X.; Wang, J.; Pan, L.; Chen, M.; Yang, Z.; He, Y.; Liang, X.; Yu, S. Identification and characterization of microRNAs from peanut (*Arachis hypogaea* L.) by high-throughput sequencing. *PLoS One* **2011**, *6*, e27530.

34. Martínez, G.; Forment, J.; Llave, C.; Pallás, V.; Gómez, G. High-throughput sequencing, characterization and detection of new and conserved cucumber miRNAs. *PLoS One* **2011**, *6*, e19523.

35. Guo, H.; Kan, Y.; Liu, W. Differential expression of miRNAs in response to topping in flue-cured tobacco (*Nicotiana tabacum*) roots. *PLoS One* **2011**, *6*, e28565.

36. Song, C.; Wang, C.; Zhang, C.; Korir, N.K.; Yu, H.; Ma, Z.; Fang, J. Deep sequencing discovery of novel and conserved microRNAs in trifoliate orange (*Citrus trifoliata*). *BMC Genomics* **2010**, *11*, 431.

37. Sunkar, R.; Zhu, J.K. Novel and stress-regulated microRNAs and other small RNAs from Arabidopsis. *Plant Cell* **2004**, *16*, 2001–2019.

38. Cai, X.; Davis, E.J.; Ballif, J.; Liang, M.; Bushman, E.; Haroldsen, V.; Torabinejad, J.; Wu, Y. Mutant identification and characterization of the laccase gene family in Arabidopsis. *J. Exp. Bot.* **2006**, *57*, 2563–2569.

39. Li, Y.; Yan, J.; Kim, I.; Liu, C.; Huo, K.; Rao, H. Rad4 regulates protein turnover at a postubiquitylation step. *Mol. Biol. Cell* **2010**, *21*, 177–185.

40. Pauluzzi, G.; Divol, F.; Puig, J.; Guiderdoni, E.; Dievart, A.; Périn, C. Surfing along the root ground tissue gene network. *Dev. Biol.* **2012**, *365*, 14–22.

41. Chung, M.Y.; Vrebalov, J.; Alba, R.; Lee, J.; McQuinn, R.; Chung, J.D.; Klein, P.; Giovannoni, J. A tomato (*Solanum lycopersicum*) *APETALA2/ERF* gene, *SlAP2a*, is a negative regulator of fruit ripening. *Plant J.* **2010**, *64*, 936–947.

42. Koyama, T.; Furutani, M.; Tasaka, M.; Ohme-Takagi, M. TCP transcription factors control the morphology of shoot lateral organs via negative regulation of the expression of boundary-specific genes in Arabidopsis. *Plant Cell* **2007**, *19*, 473–484.

43. Audic, S.; Claverie, J.M. The significance of digital gene expression profiles. *Genome Res.* **1997**, *7*, 986–995.

44. Dai, X.; Zhao, P.X. psRNATarget: A plant small RNA target analysis server. *Nucleic Acids Res.* **2011**, *39*, W155–W159.

2

Regulation of miRNA Expression by Low-Level Laser Therapy (LLLT) and Photodynamic Therapy (PDT)

Toshihiro Kushibiki *, Takeshi Hirasawa, Shinpei Okawa and Miya Ishihara

Department of Medical Engineering, National Defense Medical College 3-2 Namiki, Tokorozawa, Saitama 359-8513, Japan

* Author to whom correspondence should be addressed; E-Mail: toshi@ndmc.ac.jp

Abstract: Applications of laser therapy, including low-level laser therapy (LLLT), phototherapy and photodynamic therapy (PDT), have been proven to be beneficial and relatively less invasive therapeutic modalities for numerous diseases and disease conditions. Using specific types of laser irradiation, specific cellular activities can be induced. Because multiple cellular signaling cascades are simultaneously activated in cells exposed to lasers, understanding the molecular responses within cells will aid in the development of laser therapies. In order to understand in detail the molecular mechanisms of LLLT and PDT-related responses, it will be useful to characterize the specific expression of miRNAs and proteins. Such analyses will provide an important source for new applications of laser therapy, as well as for the development of individualized treatments. Although several miRNAs should be up- or down-regulated upon stimulation by LLLT, phototherapy and PDT, very few published studies address the effect of laser therapy on miRNA expression. In this review, we focus on LLLT, phototherapy and PDT as representative laser therapies and discuss the effects of these therapies on miRNA expression.

Keywords: low-level laser therapy (LLLT); phototherapy; photodynamic therapy (PDT); miRNA

1. Low-Level Laser Therapy (LLLT) and Its Effects on miRNA Expression

A laser (light amplification by stimulated emission of radiation) is a device that generates electromagnetic radiation that is relatively uniform in wavelength, phase and polarization. This technology was originally described by Maiman in 1960 in the form of a ruby laser [1]. The properties of lasers have allowed for numerous medical applications, including their use in surgery, activation of photodynamic agents and various ablative therapies in cosmetics, all of which are based on heat generated by the laser beam, in some cases, leading to tissue destruction [2–9]. These applications of lasers are considered "high-energy", because of their intensities, which range from about 1–100 watt (W)/cm^2.

This paper will address another type of laser application, low-level laser therapy (LLLT), which elicits its effects through non-thermal means. This field was initiated by the work of Mester *et al.*, who in 1967 reported non-thermal effects of lasers on mouse hair growth [10]. In a subsequent study, the same group reported acceleration of wound healing and improvement in the post-wounding regeneration ability of muscle fibers using a 1 J/cm^2 ruby laser [11]. Since those early days, numerous *in vitro* and *in vivo* studies of LLLT in the context of regenerative medicine have demonstrated a wide variety of therapeutic effects, including reduction of pain, anti-inflammatory effects and wound healing. According to da Silva *et al.* [12], the types of laser most frequently used for wound healing and tissue repair are helium neon (He-Ne) lasers and diode lasers, including gallium-aluminum-arsenic (Ga-Al-As), arsenic-gallium (As-Ga) and indium-gallium-aluminum-phosphide (In-Ga-Al-P) lasers.

One of the most distinctive features of LLLT relative to other modalities is that the effects are mediated not through induction of thermal effects, but rather, through a process, still not clearly defined, called "photobiostimulation". Because this effect of LLLT apparently does not depend on coherence, it is therefore possible to achieve photobiostimulation using non-laser light-generating devices, such as inexpensive light-emitting diode (LED) technology [13–17]. To date, several mechanisms of biological action have been proposed, although none have been clearly established. These include augmentation of cellular ATP levels [18–20], manipulation of inducible nitric oxide synthase (iNOS) activity [21–25], suppression of inflammatory cytokines, such as TNF-alpha [19,26–28], IL-1beta [28–30], IL-6 [28,31–34] and IL-8 [28,31,32,35], upregulation of growth factors, such as PDGF, IGF-1, NGF and FGF-2 [30,36–38], alteration of mitochondrial membrane potential [39–42], due to chromophores found in the mitochondrial respiratory chain [43–45], stimulation of protein kinase C (PKC) activation [46], manipulation of NF-kappaB activation [47], induction of reactive oxygen species (ROS) [48,49], modification of extracellular matrix components [50], inhibition of apoptosis [39], stimulation of mast cell degranulation [51] and upregulation of heat shock proteins [52]. We have also proposed that LLLT influences cell differentiation following laser stimulation [53–55].

Unfortunately, these effects have been demonstrated using a variety of laser devices in non-comparable models. To add to the confusion, dose-dependency seems to be confined to a very narrow range, and in numerous systems, the therapeutic effects disappear with increased dose. Consequently, only two studies of miRNA expression dynamics following LLLT have been reported to date, by Wang *et al.* [56] and Gu *et al.* [57]. With the exception of those studies, no data are currently available regarding the overall changes in the global expression of many hundreds of miRNAs following LLLT. Wang *et al.* [56] showed that LLLT increases the migration, proliferation and viability of rat mesenchymal stem cells (MSCs) and, also, activates the expression of various miRNAs. Using a diode laser (wavelength: 635 nm, 0.5 J/cm^2), they found that the proliferation rate and expression of cell cycle-associated genes increased in a time-dependent manner following LLLT treatment of MSCs. Microarray assays revealed subsets of miRNAs that were regulated by LLLT: 19 miRNAs were upregulated and 15 miRNAs were downregulated (Table 1); these dynamic changes were confirmed by quantitative real-time PCR.

Table 1. Aberrations in miRNA expression after low-level laser therapy (LLLT) to mesenchymal stem cells by using a diode laser (wavelength: 635 nm, 0.5 J/cm^2) [56].

Upregulation	Downregulation
miR-30e *	
miR-15b	
miR-30b-5p	miR-204 *
miR-322	miR-7a
miR-215	miR-423
miR-449a	miR-678
miR-126	miR-25 *
miR-133b	miR-327
miR-21 *	miR-351
miR-455	miR-23a
miR-759	miR-667
miR-872 *	miR-770
miR-29b	miR-324-3p
miR-192	miR-30c-2 *
miR-219-1-3p	miR-758
miR-301a	miR-320
miR-551b	miR-466c
miR-224	
miR-193	

miRNAs expression confirmed by quantitative real-time PCR are indicated by underlining. The asterisk, * indicates the star-form of miRNA.

The most highly upregulated miRNA was miR-193. Gain- and loss-of-function experiments demonstrated that miR-193 levels regulate the proliferation of MSCs of both humans and rats; in particular, blockade of miR-193 repressed the MSCs proliferation induced by LLLT. However, this miRNA apparently does not affect apoptosis or differentiation. In addition, Wang *et al.* found that miR-193 regulated expression of cyclin-dependent kinase 2 (CDK2). Bioinformatic analyses and luciferase reporter assays revealed that inhibitor of growth family, member 5 (ING5), was the most likely target of miR-193 to functionally regulate proliferation and CDK2 expression; indeed, the mRNA and protein levels of ING5 are regulated by miR-193. Furthermore, inhibition of ING5 by small interfering RNA (siRNA) upregulated both MSC proliferation and the expression of CDK2. Another miRNA, miR-335, has been shown by others to regulate the proliferation and migration of MSCs [58], so it is likely to play an important role in MSC proliferation after LLLT. Moreover, several studies have shown that LLLT also stimulates cell differentiation [53–55,59–75], and future work should reveal miRNAs specifically involved in mediating this effect.

Although some literature reported that tumor or apoptosis related miRNAs were induced by UV irradiation to cells [76–81], Gu *et al.* reported UV-phototherapy and its effect on miRNA expression [57]. They showed the effect of narrow-band ultraviolet B (NB-UVB) irradiation on miR-21 and -125b expression in psoriatic epidermis. Psoriasis is an inflammatory skin disease in which dysregulation of p63, a member of the p53 family that is crucial for skin development and maintenance, has been demonstrated [82–84]. Involvement of miR-203, miR-21 and miR-125b were implicated in the regulation of p63 or p53 in the pathogenesis of psoriasis. Skin biopsies from 12 psoriasis patients were collected before, during and after NB-UVB therapy. The p63 expression was not significantly affected, whereas NB-UVB phototherapy significantly decreased expression of miR-21and increased miR-125b levels. Since NB-UVB phototherapy is commonly used in the treatment of psoriasis [85–87], those results indicate a complex mechanism of p63 regulation, which merits further investigation in order to achieve better long-term clinical improvement.

2. Photodynamic Therapy and Its Effects on miRNA Expression

Photodynamic therapy (PDT), a class of laser therapy, is a photochemical modality approved for the treatment of various cancers and diseases in which neovascularization occurs [88,89]. The PDT process consists of injecting a photosensitizer, which selectively accumulates at the lesion site, followed by local irradiation of the tumor with light of an appropriate wavelength to activate a specific drug [90]. Irradiation leads to the generation of singlet oxygen and other reactive oxygen species (ROS) [91]. PDT is being considered not only as palliative therapy, but also as a treatment option for early-stage skin, lung, cervical and esophageal cancers, as well as basal-cell carcinomas. Currently, PDT has been approved for localized diseases and precancerous lesions, such as bladder cancers, pituitary tumors and glioblastomas [92,93]. Furthermore, numerous ongoing clinical studies have been designed to optimize the conditions for PDT; subsequently, PDT has been approved in several countries.

Upon absorption of one or more photons, the excited photosensitizer undergoes one of two possible reactions (type I or/and II) with a neighboring oxygen molecule, yielding ROS [94]. These ROS oxidize various cellular substrates, affecting cellular functions and resulting in cell death. The ROS that are produced during PDT destroy tumors by multiple mechanisms: in contrast to most conventional cytotoxic agents, which usually only trigger apoptotic cell death, PDT can cause cell death by necrosis and/or apoptosis.

The direct destruction of cancer cells (necrosis) by PDT is caused by irreversible damage to the plasma membrane and intracellular organelles, including the mitochondria, lysosomes, Golgi apparatus and endoplasmic reticulum (ER). The mechanisms of PDT-induced apoptosis have been described by many studies. Apoptosis, or programmed cell death, is one mechanism that mediates toxicity in the target tissue following PDT [95]. Apoptosis involves a cascade of molecular events leading to orderly cellular death without an inflammatory response [96–98]. The initiation of apoptosis involves a complex network of signaling pathways, both intrinsic and extrinsic to the individual cell, which are regulated, in part, by pro- and anti-apoptotic factors [96]. The initial damage can involve different molecules, ultimately leading to activation of specific death pathways. Mitochondria-localized photosensitizers can cause immediate and light-dependent photodamage to mitochondrial components, such as the anti-apoptotic Bcl-2, Bcl-xL and the other apoptosis-related proteins, prompting the release of caspase-activating molecules [99]. Photosensitizers that accumulate in the lysosomes or mitochondria and which were excited by laser light can induce Bax-mediated caspase activation (Figure 1).

Figure 1. Representative signaling pathways of apoptosis induced by photodynamic therapy (PDT). Depending on the nature of the photosensitizer and its intracellular localization, the initial photodamage can involve different molecules, with the consequent activation of specific death pathways that converge on mitochondria. Mitochondria-localized photosensitizer can cause immediate and light-dependent photodamage to the anti-apoptotic Bcl-2 and Bcl-xL proteins, prompting the release of caspase-activating molecules. Lysosomal hydrolases and ER stress also induce Bax-mediated caspase activation.

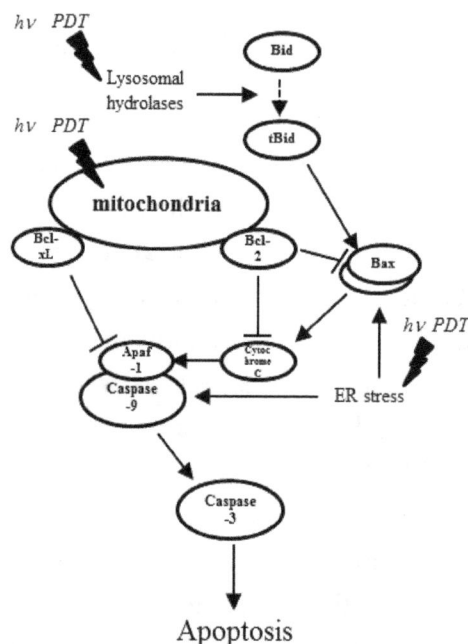

Another important cellular factor induced by PDT and released from necrotic tumor cells is heat-shock protein 70 (Hsp70) [100]. Hsp70 is significantly induced after stress; when it remains within the cell, it chaperones unfolded proteins and prevents cell death by inhibiting the aggregation of cellular proteins. Hsp70 directly binds to the caspase-recruitment domain of apoptotic-protease activating factor 1 (Apaf-1), thereby preventing the recruitment of Apaf-1 oligomerization and association of Apaf-1 with procaspase 9. These properties not only enable intracellular Hsp70 to inhibit cancer-cell death by apoptosis, but also promote the formation of stable complexes with cytoplasmic tumor antigens. These antigens can then either be expressed at the cell surface or escape intact from dying necrotic cells to interact with antigen-presenting cells, thereby stimulating an anti-tumor immune response.

The mechanisms of cell death following PDT have been thoroughly summarized in the literature [95,101–104]. A better understanding of the molecular differences between apoptosis and necrosis and identification of the crosstalk between these programs will certainly be crucial to the development of new PDT modalities aimed at increasing the efficiency of cancer-cell killing.

Another inherent consequence of PDT is local hypoxia, which can arise either directly, from oxygen consumption during treatment [105–107], or indirectly, from the destruction of tumor vasculature as a result of effective treatment [108,109]. Hypoxia is a major stimulus for angiogenesis, via its stabilization of the hypoxia-inducible factor-1α (HIF-1α) transcription factor [110,111]. HIF-1 is a heterodimeric complex of two helix-loop-helix proteins, HIF-1α and HIF-1β (ARNT). ARNT is constitutively expressed, whereas HIF-1α is rapidly degraded under normoxic conditions. Hypoxia induces the stabilization of the HIF-1α subunit, which, in turn, allows formation of the transcriptionally active protein complex. A number of HIF-1–responsive genes have been identified, including those encoding vascular endothelial growth factor (VEGF), erythropoietin and glucose transporter-1 [112,113]. Following PDT, increases in VEGF secretion and angiogenic responses stimulated via HIF-1 pathways have been documented *in vivo* [114–117]. VEGF induction can contribute to tumor survival and regrowth and, therefore, may represent one of the factors that prevent PDT from achieving its full tumoricidal potential. PDT has been considered for both palliative therapy and as an early treatment option for cancer. Numerous ongoing clinical studies have been designed to optimize PDT conditions. However, no standardized biological markers of cell death and PDT efficacy, other than cell viability itself, have been reported.

Human cancer is associated with changes in miRNA expression. The pattern of miRNA expression varies dramatically across tumor types, and miRNA profiles reflect the developmental lineage and differentiation state of a tumor [118]. miRNA is also likely to play critical roles in various aspects of hematopoiesis, including the differentiation of hematopoietic stem/progenitor cells, as well as in events that lead to hematological disorders. Nonetheless, very few miRNA expression patterns of specific diseases are available. Moreover, no profiles of miRNA expression after PDT have been reported. Cheng *et al.* found that inhibition of miR-95, -124, -125, -133, -134, -144, -150, -152, -187, -190, -191, -192, -193, -204, -211, -218, -220, -296 and -299 resulted in a decrease in cell growth, whereas inhibition of miR-21 and miR-24 profoundly increased cell growth in HeLa cells [119]. In addition, they identified miRNAs, whose expression increased levels of apoptosis (miR-7, -148, -204, -210, -216 and -296). Those data suggest that specific miRNAs are involved in the cell-death response. We have shown that a miRNA specific to apoptosis is expressed at increased levels in HeLa cells in response to PDT using talaporfin sodium as a photosensitizer [120]. Our study was the first to characterize miRNA expression levels following PDT. In our experiments, miR-210 and miR-296 expression levels increased significantly 1 h after PDT in cells treated with 50 µg/mL talaporfin sodium, relative to the control group (*i.e.*, 0 µg/mL talaporfin sodium), as shown in Figure 2. However, the expression levels of other miRNAs, e.g., miR-7, -148a, -204 and -216, were indistinguishable from those of the control group after PDT.

miR-210 is the miRNA most consistently stimulated under hypoxic conditions [121]. Because hypoxia and stabilization of intracellular HIF are inherent consequences of PDT [92], Giannakakis *et al.* investigated miR-210 expression in the context of its hypoxic effect, and they reported evidence for the involvement of the HIF signaling pathway in miR-210 regulation. To study the biological impacts of a partial or complete loss of miR-210 functions, they also identified the putative mRNA targets of miR-210. According to their report, miR-210 targets important regulators of transcription, cell metabolism, differentiation and development, *i.e.*, processes that are critically affected by hypoxia [121]. The identification of key regulators of important cellular processes among miR-210 target mRNAs, as well as the high frequency of gene copy-number aberrations in tumors, underscore the involvement of miR-210 in oncogenesis and highlight miR-210 as a potential link between hypoxia and cell-cycle control in cancer cells.

Figure 2. Expression of miR-210 and miR-296 after PDT in HeLa cells. miR-210 and miR-296 expression levels were significantly increased 1 h after PDT (60 mW/cm^2, 90 s) in cells treated with 50 μg/mL talaporfin sodium relative to levels in the control group (*i.e.*, talaporfin sodium concentration of 0 μg/mL) (1 × 10^4 cells/well). The asterisk, * indicates $p < 0.05$, a significant difference between the relative expression levels of PDT-treated cells and non-PDT-treated cells. All experiments were performed four times independently. All data are expressed as the means ± SD of four replicates from four experiments (Adapted from [120]).

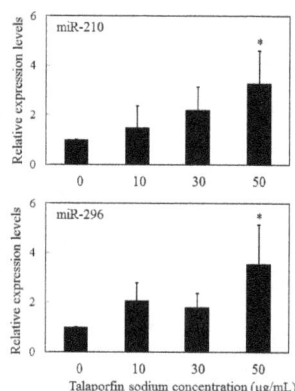

Würdinger *et al.* reported a role for miR-296 in promoting angiogenesis in tumors [122], and in particular, they showed that VEGF alone is capable of increasing miR-296 expression levels. Their results revealed a feedback loop, wherein VEGF induces miR-296 expression, which targets the hepatocyte growth factor-regulated tyrosine kinase substrate (HGS), which, in turn, results in increased levels of VEGF receptor 2 and platelet-derived growth factor (PDGF) receptor β protein and, ultimately, in an increased response to VEGF. Because increased VEGF sensitivity of cancer cells is one of the inherent consequences of PDT [115], our results suggest that inhibition of miR-296 expression should improve PDT efficacy [120]. Our study also suggested that hypoxia induced by PDT induces miR-210 expression, followed by an increased expression of both VEGF and miR-296 [120]. Hence, we reported that miR-210 and miR-296 expression levels represent markers for the efficacy of talaporfin sodium-mediated PDT in cancer cells.

Furthermore, a recently published paper by Bach *et al.* described a comprehensive analysis of changes in miRNA levels following PDT, using polyvinylpyrrolidone hypericin (PVPH) as a photosensitizer, against A431 human epidermoid carcinoma cells [123]. That study was the first comprehensive analysis of changes in miRNA induced by PDT. Using microarray analysis, Bach *et al.* identified eight miRNAs that were significantly differentially expressed 5 hr after treatment, compared with baseline levels, and three miRNAs with more than two-fold differential expression that could be detected in one or two biological replicates. The verification of these results by quantitative real-time

PCR, including a detailed time course, revealed an up to 15-fold transient upregulation of miR-634, -1246 and -1290 relative to their basal levels (Table 2).

Table 2. Aberrations in miRNA expression after PDT to human epidermoid carcinoma cells (A431) by using polyvinylpyrrolidone hypericin (PVPH) [123].

Upregulation	Downregulation
miR-1290	miR-1260b
miR-634	miR-720
miR-1246	miR-1260
	miR-1280

In silico prediction of the targets of these miRNAs yielded numerous mRNAs encoding proteins, including the apoptotic protease activating factor-1 interacting protein and the BMI1 polycomb ring finger oncogene in the apoptosis/cell death category, cyclin-dependent kinase 20 and the cell division cycle 25 homolog C in the proliferation/cell cycle category, frizzled family receptor 3 and bone morphogenetic protein 4 in the cell signaling/adhesion category and the DNA excision repair protein ERCC-8 and peroxiredoxin-6 in the cell stress category. Although several studies have investigated the PDT-induced changes in the transcriptome and proteome, no comprehensive data are currently available regarding the effect of PDT on the miRNA transcriptome. Using a comprehensive microarray platform covering 1223 mature human miRNAs, Bach *et al.* did not observe up- or down-regulation by PDT of the miRNAs reported in our study (miR-210 and -296 [120]). This difference is likely attributable to the PDT conditions, such as cell type, photosensitizer and laser dose. Furthermore, the significant increase in the apoptosis-related miRNAs (3–4-fold increase) observed in our study was measured in a mixed population of cells, consisting predominantly of surviving cells [124]. Given these discrepancies, there is a need for additional experiments that might uncover additional miRNAs that are transiently regulated following photodynamic damage. It will be also of paramount interest to study miRNA-related cellular responses under explicitly non-lethal PDT conditions, as this approach could identify possible miRNA targets, whose manipulation might increase cells' sensitivity towards PDT.

Interestingly, Bach *et al.* also found that the incubation with the photosensitizer induced a slight to moderate increase in the expression of several miRNAs (*i.e.*, miR-1260b, -1260, -1280, -3182, -1290 and -1246), particularly at later time points [123]. Conversely, several miRNAs were transiently up-regulated by light-only treatment, especially at earlier time points (miR-1260b, -1260, -1280, -3182 and -1290). They concluded that the detailed functions of the increased expression of these miRNAs following apoptosis induced by PDT remain to be elucidated [123].

3. Conclusions

In this review, we focused on miRNA expression after LLLT and PDT. As mentioned above, only a few papers have been published regarding miRNA expression in this context, and those few reports discuss only a small number of laser therapy conditions. The ability of LLLT to induce growth-factor production, inhibition of inflammation, stimulation of angiogenesis, pain reduction and direct effects on stem cells suggests that there is an urgent need to combine this modality with regenerative medicine. PDT has been employed in the treatment of many tumor types, and its effectiveness as a curative and palliative treatment is well documented, especially in the context of skin cancer. A detailed understanding of LLLT-, phototherapy- and PDT-related molecular mechanisms, including the specific effects on miRNA and protein expression, will provide an important source for new applications of laser therapy and for the development of individualized treatments.

Acknowledgments

This work was supported by Japan Society for the Promotion of Science (JSPS) KAKENHI, Grant Number 25713009.

References

1. Maiman, T.H. Stimulated optical radiation in ruby. *Nature* **1960**, *187*, 493–494.

2. Khatri, K.A.; Mahoney, D.L.; McCartney, M.J. Laser scar revision: A review. *J. Cosmet. Laser Ther.* **2011**, *13*, 54–62.

3. Chung, S.H.; Mazur, E. Surgical applications of femtosecond lasers. *J. Biophotonics* **2009**, *2*, 557–572.

4. Zhao, Z.; Wu, F. Minimally-invasive thermal ablation of early-stage breast cancer: A systemic review. *Eur. J. Surg. Oncol.* **2010**, *36*, 1149–1155.

5. Siribumrungwong, B.; Noorit, P.; Wilasrusmee, C.; Attia, J.; Thakkinstian, A. A systematic review and meta-analysis of randomised controlled trials comparing endovenous ablation and surgical intervention in patients with varicose vein. *Eur. J. Vasc. Endovasc. Surg.* **2012**, *44*, 214–223.

6. Vuylsteke, M.E.; Mordon, S.R. Endovenous laser ablation: A review of mechanisms of action. *Ann. Vasc. Surg.* **2012**, *26*, 424–433.

7. Vogel, A.; Venugopalan, V. Mechanisms of pulsed laser ablation of biological tissues. *Chem. Rev.* **2003**, *103*, 577–644.

8. Casas, A.; di Venosa, G.; Hasan, T.; Al, B. Mechanisms of resistance to photodynamic therapy. *Curr. Med. Chem.* **2011**, *18*, 2486–2515.

9. Anand, S.; Ortel, B.J.; Pereira, S.P.; Hasan, T.; Maytin, E.V. Biomodulatory approaches to photodynamic therapy for solid tumors. *Cancer Lett.* **2012**, *326*, 8–16.

10. Mester, E.; Szende, B.; Gartner, P. The effect of laser beams on the growth of hair in mice. *Radiobiol. Radiother.* **1968**, *9*, 621–626.

11. Mester, E.; Spiry, T.; Szende, B.; Tota, J.G. Effect of laser rays on wound healing. *Am. J. Surg.* **1971**, *122*, 532–535.

12. Da Silva, J.P.; da Silva, M.A.; Almeida, A.P.; Lombardi Junior, I.; Matos, A.P. Laser therapy in the tissue repair process: A literature review. *Photomed. Laser Surg.* **2010**, *28*, 17–21.

13. Buravlev, E.A.; Zhidkova, T.V.; Vladimirov, Y.A.; Osipov, A.N. Effects of laser and led radiation on mitochondrial respiration in experimental endotoxic shock. *Lasers Med. Sci.* **2012**, *28*, 785–790.

14. De Sousa, A.P.; Santos, J.N.; dos Reis, J.A., Jr.; Ramos, T.A.; de Souza, J.; Cangussu, M.C.; Pinheiro, A.L. Effect of led phototherapy of three distinct wavelengths on fibroblasts on wound healing: A histological study in a rodent model. *Photomed. Laser Surg.* **2010**, *28*, 547–552.

15.   Lev-Tov, H.; Brody, N.; Siegel, D.; Jagdeo, J. Inhibition of fibroblast proliferation *in vitro* using low-level infrared light-emitting diodes. *Dermatol. Surg.* **2012**, *39*, 422–425.

16.   Nishioka, M.A.; Pinfildi, C.E.; Sheliga, T.R.; Arias, V.E.; Gomes, H.C.; Ferreira, L.M. Led (660 nm) and laser (670 nm) use on skin flap viability: Angiogenesis and mast cells on transition line. *Lasers Med. Sci.* **2012**, *27*, 1045–1050.

17.   Pinheiro, A.L.; Soares, L.G.; Cangussu, M.C.; Santos, N.R.; Barbosa, A.F.; Silveira Junior, L. Effects of led phototherapy on bone defects grafted with mta, bone morphogenetic proteins and guided bone regeneration: A raman spectroscopic study. *Lasers Med. Sci.* **2012**, *27*, 903–916.

18.   AlGhamdi, K.M.; Kumar, A.; Moussa, N.A. Low-level laser therapy: A useful technique for enhancing the proliferation of various cultured cells. *Lasers Med. Sci.* **2012**, *27*, 237–249.

19.   Gao, X.; Xing, D. Molecular mechanisms of cell proliferation induced by low power laser irradiation. *J. Biomed. Sci.* **2009**, *16*, 4.

20.   Tafur, J.; van Wijk, E.P.; van Wijk, R.; Mills, P.J. Biophoton detection and low-intensity light therapy: A potential clinical partnership. *Photomed. Laser Surg.* **2010**, *28*, 23–30.

21.   Gavish, L.; Perez, L.S.; Reissman, P.; Gertz, S.D. Irradiation with 780 nm diode laser attenuates inflammatory cytokines but upregulates nitric oxide in lipopolysaccharide-stimulated macrophages: Implications for the prevention of aneurysm progression. *Lasers Surg. Med.* **2008**, *40*, 371–378.

22.   Lindgard, A.; Hulten, L.M.; Svensson, L.; Soussi, B. Irradiation at 634 nm releases nitric oxide from human monocytes. *Lasers Med. Sci.* **2007**, *22*, 30–36.

23.   Moriyama, Y.; Moriyama, E.H.; Blackmore, K.; Akens, M.K.; Lilge, L. *In vivo* study of the inflammatory modulating effects of low-level laser therapy on inos expression using bioluminescence imaging. *Photochem. Photobiol.* **2005**, *81*, 1351–1355.

24.   Moriyama, Y.; Nguyen, J.; Akens, M.; Moriyama, E.H.; Lilge, L. *In vivo* effects of low level laser therapy on inducible nitric oxide synthase. *Lasers Surg. Med.* **2009**, *41*, 227–231.

25.   Tuby, H.; Maltz, L.; Oron, U. Modulations of vegf and inos in the rat heart by low level laser therapy are associated with cardioprotection and enhanced angiogenesis. *Lasers Surg. Med.* **2006**, *38*, 682–688.

26.   Fukuda, T.Y.; Tanji, M.M.; Silva, S.R.; Sato, M.N.; Plapler, H. Infrared low-level diode laser on inflammatory process modulation in mice: Pro- and anti-inflammatory cytokines. *Lasers Med. Sci.* **2012**, doi:10.1007/s10103-012-1231-z.

27.   Oliveira, R.G.; Ferreira, A.P.; Cortes, A.J.; Aarestrup, B.J.; Andrade, L.C.; Aarestrup, F.M. Low-level laser reduces the production of tnf-alpha, ifn-gamma, and il-10 induced by ova. *Lasers Med. Sci.* **2013**, doi:10.1007/s10103-012-1262-5.

28.   Yamaura, M.; Yao, M.; Yaroslavsky, I.; Cohen, R.; Smotrich, M.; Kochevar, I.E. Low level light effects on inflammatory cytokine production by rheumatoid arthritis synoviocytes. *Lasers Surg. Med.* **2009**, *41*, 282–290.

29.   Aimbire, F.; Ligeiro de Oliveira, A.P.; Albertini, R.; Correa, J.C.; Ladeira de Campos, C.B.; Lyon, J.P.; Silva, J.A., Jr.; Costa, M.S. Low level laser therapy (lllt) decreases pulmonary microvascular leakage, neutrophil influx and il-1beta levels in airway and lung from rat subjected to lps-induced inflammation. *Inflammation* **2008**, *31*, 189–197.

30. Safavi, S.M.; Kazemi, B.; Esmaeili, M.; Fallah, A.; Modarresi, A.; Mir, M. Effects of low-level he-ne laser irradiation on the gene expression of il-1beta, tnf-alpha, ifn-gamma, tgf-beta, bfgf, and pdgf in rat's gingiva. *Lasers Med. Sci.* **2008**, *23*, 331–335.

31. Boschi, E.S.; Leite, C.E.; Saciura, V.C.; Caberlon, E.; Lunardelli, A.; Bitencourt, S.; Melo, D.A.; Oliveira, J.R. Anti-inflammatory effects of low-level laser therapy (660 nm) in the early phase in carrageenan-induced pleurisy in rat. *Lasers Surg. Med.* **2008**, *40*, 500–508.

32. Shiba, H.; Tsuda, H.; Kajiya, M.; Fujita, T.; Takeda, K.; Hino, T.; Kawaguchi, H.; Kurihara, H. Neodymium-doped yttrium-aluminium-garnet laser irradiation abolishes the increase in interleukin-6 levels caused by peptidoglycan through the p38 mitogen-activated protein kinase pathway in human pulp cells. *J. Endod.* **2009**, *35*, 373–376.

33. Houreld, N.N.; Sekhejane, P.R.; Abrahamse, H. Irradiation at 830 nm stimulates nitric oxide production and inhibits pro-inflammatory cytokines in diabetic wounded fibroblast cells. *Lasers Surg. Med.* **2010**, *42*, 494–502.

34. Simunovic-Soskic, M.; Pezelj-Ribaric, S.; Brumini, G.; Glazar, I.; Grzic, R.; Miletic, I. Salivary levels of tnf-alpha and il-6 in patients with denture stomatitis before and after laser phototherapy. *Photomed. Laser Surg.* **2010**, *28*, 189–193.

35. Fushimi, T.; Inui, S.; Nakajima, T.; Ogasawara, M.; Hosokawa, K.; Itami, S. Green light emitting diodes accelerate wound healing: Characterization of the effect and its molecular basis *in vitro* and *in vivo*. *Wound Repair Regen.* **2012**, *20*, 226–235.

36. Saygun, I.; Karacay, S.; Serdar, M.; Ural, A.U.; Sencimen, M.; Kurtis, B. Effects of laser irradiation on the release of basic fibroblast growth factor (bfgf), insulin like growth factor-1 (igf-1), and receptor of igf-1 (igfbp3) from gingival fibroblasts. *Lasers Med. Sci.* **2008**, *23*, 211–215.

37. Schwartz, F.; Brodie, C.; Appel, E.; Kazimirsky, G.; Shainberg, A. Effect of helium/neon laser irradiation on nerve growth factor synthesis and secretion in skeletal muscle cultures. *J. Photochem. Photobiol. B* **2002**, *66*, 195–200.

38. Yu, W.; Naim, J.O.; Lanzafame, R.J. The effect of laser irradiation on the release of bfgf from 3t3 fibroblasts. *Photochem. Photobiol.* **1994**, *59*, 167–170.

39. Hu, W.P.; Wang, J.J.; Yu, C.L.; Lan, C.C.; Chen, G.S.; Yu, H.S. Helium-neon laser irradiation stimulates cell proliferation through photostimulatory effects in mitochondria. *J. Invest. Dermatol.* **2007**, *127*, 2048–2057.

40. Lan, C.C.; Wu, C.S.; Chiou, M.H.; Chiang, T.Y.; Yu, H.S. Low-energy helium-neon laser induces melanocyte proliferation via interaction with type iv collagen: Visible light as a therapeutic option for vitiligo. *Br. J. Dermatol.* **2009**, *161*, 273–280.

41. Wu, S.; Xing, D.; Gao, X.; Chen, W.R. High fluence low-power laser irradiation induces mitochondrial permeability transition mediated by reactive oxygen species. *J. Cell. Physiol.* **2009**, *218*, 603–611.

42. Zungu, I.L.; Hawkins Evans, D.; Abrahamse, H. Mitochondrial responses of normal and injured human skin fibroblasts following low level laser irradiation—An *in vitro* study. *Photochem. Photobiol.* **2009**, *85*, 987–996.

43. Karu, T. Photobiology of low-power laser effects. *Health Phys.* **1989**, *56*, 691–704.

44. Karu, T.I. Mitochondrial signaling in mammalian cells activated by red and near-ir radiation. *Photochem. Photobiol.* **2008**, *84*, 1091–1099.

45. Tiphlova, O.; Karu, T. Role of primary photoaccepters in low-power laser effects: Action of hene laser radiation on bacteriophage t4-escherichia coli interaction. *Lasers Surg. Med.* **1989**, *9*, 67–69.

46. Zhang, L.; Xing, D.; Zhu, D.; Chen, Q. Low-power laser irradiation inhibiting abeta25-35-induced pc12 cell apoptosis via pkc activation. *Cell. Physiol. Biochem.* **2008**, *22*, 215–222.

47. Aimbire, F.; Santos, F.V.; Albertini, R.; Castro-Faria-Neto, H.C.; Mittmann, J.; Pacheco-Soares, C. Low-level laser therapy decreases levels of lung neutrophils anti-apoptotic factors by a nf-κ dependent mechanism. *Int. Immunopharmacol.* **2008**, *8*, 603–605.

48. Kushibiki, T.; Hirasawa, T.; Okawa, S.; Ishihara, M. Blue laser irradiation generates intracellular reactive oxygen species in various types of cells. *Photomed. Laser Surg.* **2013**, *31*, 95–104.

49. Lipovsky, A.; Nitzan, Y.; Lubart, R. A possible mechanism for visible light-induced wound healing. *Lasers Surg. Med.* **2008**, *40*, 509–514.

50. Ignatieva, N.; Zakharkina, O.; Andreeva, I.; Sobol, E.; Kamensky, V.; Lunin, V. Effects of laser irradiation on collagen organization in chemically induced degenerative annulus fibrosus of lumbar intervertebral disc. *Lasers Surg. Med.* **2008**, *40*, 422–432.

51. Silveira, L.B.; Prates, R.A.; Novelli, M.D.; Marigo, H.A.; Garrocho, A.A.; Amorim, J.C.; Sousa, G.R.; Pinotti, M.; Ribeiro, M.S. Investigation of mast cells in human gingiva following low-intensity laser irradiation. *Photomed. Laser Surg.* **2008**, *26*, 315–321.

52. Coombe, A.R.; Ho, C.T.; Darendeliler, M.A.; Hunter, N.; Philips, J.R.; Chapple, C.C.; Yum, L.W. The effects of low level laser irradiation on osteoblastic cells. *Clin. Orthod. Res.* **2001**, *4*, 3–14.

53. Kushibiki, T.; Awazu, K. Controlling osteogenesis and adipogenesis of mesenchymal stromal cells by regulating a circadian clock protein with laser irradiation. *Inter. J. Med. Sci.* **2008**, *5*, 319–326.

54. Kushibiki, T.; Awazu, K. Blue laser irradiation enhances extracellular calcification of primary mesenchymal stem cells. *Photomed. Laser Surg.* **2009**, *27*, 493–498.

55. Kushibiki, T.; Tajiri, T.; Ninomiya, Y.; Awazu, K. Chondrogenic mrna expression in prechondrogenic cells after blue laser irradiation. *J. Photochem. Photobiol. B* **2010**, *98*, 211–215.

56. Wang, J.; Huang, W.; Wu, Y.; Hou, J.; Nie, Y.; Gu, H.; Li, J.; Hu, S.; Zhang, H. Microrna-193 pro-proliferation effects for bone mesenchymal stem cells after low-level laser irradiation treatment through inhibitor of growth family, member 5. *Stem Cells Dev.* **2012**, *21*, 2508–2519.

57. Gu, X.; Nylander, E.; Coates, P.J.; Nylander, K. Effect of narrow-band ultraviolet b phototherapy on p63 and microrna (mir-21 and mir-125b) expression in psoriatic epidermis. *Acta Derm. Venereol.* **2011**, *91*, 392–397.

58. Tome, M.; Lopez-Romero, P.; Albo, C.; Sepulveda, J.C.; Fernandez-Gutierrez, B.; Dopazo, A.; Bernad, A.; Gonzalez, M.A. Mir-335 orchestrates cell proliferation, migration and differentiation in human mesenchymal stem cells. *Cell Death Differ.* **2011**, *18*, 985–995.

59. Bouvet-Gerbettaz, S.; Merigo, E.; Rocca, J.P.; Carle, G.F.; Rochet, N. Effects of low-level laser therapy on proliferation and differentiation of murine bone marrow cells into osteoblasts and osteoclasts. *Lasers Surg. Med.* **2009**, *41*, 291–297.

60. Da Silva, A.P.; Petri, A.D.; Crippa, G.E.; Stuani, A.S.; Stuani, A.S.; Rosa, A.L.; Stuani, M.B. Effect of low-level laser therapy after rapid maxillary expansion on proliferation and differentiation of osteoblastic cells. *Lasers Med. Sci.* **2012**, *27*, 777–783.

61. Ebrahimi, T.; Moslemi, N.; Rokn, A.; Heidari, M.; Nokhbatolfoghahaie, H.; Fekrazad, R. The influence of low-intensity laser therapy on bone healing. *J. Dent.* **2012**, *9*, 238–248.

62. Fujimoto, K.; Kiyosaki, T.; Mitsui, N.; Mayahara, K.; Omasa, S.; Suzuki, N.; Shimizu, N. Low-intensity laser irradiation stimulates mineralization via increased bmps in mc3t3-e1 cells. *Lasers Surg. Med.* **2010**, *42*, 519–526.

63. Hou, J.F.; Zhang, H.; Yuan, X.; Li, J.; Wei, Y.J.; Hu, S.S. *In vitro* effects of low-level laser irradiation for bone marrow mesenchymal stem cells: Proliferation, growth factors secretion and myogenic differentiation. *Lasers Surg. Med.* **2008**, *40*, 726–733.

64. Kim, H.; Choi, K.; Kweon, O.K.; Kim, W.H. Enhanced wound healing effect of canine adipose-derived mesenchymal stem cells with low-level laser therapy in athymic mice. *J. Dermatol. Sci.* **2012**, *68*, 149–156.

65. Lin, F.; Josephs, S.F.; Alexandrescu, D.T.; Ramos, F.; Bogin, V.; Gammill, V.; Dasanu, C.A.; de Necochea-Campion, R.; Patel, A.N.; Carrier, E.; *et al.* Lasers, stem cells, and copd. *J. Transl. Med.* **2010**, *8*, 16.

66. Luo, L.; Sun, Z.; Zhang, L.; Li, X.; Dong, Y.; Liu, T.C. Effects of low-level laser therapy on ros homeostasis and expression of igf-1 and tgf-beta1 in skeletal muscle during the repair process. *Lasers Med. Sci.* **2013**, *28*, 725–734.

67. Medrado, A.P.; Soares, A.P.; Santos, E.T.; Reis, S.R.; Andrade, Z.A. Influence of laser photobiomodulation upon connective tissue remodeling during wound healing. *J. Photochem. Photobiol. B* **2008**, *92*, 144–152.

68. Nogueira, G.T.; Mesquita-Ferrari, R.A.; Souza, N.H.; Artilheiro, P.P.; Albertini, R.; Bussadori, S.K.; Fernandes, K.P. Effect of low-level laser therapy on proliferation, differentiation, and adhesion of steroid-treated osteoblasts. *Lasers Med. Sci.* **2012**, *27*, 1189–1193.

69. Renno, A.C.; McDonnell, P.A.; Parizotto, N.A.; Laakso, E.L. The effects of laser irradiation on osteoblast and osteosarcoma cell proliferation and differentiation *in vitro*. *Photomed. Laser Surg.* **2007**, *25*, 275–280.

70. Rosa, A.P.; de Sousa, L.G.; Regalo, S.C.; Issa, J.P.; Barbosa, A.P.; Pitol, D.L.; de Oliveira, R.H.; de Vasconcelos, P.B.; Dias, F.J.; Chimello, D.T.; *et al.* Effects of the combination of low-level laser irradiation and recombinant human bone morphogenetic protein-2 in bone repair. *Lasers Med. Sci.* **2012**, *27*, 971–977.

71. Saito, K.; Hashimoto, S.; Jung, H.S.; Shimono, M.; Nakagawa, K. Effect of diode laser on proliferation and differentiation of pc12 cells. *Bull. Tokyo Dent. Coll.* **2011**, *52*, 95–102.

72. Soleimani, M.; Abbasnia, E.; Fathi, M.; Sahraei, H.; Fathi, Y.; Kaka, G. The effects of low-level laser irradiation on differentiation and proliferation of human bone marrow mesenchymal stem cells into neurons and osteoblasts—An *in vitro* study. *Lasers Med. Sci.* **2012**, *27*, 423–430.

73. Song, S.; Zhou, F.; Chen, W.R. Low-level laser therapy regulates microglial function through src-mediated signaling pathways: Implications for neurodegenerative diseases. *J. Neuroinflammation* **2012**, *9*, 219.

74. Stein, A.; Benayahu, D.; Maltz, L.; Oron, U. Low-level laser irradiation promotes proliferation and differentiation of human osteoblasts *in vitro*. *Photomed. Laser Surg.* **2005**, *23*, 161–166.

75. Stein, E.; Koehn, J.; Sutter, W.; Wendtlandt, G.; Wanschitz, F.; Thurnher, D.; Baghestanian, M.; Turhani, D. Initial effects of low-level laser therapy on growth and differentiation of human osteoblast-like cells. *Wien. Klin. Wochenschr.* **2008**, *120*, 112–117.

76. Dziunycz, P.; Iotzova-Weiss, G.; Eloranta, J.J.; Lauchli, S.; Hafner, J.; French, L.E.; Hofbauer, G.F. Squamous cell carcinoma of the skin shows a distinct microrna profile modulated by uv radiation. *J. Invest. Dermatol.* **2010**, *130*, 2686–2689.

77. Glorian, V.; Maillot, G.; Poles, S.; Iacovoni, J.S.; Favre, G.; Vagner, S. Hur-dependent loading of mirna risc to the mrna encoding the ras-related small gtpase rhob controls its translation during uv-induced apoptosis. *Cell Death Differ.* **2011**, *18*, 1692–1701.

78. Guo, L.; Huang, Z.X.; Chen, X.W.; Deng, Q.K.; Yan, W.; Zhou, M.J.; Ou, C.S.; Ding, Z.H. Differential expression profiles of micrornas in nih3t3 cells in response to UVB irradiation. *Photochem. Photobiol.* **2009**, *85*, 765–773.

79. Pothof, J.; Verkaik, N.S.; van IJcken, W.; Wiemer, E.A.; Ta, V.T.; van der Horst, G.T.; Jaspers, N.G.; van Gent, D.C.; Hoeijmakers, J.H.; Persengiev, S.P. Microrna-mediated gene silencing modulates the uv-induced DNA-damage response. *EMBO J.* **2009**, *28*, 2090–2099.

80. Tan, G.; Niu, J.; Shi, Y.; Ouyang, H.; Wu, Z.H. Nf-kappab-dependent microrna-125b up-regulation promotes cell survival by targeting p38alpha upon ultraviolet radiation. *J. Biol. Chem.* **2012**, *287*, 33036–33047.

81. Tan, G.; Shi, Y.; Wu, Z.H. Microrna-22 promotes cell survival upon uv radiation by repressing pten. *Biochem. Biophys. Res. Commun.* **2012**, *417*, 546–551.

82. Gu, X.; Lundqvist, E.N.; Coates, P.J.; Thurfjell, N.; Wettersand, E.; Nylander, K. Dysregulation of tap63 mrna and protein levels in psoriasis. *J. Invest. Dermatol.* **2006**, *126*, 137–141.

83. Okuyama, R.; Ogawa, E.; Nagoshi, H.; Yabuki, M.; Kurihara, A.; Terui, T.; Aiba, S.; Obinata, M.; Tagami, H.; Ikawa, S. P53 homologue, p51/p63, maintains the immaturity of keratinocyte stem cells by inhibiting notch1 activity. *Oncogene* **2007**, *26*, 4478–4488.

84. Shen, C.S.; Tsuda, T.; Fushiki, S.; Mizutani, H.; Yamanishi, K. The expression of p63 during epidermal remodeling in psoriasis. *J. Dermatol.* **2005**, *32*, 236–242.

85. Menter, A.; Korman, N.J.; Elmets, C.A.; Feldman, S.R.; Gelfand, J.M.; Gordon, K.B.; Gottlieb, A.; Koo, J.Y.; Lebwohl, M.; Lim, H.W.; *et al.* Guidelines of care for the management of psoriasis and psoriatic arthritis: Section 5. Guidelines of care for the treatment of psoriasis with phototherapy and photochemotherapy. *J. Am. Acad. Dermatol.* **2010**, *62*, 114–135.

86. Ozawa, M.; Ferenczi, K.; Kikuchi, T.; Cardinale, I.; Austin, L.M.; Coven, T.R.; Burack, L.H.; Krueger, J.G. 312-nanometer ultraviolet b light (narrow-band UVB) induces apoptosis of t cells within psoriatic lesions. *J. Exp. Med.* **1999**, *189*, 711–718.

87. Schneider, L.A.; Hinrichs, R.; Scharffetter-Kochanek, K. Phototherapy and photochemotherapy. *Clin. Dermatol.* **2008**, *26*, 464–476.

88. Celli, J.P.; Spring, B.Q.; Rizvi, I.; Evans, C.L.; Samkoe, K.S.; Verma, S.; Pogue, B.W.; Hasan, T. Imaging and photodynamic therapy: Mechanisms, monitoring, and optimization. *Chem. Rev.* **2010**, *110*, 2795–2838.

89. Dolmans, D.E.; Fukumura, D.; Jain, R.K. Photodynamic therapy for cancer. *Nat. Rev. Cancer* **2003**, *3*, 380–387.

90. Verma, S.; Watt, G.M.; Mai, Z.; Hasan, T. Strategies for enhanced photodynamic therapy effects. *Photochem. Photobiol.* **2007**, *83*, 996–1005.

91. Buytaert, E.; Dewaele, M.; Agostinis, P. Molecular effectors of multiple cell death pathways initiated by photodynamic therapy. *Biochim. Biophys. Acta* **2007**, *1776*, 86–107.

92. Brown, S.B.; Brown, E.A.; Walker, I. The present and future role of photodynamic therapy in cancer treatment. *Lancet Oncol.* **2004**, *5*, 497–508.

93. Dougherty, T.J. An update on photodynamic therapy applications. *J. Clin. Laser Med. Surg.* **2002**, *20*, 3–7.

94. Tomioka, Y.; Kushibiki, T.; Awazu, K. Evaluation of oxygen consumption of culture medium and *in vitro* photodynamic effect of talaporfin sodium in lung tumor cells. *Photomed. Laser Surg.* **2010**, *28*, 385–390.

95. Oleinick, N.L.; Morris, R.L.; Belichenko, I. The role of apoptosis in response to photodynamic therapy: What, where, why, and how. *Photochem. Photochem. Photobiol. Sci.* **2002**, *1*, 1–21.

96. Danial, N.N.; Korsmeyer, S.J. Cell death: Critical control points. *Cell* **2004**, *116*, 205–219.

97. Ferri, K.F.; Kroemer, G. Organelle-specific initiation of cell death pathways. *Nat. Cell Biol.* **2001**, *3*, E255–E263.

98. Hengartner, M.O. The biochemistry of apoptosis. *Nature* **2000**, *407*, 770–776.

99. Piette, J.; Volanti, C.; Vantieghem, A.; Matroule, J.Y.; Habraken, Y.; Agostinis, P. Cell death and growth arrest in response to photodynamic therapy with membrane-bound photosensitizers. *Biochem. Pharmacol.* **2003**, *66*, 1651–1659.

100. Helbig, D.; Simon, J.C.; Paasch, U. Photodynamic therapy and the role of heat shock protein 70. *Int. J. Hyperthermia* **2011**, *27*, 802–810.

101. Matroule, J.Y.; Volanti, C.; Piette, J. Nf-kappab in photodynamic therapy: Discrepancies of a master regulator. *Photochem. Photobiol.* **2006**, *82*, 1241–1246.

102. Agostinis, P.; Buytaert, E.; Breyssens, H.; Hendrickx, N. Regulatory pathways in photodynamic therapy induced apoptosis. *Photochem. Photobiol. Sci.* **2004**, *3*, 721–729.

103. Dewaele, M.; Verfaillie, T.; Martinet, W.; Agostinis, P. Death and survival signals in photodynamic therapy. *Methods Mol. Biol.* **2010**, *635*, 7–33.

104. Kessel, D.; Oleinick, N.L. Photodynamic therapy and cell death pathways. *Methods Mol. Biol.* **2010**, *635*, 35–46.

105. Chen, Q.; Huang, Z.; Chen, H.; Shapiro, H.; Beckers, J.; Hetzel, F.W. Improvement of tumor response by manipulation of tumor oxygenation during photodynamic therapy. *Photochem. Photobiol.* **2002**, *76*, 197–203.

106. Henderson, B.W.; Busch, T.M.; Vaughan, L.A.; Frawley, N.P.; Babich, D.; Sosa, T.A.; Zollo, J.D.; Dee, A.S.; Cooper, M.T.; Bellnier, D.A.; *et al.* Photofrin photodynamic therapy can significantly deplete or preserve oxygenation in human basal cell carcinomas during treatment, depending on fluence rate. *Cancer Res.* **2000**, *60*, 525–529.

107. Sitnik, T.M.; Hampton, J.A.; Henderson, B.W. Reduction of tumour oxygenation during and after photodynamic therapy *in vivo*: Effects of fluence rate. *Br. J. Cancer* **1998**, *77*, 1386–1394.

108. Engbrecht, B.W.; Menon, C.; Kachur, A.V.; Hahn, S.M.; Fraker, D.L. Photofrin-mediated photodynamic therapy induces vascular occlusion and apoptosis in a human sarcoma xenograft model. *Cancer Res.* **1999**, *59*, 4334–4342.

109. Fingar, V.H.; Kik, P.K.; Haydon, P.S.; Cerrito, P.B.; Tseng, M.; Abang, E.; Wieman, T.J. Analysis of acute vascular damage after photodynamic therapy using benzoporphyrin derivative (bpd). *Br. J. Cancer* **1999**, *79*, 1702–1708.

110. Keith, B.; Johnson, R.S.; Simon, M.C. Hif1alpha and hif2alpha: Sibling rivalry in hypoxic tumour growth and progression. *Nat. Rev. Cancer* **2012**, *12*, 9–22.

111. Semenza, G.L. Hypoxia-inducible factors in physiology and medicine. *Cell* **2012**, *148*, 399–408.

112. Forsythe, J.A.; Jiang, B.H.; Iyer, N.V.; Agani, F.; Leung, S.W.; Koos, R.D.; Semenza, G.L. Activation of vascular endothelial growth factor gene transcription by hypoxia-inducible factor 1. *Mol. Cell. Biol.* **1996**, *16*, 4604–4613.

113. Takenaga, K. Angiogenic signaling aberrantly induced by tumor hypoxia. *Front. Biosci.* **2011**, *16*, 31–48.

114. Deininger, M.H.; Weinschenk, T.; Morgalla, M.H.; Meyermann, R.; Schluesener, H.J. Release of regulators of angiogenesis following hypocrellin-a and -b photodynamic therapy of human brain tumor cells. *Biochem. Biophys. Res. Commun.* **2002**, *298*, 520–530.

115. Ferrario, A.; von Tiehl, K.F.; Rucker, N.; Schwarz, M.A.; Gill, P.S.; Gomer, C.J. Antiangiogenic treatment enhances photodynamic therapy responsiveness in a mouse mammary carcinoma. *Cancer Res.* **2000**, *60*, 4066–4069.

116. Jiang, F.; Zhang, Z.G.; Katakowski, M.; Robin, A.M.; Faber, M.; Zhang, F.; Chopp, M. Angiogenesis induced by photodynamic therapy in normal rat brains. *Photochem. Photobiol.* **2004**, *79*, 494–498.

117. Schmidt-Erfurth, U.; Schlotzer-Schrehard, U.; Cursiefen, C.; Michels, S.; Beckendorf, A.; Naumann, G.O. Influence of photodynamic therapy on expression of vascular endothelial growth factor (vegf), vegf receptor 3, and pigment epithelium-derived factor. *Invest. Ophthalmol. Vis. Sci.* **2003**, *44*, 4473–4480.

118. Lu, J.; Getz, G.; Miska, E.A.; Alvarez-Saavedra, E.; Lamb, J.; Peck, D.; Sweet-Cordero, A.; Ebert, B.L.; Mak, R.H.; Ferrando, A.A.; *et al.* Microrna expression profiles classify human cancers. *Nature* **2005**, *435*, 834–838.

119. Cheng, A.M.; Byrom, M.W.; Shelton, J.; Ford, L.P. Antisense inhibition of human mirnas and indications for an involvement of mirna in cell growth and apoptosis. *Nucleic Acids Res.* **2005**, *33*, 1290–1297.

120. Kushibiki, T. Photodynamic therapy induces microrna-210 and -296 expression in HeLa cells. *J. Biophotonics* **2010**, *3*, 368–372.

121. Giannakakis, A.; Sandaltzopoulos, R.; Greshock, J.; Liang, S.; Huang, J.; Hasegawa, K.; Li, C.; O'Brien-Jenkins, A.; Katsaros, D.; Weber, B.L.; *et al.* Mir-210 links hypoxia with cell cycle regulation and is deleted in human epithelial ovarian cancer. *Cancer Biol. Ther.* **2008**, *7*, 255–264.

122. Wurdinger, T.; Tannous, B.A.; Saydam, O.; Skog, J.; Grau, S.; Soutschek, J.; Weissleder, R.; Breakefield, X.O.; Krichevsky, A.M. Mir-296 regulates growth factor receptor overexpression in angiogenic endothelial cells. *Cancer Cell* **2008**, *14*, 382–393.

123. Bach, D.; Fuereder, J.; Karbiener, M.; Scheideler, M.; Ress, A.L.; Neureiter, D.; Kemmerling, R.; Dietze, O.; Wiederstein, M.; Berr, F.; *et al.* Comprehensive analysis of alterations in the mirnome in response to photodynamic treatment. *J. Photochem. Photobiol. B.* **2013**, *120*, 74–81.

124. Sato, M.; Kubota, N.; Inada, E.; Saitoh, I.; Ohtsuka, M.; Nakamura, S.; Sakurai, T.; Watanabe, S. Hela cells consist of two cell types, as evidenced by cytochemical staining for alkaline phosphatase activity: A possible model for cancer stem cell study. *Adv. Stem. Cell* **2013**, doi:10.5171/2013.208514.

3

Identification of MicroRNA 395a in 24-Epibrassinolide-Regulated Root Growth of *Arabidopsis thaliana* using MicroRNA Arrays

Li-Ling Lin [1], Chia-Chi Wu [2], Hsuan-Cheng Huang [3,*], Huai-Ju Chen [4], Hsu-Liang Hsieh [4,*] and Hsueh-Fen Juan [1,2,5,*]

[1] Department of Life Science, National Taiwan University, Taipei 106, Taiwan;
E-Mail: f94b43019@ntu.edu.tw

[2] Institute of Molecular and Cellular Biology, National Taiwan University, Taipei 106, Taiwan;
E-Mail: jcwu0417@hotmail.com

[3] Institute of Biomedical Informatics, Center for Systems and Synthetic Biology,
National Yang-Ming University, Taipei 112, Taiwan

[4] Institute of Plant Biology, National Taiwan University, Taipei 106, Taiwan;
E-Mail: d92621103@ntu.edu.tw

[5] Graduate Institute of Biomedical Electronic and Bioinformatics, National Taiwan University,
Taipei 106, Taiwan

* Authors to whom correspondence should be addressed; E-Mails: hsuancheng@ym.edu.tw (H.-C.H.);
hlhsieh@ntu.edu.tw (H.-L.H.); yukijuan@ntu.edu.tw (H.-F.J.)

Abstract: Brassinosteroids (BRs) are endogenous plant hormones and are essential for normal plant growth and development. MicroRNAs (miRNAs) of *Arabidopsis thaliana* are involved in mediating cell proliferation in leaves, stress tolerance, and root development. The specifics of BR mechanisms involving miRNAs are unknown. Using customized miRNA array analysis, we identified miRNAs from *A. thaliana* ecotype Columbia (Col-0) regulated by 24-epibrassinolide (EBR, a highly active BR). We found that miR395a was significantly up-regulated by EBR treatment and validated its expression under these conditions. miR395a was over expressed in leaf veins and root tissues in EBR-treated miR395a promoter::GUS plants. We integrated bioinformatics methods and publicly available DNA microarray data to predict potential targets of miR395a. GUN5—a multifunctional protein involved in plant metabolic functions such as chlorophyll synthesis and the abscisic acid

(ABA) pathway—was identified as a possible target. ABI4 and ABI5, both genes positively regulated by ABA, were down-regulated by EBR treatment. In summary, our results suggest that EBR regulates seedling development and root growth of *A. thaliana* through miR395a by suppressing GUN5 expression and its downstream signal transduction.

Keywords: brassinosteroids; miR395a; root growth; *Arabidopsis thaliana*; microRNA array

1. Introduction

In 1970, a new family of plant hormones, brassins, was reported but later found to be a mixture of multiple compounds [1]. In 1979, another steroid hormone named brassinolide was identified from rape pollen of *Brassica napus* and its structure determined [2,3]. A number of related steroid hormones have since been isolated and collectively classified under the general term brassinosteroids (BRs). To date, more than 50 BR forms including 24-epibrassinolide (EBR) have been identified in a wide variety of plant species [4]. In an attempt to understand how BRs act on plant growth and in what mechanisms they are involved, numerous studies have been and are being conducted in wide-ranging fields, including structural biology, plant physiology, molecular biology, and genetics [5]. Concurrent with biosynthetic research, a large number of BR-deficient or -insensitive mutants have been investigated, among them the *bri*1 mutant, which enabled the exploration of the affected gene's role in BR receptor expression [6]. The components of the BRs signal transduction pathway have subsequently been studied in an effort to discriminate the relevant mechanisms [7].

BRs are structurally similar to animal and insect steroid hormones [8] and are products of the isoprenoid biosynthetic pathway; however, they differ in their subsequent metabolism of squalene-2,3-epoxide. While in animals the compound is converted to the precursor of cholesterol and steroid hormones, lanosterol, it is metabolized to cycloartenol in plants, which is the parent compound of all plant sterols [3,5,7].

In 2001, a study comparing *A. thaliana* ecotype Columbia (Col-0) with BR-deficient mutants demonstrated that BR stimulates seed germination by reversing ABA-induced dormancy [9]. A recent study showed that 2 µM of exogenous BR reduced the inhibitory effect of high salt concentrations on seed germination and promoted early stages of seedling growth in *Brassica napus* [10]. Another study indicated that overexpression of the gene *AtDWF4*, essential for BR biosynthesis, was able to overcome ABA-induced inhibition of seed germination [11]. It has been proposed that exogenous BR can regulate other endogenous hormones, and the effect of BR on other plant hormones has been explored in several studies [12].

Plant miRNAs occupy only a small number of functional genes. Currently, 299 *A. thaliana* miRNAs are recorded in miRBASE (release 19) [13]. miRNAs bind to complementary sequences on target mRNAs and in plants mostly act to degrade them [14]. Complementary features of plant miRNAs target their mRNAs by an almost perfect match; most miRNA binding sites exist in coding exons [14,15]. Recently, miRNAs have been reported to be hypersensitive ubiquitous stress regulators: *i.e.*, they function to mediate expression of their target genes when unbalanced nutrient conditions are encountered [15]. miR399 and miR395 have been identified as being involved in sulfate- and phosphate-starvation responses [16,17].

While recent studies with a structural, genetic, molecular, transcriptomic and proteomic focus have helped elucidate the regulatory mechanisms of the BRs signaling pathway [18–21], the mechanisms of miRNA involvement with BRs are unknown. To gain insights into the mechanism of BR actions at the molecular level, we carried out global screening of miRNAs in *A. thaliana*, which responds rapidly to EBR treatment, successfully investigating potential targets of miRNA and their interaction in plant development.

2. Results and Discussion

2.1. EBR Regulates the Root Development of Arabidopsis

We monitored the morphology of EBR-treated seedlings for root length measurement and germination analysis. It has previously been shown that *Arabidopsis* with different levels of BRs display differences in root development [22,23]. Low concentrations (0.1 and 0.5 nM) of exogenous BRs promoted root elongation in wild-type strains and BR-deficient mutants [22]. In contrast, higher concentrations (1–100 nM) were inhibitory for primary root elongation, instead promoting lateral root formation [23,24]. In this study, we treated *Arabidopsis* with 10 nM EBR. Our results show that primary root length was significantly decreased ($p < 0.01$) and the number of lateral roots was significantly increased ($p < 0.05$) (Figure 1). Expression levels of *BRU6* and *SAURAC-1* genes in *Arabidopsis* were regulated by EBR stimulation (Figure S1), results that corroborate previous studies [21,24]. The development of germination was maintained after EBR treatment, but the root phenotype appeared obviously curved in our germination analysis (Figure 2). Based on these results, we confirm that the root development of *Arabidopsis* can be regulated by EBR at concentrations like those used in our treatments.

Figure 1. 24-epibrassinolide (EBR) regulates root development. (**A**) The development of lateral roots was enhanced in EBR-treated seedlings. Each plate contained 10 nM EBR or mock solution (control). The red line represents initial length before treatment; (**B**) Differences in primary root length between day 5 and day 11. Primary root length was significantly shorter under EBR treatment; and (**C**) Number of lateral roots on day 11. Lateral root number was significantly increased in the EBR supplement plate. Representative data from three independent experiments are presented as mean ± SD. * $p < 0.05$, ** $p < 0.01$ *vs.* control treatment values.

A

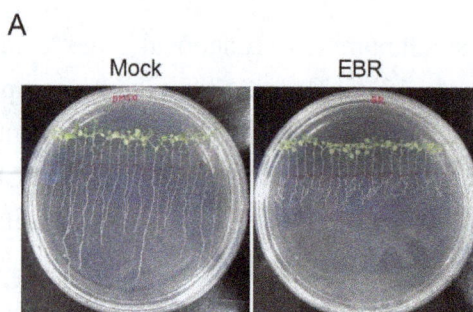

Figure 1. *Cont.*

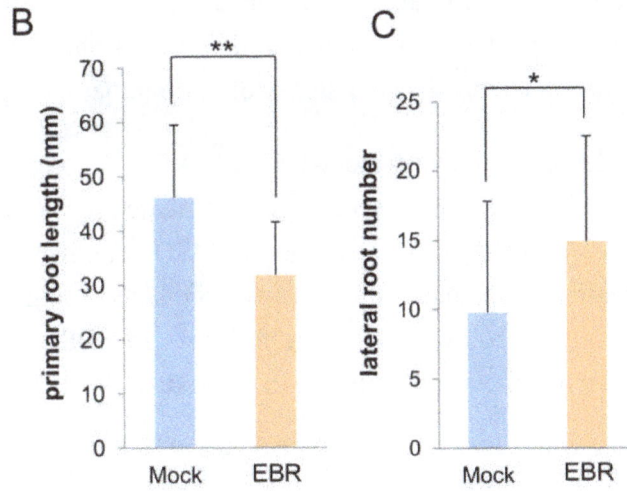

Figure 2. EBR has no effect on germination. There was no significant difference in germination rate between control and EBR-treated seeds; however roots were shorter and more strongly curved in EBR-supplemented plates. Germination was recorded on day 3 and day 13 after imbibition.

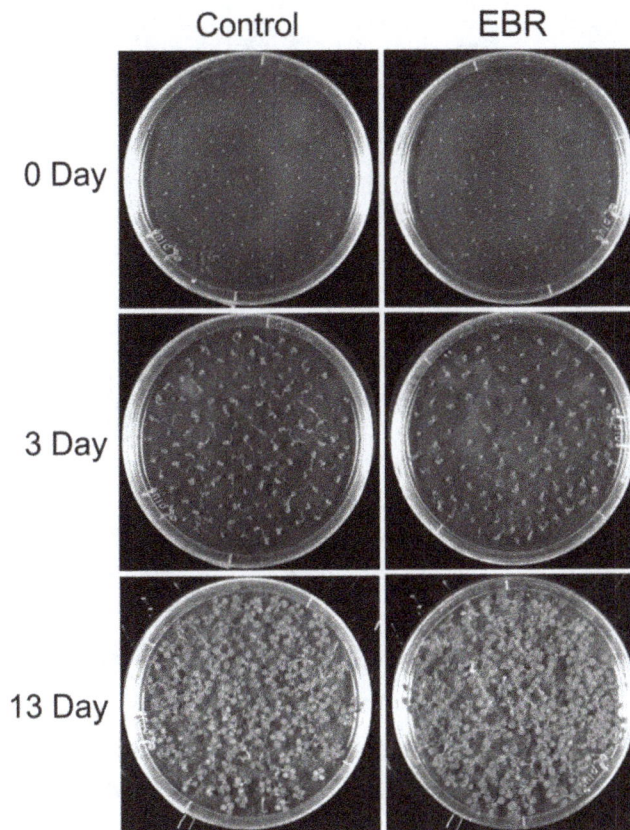

2.2. Identification of EBR-Regulated miRNAs in Arabidopsis

To explore the role of miRNAs in BR-mediated pathways, we analyzed differences in miRNA profiles between control (mock solution) and EBR treatments from customized miRNA microarrays.

Seeds were separately cultured under exogenous 10 nM EBR treatments for 30 (EBR30) or 180 (EBR180) minutes (Figure 3), and total RNA of all seedlings was extracted after seven days of growth. A scatter plot of probe intensities of duplicate microarrays shows no difference ($R^2 > 0.99$) among the duplicates, therefore validating the consistency of our microarray experiments (Figure S2).

Figure 3. Schematic flowchart of experimental design. After *Arabidopsis* Col-0 seeds had grown in MS liquid medium for 7 days, seedlings were treated for 30 or 180 min with MS medium supplemented with EBR or mock solution (DMSO), followed by RNA extraction, labeling, and hybridization. Candidate miRNAs were predicted using miRU, WMD3, and psRNATarget databases. The roles of candidate miRNAs in EBR-treated seedlings were investigated by further experiments as described in the text.

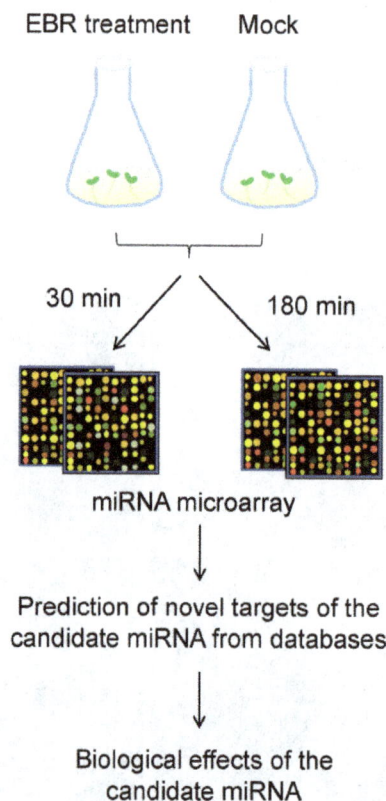

The expressed fold changes of miRNAs from EBR-treated seedlings were normalized to a DMSO-treated control. Fourteen miRNAs with significantly different expression ratios ($p < 0.05$) (Table 1) from both EBR30 and EBR180 treatments were selected for hierarchical clustering (Figure 4A). Among these, 11 miRNAs were up-regulated and three down-regulated for EBR30, and six up-regulated and eight down-regulated for EBR180. Of these, miR395a exhibited the highest fold change (1.6-fold) from microarray data at EBR180. Similarly, EBR-treated seedlings showed a higher expression of miR395a (4.3-fold) than control seedlings in real-time PCR (qPCR) analysis (Figure 4B). These results validate that miR395a expression is up-regulated by EBR.

Table 1. EBR regulates the expression of several microRNAs (miRNAs).

microRNAs	Fold change (EBR/DMSO)	
	30 min	180 min
ath-miR824	1.27	0.93
ath-miR169h	1.37	0.86
ath-miR173	1.15	0.95
ath-miR158a	1.16	0.92
ath-miR157d	1.18	0.87
ath-miR160a	1.25	1.06
ath-miR156h	1.25	1.04
ath-miR159a	1.07	1.05
ath-miR169a	1.26	1.04
ath-miR400	1.21	0.89
ath-miR161.2	1.31	0.71
ath-miR854a	0.98	0.78
ath-miR395a	0.97	1.60
ath-miR397a	0.53	1.25

Figure 4. EBR up-regulates miR395a in miRNA microarray analysis. (**A**) Hierarchical clustering of selected miRNAs expression regulated by EBR. Seedlings were treated with EBR for 30 or 180 min. miRNA expression was assessed with miRNA microarrays. Fourteen miRNAs had significantly different expression levels after EBR treatment ($p < 0.05$) and were further analyzed by a hierarchical clustering algorithm; and (**B**) Fold changes of miR395a in miRNA microarrays and qPCR analysis. miR395a was up-regulated after EBR treatment for 180 min. snoR85 was used as an internal control for normalization.

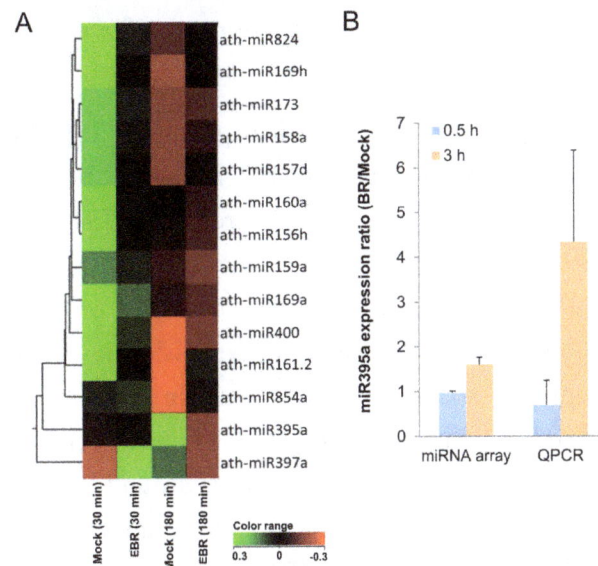

2.3. GUN5 Is a Novel Target of miR395a

To investigate the role of miR395a in *Arabidopsis* development, we explored its potential targets by the bioinformatics approach of complementary base-pairing. We obtained potential target genes of miR395a from three databases: miRU [25] (Table S1), which integrates most known plant miRNAs and target genes and can be employed to search potential plant miRNA targets and target-sites within mismatch and miRNA conservation thresholds in target recognition; WMD3 [26] (Table S2), which uses principles of artificial miRNA design to mimic natural plant miRNAs; and psRNATarget [27] (Table S3), which provides scoring schemata and evaluates target-site accessibility in miRNA target recognition. The *Arabidopsis* TAIR9 cDNA library and default parameters were used in predicting target sequences from these databases. Among the candidate targets, *GUN5* (At5g13630), the CHLH subunit of Mg-chelatase, has been reported as possibly involved in the abscisic acid (ABA) pathway [28,29]. The interaction between GUN5 and miR395a is however still unknown. ABA, a plant hormone, mediates the development of plants by inhibiting seedling germination, maintaining primary root growth and reducing lateral root density [30–32]. We therefore decided to focus our investigation on whether *GUN5* is a target gene of miR395a; the target sequence of *GUN5* is shown in Figure 5A. We furthermore analyzed *GUN5* expressions in EBR- and DMSO-treated seedlings by qPCR. As shown in Figure 5B, these gene expressions were decreased in EBR-treated seedlings compared with DMSO-treated seedlings, indicating that *GUN5* might be associated with miR395a.

To further validate the interaction between miR395a and *GUN5* in *Arabidopsis*, we constructed pRTL2-miR395a and smGFP/pRTL2-*GUN5*. We co-transformed the plasmids into PSB-D cells and detected the fluorescence intensity of smGFP/pRTL2-*GUN5* and the internal control RFP. We also analyzed the levels of GFP-*GUN5* expression in transformed PSB-D cells. As shown in Figure 5C, lower *GUN5* expression was detected in PSB-D cells with miR395a overexpression than in PSB-D cells with control vectors. In plants with miR395a knockout, the gene expression of *GUN5* was also significantly increased (Figure 5D). These results suggest that *GUN5* is a target gene of miR395a.

2.4. Distribution of miR395a in Vascular Bundles, Leaf Veins and Roots of Arabidopsis

GUN5 has been reported as being instrumental in leaf greening [28], and a decrease in chlorophyll accumulation has been found in *gun5* mutants [33]. BR is also a crucial factor in the regulation of chloroplast development, playing a role as a negative regulator [34]. Based on these similarities, we explored the role of miR395a in EBR-treated *Arabidopsis*, following *GUN5* suppression. To clarify the expression sites of miR395a, we used T2 seeds from miR395a promoter::GUS plants to examine the miR395a expression pattern under EBR treatment (expression levels of miR395a promoter::GUS were also found to be up-regulated under EBR treatment). In leaf and root development, miR395a specifically was concentrated in leaf veins of the cotyledon (Figure 6A) and in partial vascular bundles of roots (Figure 6B) and was also distributed in chloroplasts around leaf veins (Figure S3). Figure 6B also shows that root diameter under EBR treatment was larger under mock treatment, indicating that miR395a might regulate root development through EBR signaling.

Figure 5. GUN5 is a target gene of miR395a. (**A**) The prediction was based on complementary base-pairing between miR395a and mRNA. Putative target genes for miR395a were predicted using web-based databases (miRU, psRNATarget, WMD3) and simultaneous comparisons to gene microarray data. The putative target genes of miR395a were down-regulated by EBR treatment; (**B**) Expression levels of GUN5 were analyzed by Q-PCR and normalized to 18S rRNA; (**C**) Fluorescence assay of miR395a and potential target GUN5 in the *Arabidopsis* PSB-D cell line. The fluorescent expression levels of GUN5 revealed significant down-regulation by miR395a; and (**D**) Gene expression of GUN5 measured by qPCR in miR395a knockout plants. GUN5 expression was significantly down-regulated by miR395a. * $p < 0.05$.

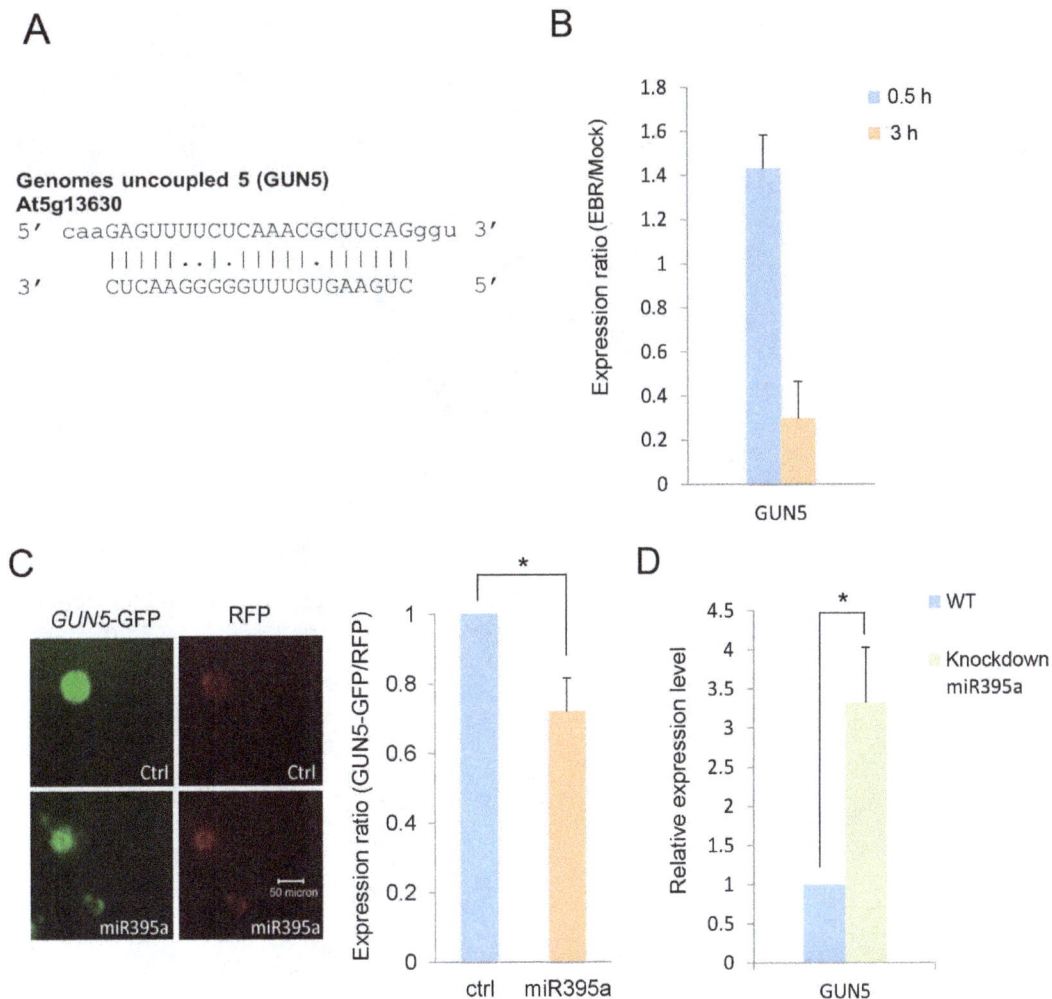

A previous study has shown that the transcription factors *ABI4* and *ABI5* are positive regulators of ABA signaling and can be considered downstream genes of *GUN5* [19]. We found that these genes could be suppressed by EBR treatment (Figure 7A). It has also been reported that a mutation of *ABI4* can increase the number of lateral roots [35] and that *ABI5* activity inhibits seedling germination and promotes primary root growth [36,37]. These results lead us to propose that EBR may maintain seedling germination, inhibit primary root growth, and increase the number of lateral roots through regulation of miR395a effects on *ABI4* and *ABI5* via *GUN5* (Figure 7B).

Figure 6. Histochemical GUS staining of miR395a expression in *A. thaliana*. Expression patterns of miR395a promoter::GUS plants in (**A**) leaf and (**B**) root tissue. After growing for 7 days, seedlings were grown under EBR treatment (10 nM EBR) or mock control for 3 h. The arrow indicates a high concentration of miR395a distributed in the vascular bundle compared with the mock treatment.

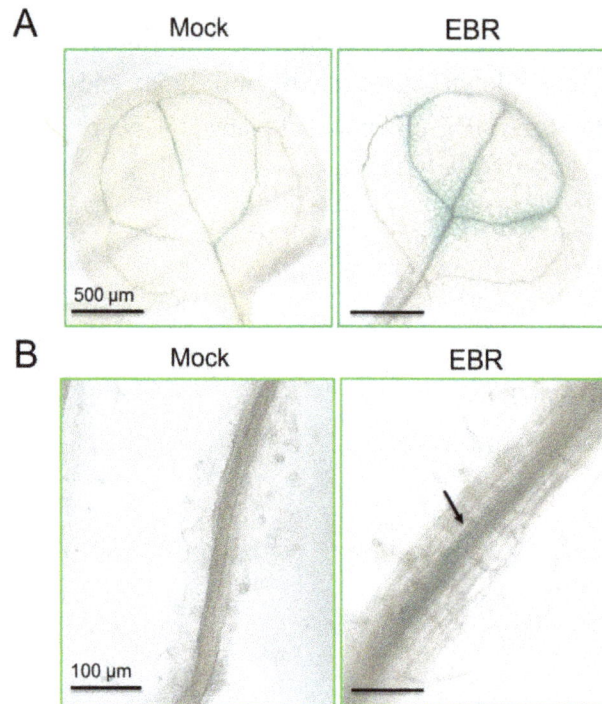

Figure 7. Expression of GUN5 downstream genes in the ABA pathway. (**A**) Relative expression ratio of GUN5 downstream genes; and (**B**) Diagram of GUN5-dependent ABA pathway.

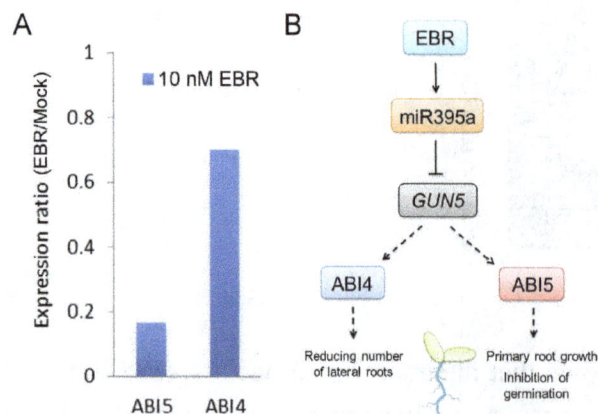

2.5. Discussion

BRs can induce a wide range of physiological effects in cell elongation and division, photosynthesis, photomorphogenesis, flowering, senescence, seed germination, root development, male fertility, and abiotic and biotic stress resistance [8,19,38,39]. They are active at low concentrations throughout the plant kingdom and widely distributed in plants at varying levels of complexity [7,8]. Higher concentrations of BRs are seen primarily in young growing tissues rather than in mature tissues [7]. BR-insensitive mutants in *Arabidopsis* exhibit phenotypes such as dwarfism, dark-green leaves, reduced fertility, prolonged life span, and abnormal skotomorphogenesis [3,8,39].

miRNAs also provide examples of regulation at various stages of plant development. Some miRNAs, such as miR159 and miR160, play roles during early development stages including seed germination. During post-germination stages, miR156 and miR172 mediate the emergence of vegetative leaves, a stage of transition to autotrophic growth [40,41]. miRNA-mediated signaling is also involved in the development of various tissues; several miRNA families such as miR160, miR164, miR167, and miR390 have been demonstrated to be involved in root cap formation and lateral root development [42]. However, the relationship between BRs and miRNAs is unknown.

In the present study, we screened different miRNA expression profiles in *Arabidopsis* with 10 nM EBR for periods of 30 and 180 min. The results show that in both cases the expression of miR395a was significantly up-regulated by EBR (Figure 4). Recent studies have indicated that miR395a is up-regulated in roots and expressed in cortex, phloem companion cells and epidermis under low-sulfur conditions [43]. miR395a is mostly expressed in roots when playing a role in homeostasis regulation [43]. As discussed above, the morphology of BR-treated plants showed a decrease in taproot length and an increase in lateral root formation [23]. These effects might be caused by miR395a-involved mechanisms, and miR395a might be among the factors affecting root growth and development. miR395 and miR397 play roles in sulfate metabolism and copper homeostasis, respectively [43–45]. The function of these miRNAs lies mostly in adapting to unbalanced conditions, which implies that the experimental concentration of 10 nM EBR might have been in excess of physiological levels and affected the homeostasis of the seedlings.

In addition to miR395a, several significantly different expressions of miRNAs may have potential functions relevant to BR-treated seedlings:

(a) miR824 was down-regulated in BR-treated seedlings. It is involved in stomatal development by targeting *AGL16*, through which it causes a decrease in the number of stomata [46]. This suggests an increased stomata number in BR-treated plants [46,47]. Proper amounts and distributions of stomata are essential for successful gas exchange [46], and so an increase in the stomata number might therefore contribute to greater metabolic efficiency in plants.

(b) miR169a, which can regulate adaptive responses to nutrient deprivation [48], was also up-regulated in our miRNA profiles. This suggests that miR169a might have acted in this capacity of adaptation to environmental change when we supplied exogenous BR.

(c) miR160 mediates agravitropic roots with disorganized root caps as well as lateral root development, primary root growth, floral organs in carpels, and germination [40,42,49]. Our miRNA arrays indicated that the up-regulation of miR160a might have resulted in the expression of the phenotype observed in the present study. Since the lateral root formation caused by miR160 was similar to the morphology of BR-treated seedlings, we suspect miR160 might play an important role in lateral root development in BR-supplied plants.

(d) miR156 has been shown in recent studies to increase leaf initiation, phase change, floral induction, and phosphate homeostasis, to decrease apical dominance, and to delay flowering time [40,42,49]. As suspected, miR156h was up-regulated in the miRNA profiles, suggesting a crucial function in promoting growth and development.

(e) miR159 regulates germination, anthers, and flowering time by targeting the MYB transcription factor [49,50]. Overexpression of miR159 results in male sterility and delayed flowering time.

To further explore the role of miR395a, we predicted target genes of miR395a from several different databases and identified *GUN5* as a novel potential target of miR395a in *Arabidopsis*. We were able to show that the expression of *GUN5* was suppressible by miR395a (Figure 5C,D). Similar to the phenotype of the *gun4* mutant, the *gun5* mutant showed a decrease in chlorophyll accumulation, while the *gun4gun5* double mutant displayed the even more noticeable characteristic of albino leaves [28]. *GUN2/3/4/5* are also involved in communicating along plastid-to-nucleus retrograde signaling pathways with Mg-ProtoIX acting as a signaling molecule between chloroplast and nucleus [51]. In contrast, BR inhibits chloroplast development [34], and down-regulates *GUN5* expression (Figure 5B). After EBR treatment, miR395a was up-regulated and strongly expressed in cotyledon leaf veins and root vascular bundles (Figures 4B and 6). These results suggest that BR might enhance miR395a to suppress *GUN5* expression during plant development. However, the exact relationship between BR, miR395a and *GUN5* remains unknown. Recent studies have indicated that *ABI4* is a downstream regulator between chloroplast and nucleus that connects to ABA via retrograde signaling [52]. Hence, *GUN5* is likely to play a role in chlorophyll synthesis by connecting ABA to different pathways [33]. Additionally, we found that ABA regulatory genes were suppressed by EBR (Figure 7). These outcomes indicate that the interaction between miR395a and *GUN5* may regulate chlorophyll synthesis through the ABA signaling pathway.

3. Experimental Section

3.1. Plant Material and Growth Conditions

A. thaliana ecotype Columbia (Col-0) was used as plant material in this study. Before sowing, seeds were surface sterilized by rinsing them in 1% bleach (sodium hypochlorite) with 0.5% Tween 20 and vortexing for eight minutes, washed 5–6 times and then cold-treated for two days at 4 °C under dark conditions. Plants were sown in pots (containing 50% vermiculite and 50% soil mixture), medium, or agar plates and kept in a growth chamber operating at photoperiod conditions of 14 h light and 10 h darkness at 22 °C after stratification.

3.2. Germination Assay

Seeds were grown on half-strength Murashige and Skoog medium (1/2 MS medium; Duchefa Biochemie B.V., Haarlem, Netherlands) with 1.5% (*w/v*) sucrose (Sigma-Aldrich Co. LLC., Dorset, UK) and 0.8% (*w/v*) plant agar (Sigma-Aldrich, St. Louis, MO, USA) containing 10 nM 24-epibrassinolide (EBR, a highly active BR; Sigma-Aldrich) or mock solution (dimethyl sulfoxide, DMSO). Images were taken at zero, three, and thirteen days after sowing.

3.3. Root Growth Assay

For root elongation analysis, seedlings were grown vertically on 1/2 MS medium with 1.5% sucrose and 0.8% plant agar for five days after germination. Seedlings were then transferred to new plates containing MS medium supplemented with 10 nM EBR or mock solution for another six days, with images taken after five and 11 days. Differences in primary root length between the two images were measured, and number of lateral roots was calculated after 11 days, using ImageJ software [53].

3.4. MicroRNA Microarray Hybridization and Analysis

For the miRNA array experiments, seedlings were grown in 1/2 MS medium with 1.5% sucrose. After stratification, seeds were transferred into 50 mL flasks with 10 mL liquid medium and incubated for seven days at 50 rpm and 22 °C under continuous light conditions. Seedlings were then treated with medium containing 10 nM EBR or mock solution for 0.5 and 3 h, respectively. Total RNAs were extracted from complete frozen seedlings using TRIzol® Reagent (Invitrogen, Carlsbad, CA, USA), and RNA purity was confirmed by spectrophotometry (A_{260}/A_{280} ratio) and capillary electrophoresis (Agilent 2100 Bioanalyzer, Agilent Technologies, Palo Alto, CA, USA). Then, 100 ng total RNAs of each sample were prepared for labeling with Cyanine 3-pCp. RNA processing and hybridization were performed using miRBASE V14 arrays (Agilent Technologies, Palo Alto, CA, USA) according to the manufacturer's protocol; this version contains 161 *Arabidopsis thaliana* miRNA genes. Each plex on these customized eight-plex microarrays contained duplicate or triplicate probes for each miRNA, with 20 replicates for each probe. Microarray analysis was carried out in GeneSpring GX version 11 (Agilent, city, state, country). The data (covering the four conditions) were classified into groups by the averages of duplicates, and the median of all samples was set as a baseline. Differences in miRNA expression were tested using a one-way ANOVA. miRNAs with significant differences ($p < 0.05$) between mock control and EBR-treated seedlings were selected for clustering and those with the highest fold change were subjected to further analysis. Array data were submitted to the GEO database (series record number GSE46377).

3.5. Real-Time RT-PCR

All cDNA synthesis was carried out on total RNAs using the RevertAid H Minus Reverse Transcriptase Kit (Fermentas, Maryland, NY, USA) according to the manufacturer's instructions. Reactions for expression analysis of Col-0 genes treated with EBR or mock control were performed in triplicate and monitored using the iQ5 Real-time PCR Detection System (Bio-Rad, Philadelphia, PA, USA). Investigated genes and corresponding primers are listed in Table S5. Relative abundance of transcripts was normalized to the constitutive expression levels of 18S rRNA (At3g41768). For miRNA expression analysis, specific miRNAs were measured with TaqMan microRNA assays (Applied Biosystems, Foster City, CA, USA) according to the manufacturer's instructions. All reactions were run in triplicate and snoR85 was used as the internal control for normalization.

3.6. Prediction of Novel miRNA Target Genes

We obtained *Arabidopsis* miRNA sequences from the miRBASE dataset [13]. The mature sequences of all miRNA genes were used in this study. For predictions of miRNA targets, the programs miRU [25], psRNATarget [27] and WMD3 [26] were employed. The *Arabidopsis thaliana* full genome (TAIR9) was selected in the psRNATarget and WMD3 databases (other parameters were left at default).

3.7. Vector Construction

The genomic DNA of *Arabidopsis* leaves was extracted using QuickExtract™ Plant DNA Extraction Solution (Epicentre, Madison, WI, USA), following the manufacturer's instructions. Different vectors were used for specific purposes. For validating the interaction of miRNA and genes, vector pRTL2-mGFP (Biovector Co., LTD, Beijing, China) was used to construct miR395a and control. Vector pRTL2 was used to delete the mGFP gene from pRTL2-mGFP via the restriction enzymes *EcoR* I and *Xba* I of miR395a. Vector 326-RFP is an internal control for cell numbers in *Arabidopsis* cell lines. For cloning *GUN5* in translational fusion, the coding region of *GUN5* was inserted into smGFP/pRTL2 using the restriction site *Spe I*. Transcriptional fusions were created with GUN5 and smGFP for activity analysis in protoplast system. Protoplasts were prepared following the protocol of Miao and Jiang [54]. Vector pZP221 was used in transgenic plant construction for the miR395a-overexpressing line; the insertion containing miR395a with CaMV 35S promoter and terminator in the pRTL2-miR395a vector was cloned into the *Pst I* site of pZP221. The binary vector pBI101 with the reporter gene β-glucuronidase (GUS) was used for promoter activity analysis of the miR395a promoter line. Transcriptional fusions for analysis of promoter activity in plants were generated using the miR395a promoter with built-in cloning sites *Sal I* and *Xba I*, and GUS.

The miR395a knockout line was purchased from the *Arabidopsis* Information Resource (TAIR). Transformations were performed with the Gene Pulser Xcell™ Electroporation System (Bio-Rad, Richmond, CA, USA) at pulse settings of 130 V at 1000 μF.

3.8. Fluorescence Assay for Validating miR395a and GUN5

For fluorescence assays, 200 μL of transformed protoplast cells were transferred to black opaque 96-well microplates (Greiner Bio-One, Wemmel, Belgium) and immediately measured in a multimode microplate reader (FlexStation 3 microplate reader; MDS Analytical Technologies, Sunnyvale, CA, USA). Excitation and emission wavelengths were 488 and 508 nm for green light (smGFP) and 558 and 583 nm for red light (DsRed).

3.9. Detection of the Expression Pattern of miR395a in Arabidopsis thaliana

When *Arabidopsis* plants had grown for four to six weeks, the first bolt was cut to induce the emergence of further bolts. About one week after clipping, plants containing numerous unopened floral buds were immersed in a buffer of *Agrobacterium tumefaciens*. The buffer was prepared as follows: Transformed *A. tumefaciens* cells were grown at 28 °C and 180 rpm shaking in LB medium with the appropriate antibiotics. A 10-mL pre-culture was grown for two days and then transferred to the 200-mL main culture. This was incubated until an OD600 value of 0.8 was reached, and then was centrifuged at 4000× *g* for 15 min at 4 °C. The supernatant was discarded and sucrose and Silwet L-77 (Sigma, St. Louis, MO, USA) were added to the culture to obtain final concentrations of 5% and 0.05%, respectively.

Pots of plants were inverted and the inflorescence shoots dipped into suspension, then laid on a flat plastic surface and left covered and dark for the next 24 h, and afterwards returned to normal growing conditions. T1 plants were grown from selected transformants. The transgenic character of plants was confirmed by PCR and GUS staining.

When T2 trangenic lines were obtained, we used a GUS staining kit (GUSS; Sigma) to detect the expression pattern of miR395a. Seedlings were incubated at room temperature for 45 min with a fixation solution, which was then poured off. They were washed three times with wash solution for one minute, then left to incubate with staining solution for up to 24 h at 37 °C. Finally, the chlorophyll was removed by distaining the samples with ethanol. Tissues were stored in ethanol. Manufacturer's instructions were followed in performing the assay.

3.10. Statistical Analysis

Data were represented as mean ± standard deviation (SD). Differences between independent groups were analyzed using a two-tailed Student's *t*-test. MicroRNA microarrays for miRNA expression were analyzed using a one-way ANOVA (GeneSpring 7.3.1, Agilent Technologies, Palo Alto, CA, USA). A *p* value < 0.05 was taken to indicate statistical significance.

4. Conclusions

Our results show that miR395a was significantly up-regulated by EBR in *Arabidopsis*, was expressed more strongly in leaf veins and roots of EBR-treated miR395a promoter::GUS plants, and targeted GUN5 with the effect of suppressing its expression. EBR was able to suppress GUN5 downstream genes to regulate seedling germination and the formation of primary and lateral roots. These results suggest that the reduced amount of chlorophyll in leaf veins and root growth of *Arabidopsis* might be attributable to the interaction between miR395a and GUN5. This study provides new insights into the function of miRNAs that will be useful in further research into the roles miRNAs play in the molecular mechanisms of plant development.

Acknowledgments

This work was supported by the National Science Council, Taiwan (NSC 99-2621-B-002-005-MY3, NSC 99-2621-B-010-001-MY3, NSC 101-2120-M-002-010) and the National Taiwan University Cutting-Edge Steering Research Project (NTU-CESRP-102R7602C3).

References

1.   Mitohell, J.W.; Mandava, N.; Worley, J.F.; Plimmer, J.R.; Smith, M.V. Brassins-a new family of plant hormone from rape pollen. *Nature* **1970**, *225*, 1065–1066.

2.   Grove, M.D.; Spencer, G.F.; Rohwedder, W.K. Brassinolide, a plant growth-promoting steroid isolated. *Nature* **1979**, *281*, 216–217.

3.   Altmann, T. Molecular physiology of brassinosteroids revealed by the analysis of mutants. *Planta* **1999**, *208*, 1–11.

4.   Hategan, L.; Godza, B.; Russinova, M. Regulation of Brassinoteroid Metabolism. In *Brassinosteroids: A Class of Plant Hormone*; Hayat, S., Ahmad, A., Eds.; Springer: Heidelberg, Germany, 2011; p. 67.

5.   Yokota, T. The structure, biosynthesis and function of brassinosteroids. *Trends Plant Sci.* **1997**, *2*, 137–143.

6.   Clouse, S.D.; Langford, M.; McMorris, T.C. A brassinosteroid-insensitive mutant in *Arabidopsis thaliana* exhibits multiple defects in growth and development. *Plant Physiol.* **1996**, *111*, 671–678.

7.   Clouse, S.D.; Sasse, J.M. Brassinosteroids: Essential regulators of plant growth and development. *Annu. Rev. Plant Physiol. Plant Mol. Biol.* **1998**, *49*, 427–251.

8.   Clouse, S.D. Brassinosteroid signal transduction: Clarifying the pathway from ligand perception to gene expression. *Mol. Cell* **2002**, *10*, 973–982.

9.   Steber, C.M.; McCourt, P. A role for brassinosteroids in germination in *Arabidopsis*. *Plant Physiol.* **2001**, *125*, 763–769.

10. Kagale, S.; Divi, U.K.; Krochko, J.E.; Keller, W.A.; Krishna, P. Brassinosteroid confers tolerance in *Arabidopsis thaliana* and *Brassica napus* to a range of abiotic stresses. *Planta* **2007**, *225*, 353–364.

11. Divi, U.K.; Krishna, P. Overexpression of the brassinosteroid biosynthetic gene AtDWF4 in *Arabidopsis* seeds overcomes abscisic acid-induced inhibition of germination and increases cold tolerance in transgenic seedlings. *J. Plant Growth Regul.* **2010**, *29*, 385–393.

12. Gomes, M.M.A. Physiological Effects Regulated to Brassinoteroid Application in Plants. In *Brassinoteroids: A class of Plant Hormone*; Hayat, S., Ahmad, A., Eds.; Springer: Heidelberg, Germany, 2011; pp. 193–204.

13. Kozomara, A.; Griffiths-Jones, S. miRBase: Integrating microRNA annotation and deep-sequencing data. *Nucleic Acids Res.* **2011**, *39*, D152–D157.

14. Voinnet, O. Origin, biogenesis, and activity of plant microRNAs. *Cell* **2009**, *136*, 669–687.

15. Bartel, D.P. microRNAs: Target recognition and regulatory functions. *Cell* **2009**, *136*, 215–233.

16. Liu, H.H.; Tian, X.; Li, Y.J.; Wu, C.A.; Zheng, C.C. Microarray-based analysis of stress-regulated microRNAs in *Arabidopsis thaliana*. *RNA* **2008**, *14*, 836–843.

17. Fujii, H.; Chiou, T.J.; Lin, S.I.; Aung, K.; Zhu, J.K. A miRNA involved in phosphate-starvation response in *Arabidopsis*. *Curr. Biol.* **2005**, *15*, 2038–2043.

18. Deng, Z.; Zhang, X.; Tang, W.; Oses-Prieto, J.A.; Suzuki, N.; Gendron, J.M.; Chen, H.; Guan, S.; Chalkley, R.J.; Peterman, T.K.; *et al.* A proteomics study of brassinosteroid response in *Arabidopsis*. *Mol. Cell. Proteomics* **2007**, *6*, 2058–2071.

19. Tang, W.; Deng, Z.; Oses-Prieto, J.A.; Suzuki, N.; Zhu, S.; Zhang, X.; Burlingame, A.L.; Wang, Z.-Y. Proteomics studies of brassinosteroid signal transduction using prefractionation and two-dimensional DIGE. *Mol. Cell. Proteomics* **2008**, *7*, 728–738.

20. Nemhauser, J.L.; Hong, F.; Chory, J. Different plant hormones regulate similar processes through largely nonoverlapping transcriptional responses. *Cell* **2006**, *126*, 467–475.

21. Goda, H.; Shimada, Y.; Asami, T.; Fujioka, S.; Yoshida, S. Microarray analysis of brassinosteroid-regulated genes in *Arabidopsis*. *Plant Physiol.* **2002**, *130*, 1319–1334.

22. Mussig, C.; Shin, G.H.; Altmann, T. Brassinosteroids promote root growth in *Arabidopsis*. *Plant Physiol.* **2003**, *133*, 1261–1271.

23. Bao, F.; Shen, J.; Brady, S.R.; Muday, G.K.; Asami, T.; Yang, Z. Brassinosteroids interact with auxin to promote lateral root development in *Arabidopsis*. *Plant Physiol.* **2004**, *134*, 1624–1631.

24. Fukaki, H.; Tasaka, M. Hormone interactions during lateral root formation. *Plant Mol. Biol.* **2009**, *69*, 437–499.

25. Zhang, Y. miRU: An automated plant miRNA target prediction server. *Nucleic Acids Res.* **2005**, *33*, W701–W704.

26. Schwab, R.; Ossowski, S.; Riester, M.; Warthmann, N.; Weigel, D. Highly specific gene silencing by artificial microRNAs in *Arabidopsis*. *Plant Cell* **2006**, *18*, 1121–1133.

27. Dai, X.; Zhao, P.X. psRNATarget: A plant small RNA target analysis server. *Nucleic Acids Res.* **2011**, *39*, W155–W159.

28. Mochizuki, N.; Brusslan, J.A.; Larkin, R.; Nagatani, A.; Chory, J. *Arabidopsis* genomes uncoupled 5 (GUN5) mutant reveals the involvement of Mg-chelatase H subunit in plastid-to-nucleus signal transduction. *Proc. Natl. Acad. Sci. USA* **2000**, *98*, 2053–2058.

29. Rodermel, S.; Park, S. Pathways of intracellular communication: Tetrapyrroles and plastid-to-nucleus signaling. *BioEssays* **2003**, *25*, 631–636.

30. Finkelstein, R.R.; Gampala, S.S.; Rock, C.D. Abscisic acid signaling in seeds and seedlings. *Plant Cell* **2002**, *14*, S15–S45.

31. Zhang, H.; Han, W.; de Smet, I.; Talboys, P.; Loya, R.; Hassan, A.; Rong, H.; Jurgens, G.; Paul Knox, J.; Wang, M.H. ABA promotes quiescence of the quiescent centre and suppresses stem cell differentiation in the *Arabidopsis* primary root meristem. *Plant J.* **2010**, *64*, 764–774.

32. De Smet, I.; Signora, L.; Beeckman, T.; Inze, D.; Foyer, C.H.; Zhang, H. An abscisic acid-sensitive checkpoint in lateral root development of *Arabidopsis*. *Plant J.* **2003**, *33*, 543–555.

33. Jenks, M.A.; Wood, A.J. *Genes for Plant Abiotic Stress*, 1st ed.; John Wiley & Sons, Ltd.: Hoboken, NJ, USA, 2009.

34. Nakano, T.; Kimura, T.; Kaneko, I.; Nagata, N.; Matsuyama, T.; Asami, T.; Yoshida, S. Molecular mechanism of chloroplast development regulated by plant hormones. *RIKEN Rev.* **2001**, *41*, 86–87.

35. Shkolnik-Inbar, D.; Bar-Zvi, D. ABI4 mediates abscisic acid and cytokinin inhibition of lateral root formation by reducing polar auxin transport in *Arabidopsis*. *Plant Cell* **2010**, *22*, 3560–3573.

36. Lopez-Molina, L.; Mongrand, S.; McLachlin, D.T.; Chait, B.T.; Chua, N.H. ABI5 acts downstream of ABI3 to execute an ABA-dependent growth arrest during germination. *Plant J.* **2002**, *32*, 317–328.

37. Miura, K.; Lee, J.; Jin, J.B.; Yoo, C.Y.; Miura, T.; Hasegawa, P.M. Sumoylation of ABI5 by the *Arabidopsis* SUMO E3 ligase SIZ1 negatively regulates abscisic acid signaling. *Proc. Natl. Acad. Sci. USA* **2009**, *106*, 5418–5423.

38. Ye, Q.; Zhu, W.; Li, L.; Zhang, S.; Yin, Y.; Ma, H.; Wang, X. Brassinosteroids control male fertility by regulating the expression of key genes involved in *Arabidopsis* anther and pollen development. *Proc. Natl. Acad. Sci. USA* **2010**, *107*, 6100–6105.

39. Clouse, S.D. Molecular genetic analysis of brassinosteroid action. *Physiol. Plant* **1997**, *100*, 702–709.

40. Nonogaki, H. microRNA gene regulation cascades during early stages of plant development. *Plant Cell Physiol.* **2010**, *51*, 1840–1846.

41. Poethig, R.S. Small RNAs and developmental timing in plants. *Curr. Opin. Genet. Dev.* **2009**, *19*, 374–378.

42. Meng, Y.; Ma, X.; Chen, D.; Wu, P.; Chen, M. microRNA-mediated signaling involved in plant root development. *Biochem. Biophys. Res. Commun.* **2010**, *393*, 345–349.

43. Kawashima, C.G.; Yoshimoto, N.; Maruyama-Nakashita, A.; Tsuchiya, Y.N.; Saito, K.; Takahashi, H.; Dalmay, T. Sulphur starvation induces the expression of microRNA-395 and one of its target genes but in different cell types. *Plant J.* **2009**, *57*, 313–321.

44. Liang, G.; Yang, F.; Yu, D. microRNA395 mediates regulation of sulfate accumulation and allocation in *Arabidopsis thaliana*. *Plant J.* **2010**, *62*, 1046–1057.

45. Yamasaki, H.; Hayashi, M.; Fukazawa, M.; Kobayashi, Y.; Shikanai, T. SQUAMOSA promoter binding protein-like7 is a central regulator for copper homeostasis in *Arabidopsis*. *Plant Cell* **2009**, *21*, 347–361.

46. Kutter, C.; Schob, H.; Stadler, M.; Meins, F., Jr.; Si-Ammour, A. microRNA-mediated regulation of stomatal development in *Arabidopsis*. *Plant Cell* **2007**, *19*, 2417–2429.

47. Petti, F.B.; Liguori, A.; Ippoliti, F. Study on cytokines IL-2, IL-6, IL-10 in patients of chronic allergic rhinitis treated with acupuncture. *J. Tradit. Chin. Med.* **2002**, *22*, 104–111.

48. Zhao, M.; Ding, H.; Zhu, J.-K.; Zhang, F.; Li, W.-X. Involvement of miR169 in the nitrogen-starvation responses in *Arabidopsis*. *New Phytol.* **2011**, *190*, 906–915.

49. Jones-Rhoades, M.W.; Bartel, D.P.; Bartel, B. microRNAs and their regulatory roles in plants. *Annu. Rev. Plant Biol.* **2006**, *57*, 19–53.

50. Achard, P.; Herr, A.; Baulcombe, D.C.; Harberd, N.P. Modulation of floral development by a gibberellin-regulated microRNA. *Development* **2004**, *131*, 3357–3365.

51. Strand, A.; Asami, T.; Alonso, J.; Ecker, J.R.; Chory, J. Chloroplast to nucleus communication triggered by accumulation of Mg-protoporphyrinIX. *Nature* **2003**, *421*, 79–83.

52. McCourt, P.; Creelman, R. The ABA receptors—We report you decide. *Curr. Opin. Plant Biol.* **2008**, *11*, 474–478.

53. Rasband, W.S.; ImageJ, U.S. ImageJ software; National Institutes of Health: Bethesda, MD, USA, 1997–2012.

54. Miao, Y.; Jiang, L. Transient expression of fluorescent fusion proteins in protoplasts of suspension cultured cells. *Nat. Protoc.* **2007**, *2*, 2348–2353.

4

Systems Biology Approach to the Dissection of the Complexity of Regulatory Networks in the *S. scrofa* Cardiocirculatory System

Paolo Martini [1], Gabriele Sales [1], Enrica Calura [1], Mattia Brugiolo [2], Gerolamo Lanfranchi [1,2], Chiara Romualdi [1],* and Stefano Cagnin [1,2],*

[1] Department of Biology, University of Padova, Via G. Colombo 3, Padova 35121, Italy;
    E-Mails: paolo.martini@unipd.it (P.M.); gabriele.sales@unipd.it (G.S.);
    enrica.calura@unipd.it (E.C.); gerolamo.lanfranchi@unipd.it (G.L.)

[2] C.R.I.B.I. Biotechnology Centre, University of Padova, Via U. Bassi 58/B, Padova 35121, Italy;
    E-Mail: brugiolo@mpi-cbg.de

* Authors to whom correspondence should be addressed; E-Mails: chiara.romualdi@unipd.it (C.R.);
    stefano.cagnin@unipd.it (S.C.);

Abstract: Genome-wide experiments are routinely used to increase the understanding of the biological processes involved in the development and maintenance of a variety of pathologies. Although the technical feasibility of this type of experiment has improved in recent years, data analysis remains challenging. In this context, gene set analysis has emerged as a fundamental tool for the interpretation of the results. Here, we review strategies used in the gene set approach, and using datasets for the pig cardiocirculatory system as a case study, we demonstrate how the use of a combination of these strategies can enhance the interpretation of results. Gene set analyses are able to distinguish vessels from the heart and arteries from veins in a manner that is consistent with the different cellular composition of smooth muscle cells. By integrating microRNA elements in the regulatory circuits identified, we find that vessel specificity is maintained through specific miRNAs, such as miR-133a and miR-143, which show anti-correlated expression with their mRNA targets.

Keywords: pathway analysis; miRNA; cardiocirculatory; network reconstruction; integrative analysis; pig; artery; vein; vessel

---

1. Introduction

Genome-wide experiments on RNA expression typically provide lists of differentially expressed genes (DEGs) [1,2] that represent the starting point of a highly challenging process of result interpretation in which the gene-by-gene approach is often used. The lists obtained are highly dependent on the statistical tests adopted and on the threshold used to declare a gene significant. This variability has raised substantial criticism concerning the reproducibility of array experiments. Several studies have demonstrated greater consistency of array results using gene set approaches, rather than single gene approaches [3], indicating that there is greater reproducibility of the main biological themes than of their single elements. A gene set is defined as a set of genes that are functionally related. Gene sets are usually identified based on a priori biological knowledge (see, for example, Gene Ontology "GO" (http://www.geneontology.org/ (accessed on 13 November 2013)) and the Kyoto Encyclopedia of Genes and Genomes "KEGG" (http://www.genome.jp/kegg/ (accessed on 13 November 2013))). In this regard, several new bioinformatics tools have been developed that allow the integration of information such as gene location [4–6], ontological annotations [7–10], or sequence features [11]. These methods can be broadly divided into supervised and unsupervised approaches. Supervised methods use *a priori* information on the functional relationships among genes to identify the processes involved in an experimental condition, while unsupervised approaches attempt to reconstruct functional associations among genes without relying on external information. In the following, we will briefly review these strategies, focusing specifically on their pros and cons; in addition, we will apply these strategies to a case study.

1.1. Supervised Approaches: Pathway Analysis

The integration of gene expression profiles with additional information on pathway annotations is called pathway analysis. The pathway analysis approach evaluates gene expression profiles among related genes, looking for coordinated changes in their expression levels. Several implementations of pathway analysis are now available, from the widely used algorithm developed by Subramanian and colleagues (Gene Set Enrichment Analysis; GSEA) [9], with its improvements [10,12], to more sophisticated implementations that exploit the topology of the pathway [13,14] (for a comprehensive review of existing methods, see [15]). Pathway analysis methods can be divided into (i) methods based on enrichment analysis and performed on a list of genes selected through a gene-level test; and (ii) methods based on global and multivariate approaches that define a model based on the whole gene set. With the first class of methods, the primary concerns are the assumption that genes are independent and the use of a threshold value for the selection of differentially expressed genes. Due to the latter, many genes with moderate but meaningful expression changes are discarded based on the strict cut-off value, leading to a reduction in statistical power. On the other hand, global and multivariate approaches relax the assumption of independence among genes belonging to the same gene sets and identify moderate but coordinated expression changes that cannot be detected by the enrichment analysis approach [16].

From this perspective, we recently developed three novel algorithms that can be used to perform gene set and pathway analysis. Graphite, a Bioconductor package [17], is a computational framework that can be used to manage, interpret, and convert pathway annotations to gene-gene networks, while STEPath [18] integrates expression levels and chromosome positioning to identify regional gene activation and CliPPER [14,19] explores the topology of a pathway, highlighting the portions most involved in its deregulation. We have implemented most of these analyses in a new web tool called GraphiteWeb [20].

One of the major drawbacks associated with these approaches is the limitation of pathway annotation. Pathway annotation is a highly challenging procedure that exploits the efforts of many researchers, who manually curate each single pathway based on information available in the literature. Pathways are often thought of as the elementary functional and evolutionary building blocks of the complete metabolic network, with each pathway representing a "self-contained" elementary biochemical process. To partition the reaction network of an organism into a set of (possibly overlapping) metabolic pathways requires arbitrary decisions as to where such partitions should be made and how pathway variants should be described [21]. For these reasons, only a portion (in humans, approximately one-third) of known genes are currently annotated in at least one pathway.

In KEGG [22], the metabolic pathways—called "maps"—are subparts of the overall reaction graph. Reactions within a map are connected by their constituent metabolites, which also provide links to reactions in other maps. KEGG metabolic maps are described without reference to a particular species, and each map includes the reactions belonging to all known variants of a particular pathway. MetaCyc is a database of non-redundant, experimentally elucidated metabolic pathways that are found in many species [23] while, in the smaller Reactome database [24], the human database is used as the reference for predicting reactions and pathways in other organisms.

1.2. Unsupervised Approaches: Reverse Engineering Approach

A different approach to dealing with biological networks is the *ab initio* strategy: using genome-wide expression values, these algorithms try to infer the best network of interactions satisfying specific conditions. Unlike the pathway analysis approach, here, all known genes can be taken into consideration. Several methods have been proposed for the reconstruction of gene regulatory networks (GRNs) from experimental data; these include Bayesian Networks (BN) [25], Relevance Networks (RN) [26], and Graphical Gaussian Models (GGM) [27,28]. While BN and GGM distinguish between direct and indirect edges, RN does not. It is worth noting that although BN and GGM are able to infer edge direction this does not necessarily imply an ability to identify biological causality.

BN and GGM function poorly in cases involving thousands of genes and a small number of replicates, while RN has the ability to address such cases. RN uses association measures between two expression profiles, such as correlation and mutual information, to rank gene-gene interactions according to their strengths; the higher the association measure, the greater the probability of a functional interaction between the two genes. All of these approaches produce a large number of false positives (false interactions). The seminal paper of Basso *et al.*, 2005 [29], extends RN, introducing an algorithm based on Data Processing Inequality (DPI) for removing indirect edges. Their approach, called ARACNE (Algorithm for the Reconstruction of Accurate Cellular Networks) [30], has been successfully used to reconstruct the sub-network of the MYC gene in human B cells.

In this context, we developed a new R package, *parmigene*, that performs network inference by implementing an unbiased estimation of the mutual information between expression profiles, thus yielding more precise results than existing software at strikingly less computational cost [31].

Apart from their low specificity, a significant issue raised by the last network inference challenge (DREAM 5) is that no single network inference method performs optimally across all data sets. In contrast, integration of predictions from multiple inference methods through a consensus network shows robust and high performance across diverse data sets [32].

Apart from the algorithm used, once the whole network has been inferred, the classical approach to dealing with large amounts of interactions is identifying small-connected components as a means of testing their enrichment in specific biological processes.

1.3. The Missing Element: MicroRNAs (miRNAs)

Although highly innovative, the supervised and unsupervised approaches described so far do not take miRNAs into consideration. Many efforts have been made to predict miRNA/mRNA interactions, first by developing various target prediction algorithms and then by introducing new experimental techniques to isolate miRNA/mRNA complexes [33–36]. Computational target prediction is still widely used, although it is characterized by many false positives. For exhaustive reviews on miRNA discovery algorithms and *in silico* target prediction [37,38].

The integration of target predictions with miRNA and gene expression profiles has recently been proposed as a means of computationally improving and refining miRNA-target predictions. As miRNAs act predominantly through target degradation, the expression profiles of miRNAs and those of their target genes are expected to be inversely correlated [39,40].

Although the key role of miRNA in post-transcriptional regulation is universally recognized, few attempts have been made to use combinations of miRNA elements in developing gene set approaches. The only such attempt was described by Nam and colleagues [41], who performed GSEA on the mRNA targets of de-regulated miRNAs.

1.4. Case Study: The Pig as a Model Organism

Considering the advantages and disadvantages of the approaches described above, here we propose a consensus strategy based on the integration of pathway analysis, relevance networks and miRNA expression using as a model organism the pig and its cardiocirculatory system.

The size of organs, as well as various anatomical features, general physiology, and features of organ development, are very similar in pigs and humans. This permits the use of the pig as a model in the study of a number of pathologies, such as those affecting eyes [42], muscle [43], organ transplantation [44,45], and the gastrointestinal [46], nervous [47], and cardiovascular [48] systems. The coronary artery distribution in the pig is more similar to that of humans than is that of other animals. In addition, pigs present very similar cardiac output to humans; they possess a vaso vasorum in the aorta, and the left azygous vein empties into the coronary sinus instead of into the precava. Blood pressure (145–160/105 BP), heart rate (100–150 BPM) and pulmonary pressure are higher in pigs than in humans.

Despite the medical importance of the pig as a species for study, our knowledge of the genome organization, gene expression regulation, and the molecular mechanisms underlying the pathophysiological processes of the pig is far less than the knowledge we have acquired of the mouse and rat. More than 90% of the porcine genome has been sequenced by the Swine Genome Sequencing Consortium [49]. The availability of detailed information on the porcine genome, together with emerging transgenic technologies, will enhance our ability to create specific and useful pig models. Recently, an atlas of DNA methylomes in porcine adipose and muscle tissues was published [50], and a great effort was made to combine genome sequence information with our knowledge of gene expression. Many of these studies focused on the swine immune system [51–54], while a genome-wide expression analysis in different tissues was described in Freeman's paper [55]. Recently, using sequencing approaches, a compendium of small non-coding RNAs was identified in various pig tissues (e.g., skeletal muscle [56–62], kidney [63], tooth [64], intestinal tract [65], brain [66], testis, ovary, sperm, and embryo [67–71] and pituitary gland [72]). Li and colleagues demonstrated that a complex regulatory network of porcine subcutaneous fat development is reflected in a great diversity of miRNA composition and expression between muscle and adipose tissue [73].

Here, we generate new custom mRNA and miRNA platforms that can be used to dissect the transcriptomic changes and regulatory circuits that are involved in the maintenance of veins and arteries in the pig. An integrative approach, combining pathway analysis and *de novo* network reconstruction, was used to expand our current knowledge of these regulatory circuits and to integrate miRNA activity into these circuits demonstrating their role in vessel specification. We show that vessel specificity can be maintained through different miRNAs (e.g., miR-133a and miR-143), the expression of which is inversely correlated with that of their mRNA targets.

2. Results and Discussion

The integration and analysis of gene and miRNA expression profiles across different tissues is fundamental to our understanding of tissue-specific processes. Here, we focus our analysis on differences in gene and miRNA expression among different tracts of the circulatory system: the two largest veins of the body (superior and inferior vena cava), the aorta (ascending and descending), the pulmonary artery, and the coronary artery. To achieve this goal, we created mRNA and miRNA [74] platforms, the latter based on the RAKE (RNA primed–array-based Klenow enzyme assay) method [75,76], to quantify coding and non-coding gene expression in pig tissues. After quantifying miRNA and mRNA expression, we used a combination of supervised and unsupervised approaches to detect transcriptional and post-transcriptional differences among different tracts of the circulatory system.

Ensembl transcripts (Ver. 56; EMBL-EBI, Wellcome Trust Genome Campus, Hinxton, Cambridgeshire, UK) and UniGene (Ver. 38; National Center for Biotechnology Information, U.S. National Library of Medicine, Bethesda, MD, USA) pig sequences were used to produce a dedicated microarray platform for monitoring mRNA expression. On the basis of sequence similarity, UniGene features that overlapped more than 40% with an Ensembl transcript were discarded. After this filter, we obtained 40,267 UniGene clusters and 19,603 Ensembl transcripts (protein coding + pseudogenes + retrotransposed elements). For this selected collection of sequences, we designed microarray probes with different specificities and located at different distances from the 3' ends of specific transcripts using six different algorithms. The two best probes for each sequence, as determined by the reliability of the prediction algorithm and by the probe's vicinity to the 3'-end, were experimentally tested in a hybridization trial performed with a pool of mRNA populations independently prepared from 20 pig tissues (GEO: GSE28636). For each transcript with a replicated probe, we selected the probe that was the most responsive and specific on the basis of the intensity of fluorescence in the hybridization test, as suggested by Kronick [77]. The resulting pig whole-genome microarray, which was used in the gene expression analysis, is composed of: (i) 17,048 replicated probes and 963 single probes specific for the Ensembl transcripts; (ii) 11,363 replicated probes specific for the UniGene clusters of lengths between 778 nt and 1348 nt; and (iii) 28,790 single probes specific for the remaining UniGene clusters. Our analysis was not able to identify specific probes for 114 UniGene clusters and 1592 Ensembl transcripts. A limitation we faced in working with gene expression in pig was the poor gene annotation available. The number of annotated features on the array was increased by mining description and protein annotations to associate gene names with our probe symbols. Basically, for genes for which the HUGO (Human Genome Organisation) symbol was not present, we mined the description available from the Unigene database and retrieved additional gene or protein IDs, if present. All IDs were manually curated (ArrayExpress ID: A-MEXP-2351).

Recently, a new microarray platform based on 52,355 expressed sequences comprising miRNAs in miRBase Ver. 15 (Wellcome Trust Sanger Institute, Cambridge, UK) for pigs, cows, humans, and mice was described [55]. Unlike this new platform, which was constructed by spanning 22 probes along the transcripts, the platform we developed detects the 3'-UTR of each transcript; therefore, we are able to distinguish mRNA isoforms. This feature is fundamental because the activity of miRNAs is predominantly based on their interactions with the 3'-UTR region of mRNAs.

The identification of miRNAs was described in [74]. Briefly, bioinformatic analyses were performed on the pig genome for the identification of putative pre-miRNAs. These were experimentally tested using six independent RAKE experiments to identify 5' and 3' miRNA boundaries. After this experimental confirmation, all the pre-miRNAs identified as responsive (1235 hairpins) were tested for the presence of mature miRNA through RNA sequencing experiments. RNA sequencing experiments identified 343 hairpins coding for miRNAs. However, using PCR we were able to validate several miRNAs that were not confirmed by RNA sequencing. Therefore, we decided to produce an miRNA microarray platform (Array Express ID: A-MEXP-2348) containing all miRNAs detected by RAKE experiments. In the following analysis, we will discuss only miRNAs that were confirmed in sequencing experiments. Each specific probe is flanked by a background probe that was used to subtract the corresponding background fluorescence signal in the analysis (Figure 1).

Figure 1. Explicative scan portion of miRNA microarray after the RAKE and labeling reactions (**A**) and before hybridization (**B**). Spike-in spots are indicated by red lines; the blue arrow indicates a specific probe, and the orange arrow indicates its background probe. Each background probe was positioned to the right of its probe.

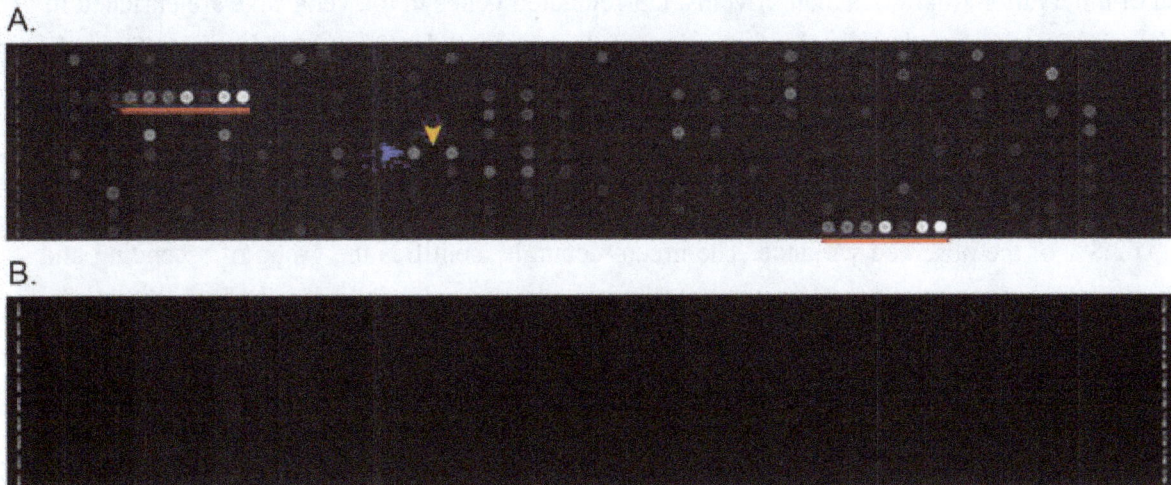

The short length of miRNAs makes complementary probe selection and the identification of optimized PCR primers a challenging task. While miRNA microarrays permit massive parallel and accurate relative measurement of all known miRNAs, they have been less useful for absolute quantification. We developed a new method that integrates the hybridization of miRNAs with an enzymatic elongation reaction that can take place only following a perfect match between the miRNA and the probe. Moreover, we introduced oligonucleotide spikes into the hybridization-enzymatic reaction, permitting the quantification of miRNAs over the linear dynamic range of 10^{-18} moles to 10^{-14} moles and avoiding biases related to sequence, labeling, or hybridization [74].

2.1. Differences between Arteries and Veins

We compared different tracts of the circulatory system: the two largest veins (the superior and inferior vena cava), the aorta (ascending and descending tracts), the pulmonary artery, and the coronary artery. As expected, the ascending and descending aorta and the coronary artery display similar gene expression profiles that are distinct from those of the superior and inferior vena cava (Figure 2A), while the pulmonary artery has an intermediate expression profile (Figure 2A). Arteries and veins are structurally different in terms of their relationship to the heart. Arteries receive blood directly from the heart and are therefore characterized by high pressure; in contrast, veins receive blood from peripheral body regions, and low pressure characterizes them. For this reason, some of the blood in the veins may not return to the heart but instead may back up or collect in these vessels. Veins transport de-oxygenated blood, while arteries transport oxygenated blood (with the exception of the pulmonary artery, which transports de-oxygenated blood to the lungs for oxygenation). The difference in blood pressure in arteries and veins is reflected in the different structures of these vessels. Arteries and arterioles have thicker walls than veins and venules; specifically, they possess an increased amount of smooth muscle that provides extra strength and elasticity to withstand surges of blood from the heart. Moreover, the thinner the vessel, the lower its innervation.

In accordance with the increased number of smooth muscle cells in arteries, the aorta expresses more smooth muscle-specific transcripts than the vena cava (Figure 2B). Genes that are up-regulated in the aorta include genes related to biological structures such as adherence junctions and processes such as nerve function and blood circulation (Table S1). This is consistent with the significantly higher level of innervation of arteries than of veins. Up-regulated genes in the vena cava are enriched in genes coding for proteins involved in the formation of the extracellular matrix (Table S1). These findings may be associated with the differences in elasticity between veins and arteries (veins have less elastic tissue than arteries).

Figure 2. (A) Principal component analysis (PCA). The first three components account for 62.8% of the observed variance. The green rectangle identifies the group of ascending and descending aorta samples (green dots); the coronary artery is indicated by a black dot, the red rectangle highlights pulmonary artery samples (red dots), and the blue rectangle surrounds superior and inferior vena cava samples (blue dots). On the right, separated from other samples, are heart samples; **(B)** Heat map of muscle transcripts. Transcripts coding for muscle proteins are up-regulated in arteries with respect to veins. The red squares indicate up-regulated genes, and the green squares indicate down-regulated genes. The grey squares indicate genes for which no expression was detected. L.P.V. = leaflet of pulmonary valve; Inf. Vena Cava = inferior vena cava; Sup. Vena Cava = superior vena cava. The numbers following the sample names indicate the number of experimental replicates.

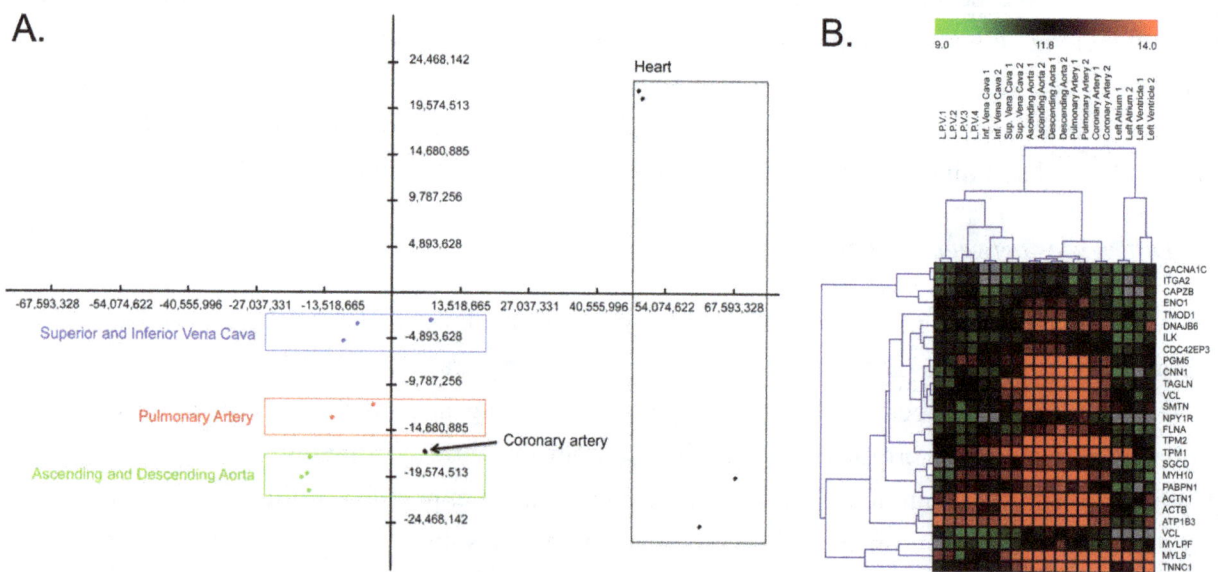

A major component of the vessel walls of large arteries and veins is the extracellular matrix (ECM), which consists of collagens, elastin, and proteoglycans. The smooth muscle cells of the aorta and vena cava synthezise different amounts of collagen. As expected, our data show that collagen synthesis is four-fold higher in venous than in arterial [78]; collagen type I (COL1A2) is the most highly expressed extracellular matrix component.

Procollagen C-endopeptidase enhancer 2 (*PCOLCE2*) and *P4HA1 prolyl 4-hydroxylase, α polypeptide I* (*P4H*) genes were found to be up-regulated in the vena cava. PCOLCE2 binds to the C-terminal propeptide of type I and II procollagens and may enhance the cleavage of their propeptides, while P4H4 is a key enzyme in collagen synthesis. Moreover, we found *type VIII collagen* (*COL8A1*), which is typical of the endothelium lining vessels, and *type VI collagen* (*COL6A3*), a subendothelial constituent [79], to be highly expressed in the vena cava.

2.2. Pathway Analysis

Using multivariate pathway analysis methods such as GSEA, we overcame the major limitation of the classical enrichment approach, cut-off-based gene selection, focusing instead on coordinated changes in gene expression. Using this method, we were able to identify gene pathways that are specifically expressed in arteries and veins (Table 1). Among the activated pathways in arteries are those associated with smooth muscle contraction, calcium-calmodulin-dependent events, genome stability and regulation of intracellular signaling cascades. This finding is consistent with the presence of a thicker smooth muscle ring in arteries than in veins. Among the activated pathways in veins, we find the complement cascade, arachidonic acid metabolism, cell surface interactions at the vascular wall, and extracellular matrix metabolism (glycosaminoglycan metabolism and keratin/keratan sulphate metabolism). Arachidonic acid metabolism is involved in the control of various processes within the cardiocirculatory system, including vasoconstriction [80] and vasodilation [81,82]. The two most highly expressed genes related to arachidonic acid metabolism were *prostaglandin-endoperoxide synthase 2* (*PTGS2 or COX-2*) and *γ-glutamyltransferase 5* (*GGT5*). *COX-2* and *endothelial nitric oxide synthase* (*eNOS*) are primarily expressed in endothelial cells and are considered important regulators of vascular function. Under normal conditions, laminar flow induces COX-2 expression and synthesis of PGI$_2$, which in turn stimulates eNOS activity [83]. GGT expression was also localized in the endothelium [84]. As blood normally flows more slowly through veins than through arteries, thromboses are more common in veins than in arteries. This could be the reason for the control of vasodilation and vasoconstriction through metabolites of arachidonic acid.

In support of the up-regulation of elements of the complement cascade in veins, it is known that inflammation is more readily induced in venous than in arterial epithelium due to the conditions of the venous circulation. We checked for the presence of an inflammatory process by analyzing the expression of complement components in 19 tissues (Figure 3). We find that not all complement components are up-regulated in veins, while most are highly expressed in lymph nodes, spleen, and liver. This is in accordance with complement system synthesis and laundering. The complement system consists of a dozen circulating proteins, most of which are synthesized by the liver, that have the ability to bind to cellular membranes. The spleen and the liver are able to remove immune complexes composed of complement elements linked to erythrocyte membranes [85].

Finally, it is worth noting that pathways describing mucopolysaccharidosis syndromes such as Hurler, Sanfilippo, and Morquio syndromes were found to be significantly expressed in veins. Altered glycosaminoglycan metabolism is a key feature of these pathologies. Glycosaminoglycans are proteoglycans that bind to a varying degree water, electrolytes and macromolecules, such as collagen, within the connective tissue. The lining of veins and arteries comprises a substantial amount of the body's connective tissue. The outer layer of vessels (tunica adventitia) consists chiefly of connective tissue and is the thickest layer of the vein.

Table 1. Summary of Gene Set Enrichment Analysis (GSEA) analysis based on the Reactome database (http://www.reactome.org/ (accessed on 13 Novembre 2013)). Set size refers to the dimension of the pathway, and NTK (Normalized T-test of the kth gene set) is the observed value of the statistic as defined in the Graphite web tool [20]. Negative NTK values indicate pathways activated in veins, while positive values indicate pathways activated in arteries. It is worth noting that GSEA is known to have low statistical power; the suggested Q-value cut-off for identification of significant pathways is 0.25.

Pathway	Set size	NTk	Q-Value
Complement cascade	18	−5.29	0
Arachidonic acid metabolism	11	−3.09	0.044912281
Glycosaminoglycan metabolism	54	−3.09	0.044912281
MPS I—Hurler syndrome	54	−3.09	0.044912281
MPS II—Hunter syndrome	54	−3.09	0.044912281
MPS IIIA—Sanfilippo syndrome A	54	−3.09	0.044912281
MPS IIIB—Sanfilippo syndrome B	54	−3.09	0.044912281
MPS IIIC—Sanfilippo syndrome C	54	−3.09	0.044912281
MPS IIID—Sanfilippo syndrome D	54	−3.09	0.044912281
MPS IV—Morquio syndrome A	54	−3.09	0.044912281
MPS IV—Morquio syndrome B	54	−3.09	0.044912281
Biological oxidations	56	−2.75	0.106666667
Cell surface interactions at the vascular wall	54	−2.75	0.106666667
Keratan sulfate/keratin metabolism	20	−2.46	0.205977011
G α (12/13) signaling events	35	−2.37	0.24
Antigen presentation: Folding, assembly and peptide loading of class I MHC	11	−2.33	0.250980392
Golgi associated vesicle biogenesis	29	−2.29	0.247017544
Glutathione conjugation	10	−2.26	0.249756098
Phase II conjugation	23	−2.26	0.249756098
EGFR interacts with phospholipase C-γ	17	2.12	0.273710692
Ca-dependent events	14	2.14	0.262564103
Calmodulin induced events	14	2.14	0.262564103
CaM pathway	14	2.14	0.262564103
Cell-extracellular matrix interactions	15	2.2	0.254184397
PLCG1 events in ERBB2 signaling	18	2.23	0.252121212
DARPP-32 events	12	2.26	0.249756098
DAG and IP3 signaling	15	2.29	0.247017544
PLC-γ1 signaling	15	2.29	0.247017544
Amyloids	18	2.33	0.250980392
Telomere Maintenance	31	2.46	0.192688172
RNA polymerase I promoter opening	18	2.65	0.131282051
Chromosome maintenance	53	2.75	0.1024
Meiotic synapsis	24	2.88	0.077575758
Deposition of new CENPA-containing nucleosomes at the centromere	21	2.88	0.077575758
Nucleosome assembly	21	2.88	0.077575758
Packaging of telomere ends	12	3.09	0.044912281
Striated muscle contraction	21	4.76	0
Smooth muscle contraction	19	6.13	0
Muscle contraction	36	7.25	0

Figure 3. Expression of genes involved in the complement response. The numbers represent gene expression levels normalized to the average expression of the same gene across all tissues. Down-regulated genes are shown in green, and up-regulated genes are shown in red. Most of the up-regulated genes are expressed in the liver, which is responsible for the synthesis of most of the proteins of the complement system, in the spleen and in lymph nodes (lymphoid organs). NA = Expression not detected; L.P.V. = leaflet pulmonary valve; WBC.A = white blood cells from arterial blood; WBC.V = white blood cells from venous blood.

	Ascending Aorta	Descending Aorta	Conoray Artery	Pulmonary Artery	Inf. Vena Cava	Sup. Vena Cava	Left Atrium	Left Ventr	L.P.V. (animal 1)	L.P.V. (animal 2)	Liver	WBC.A	WBC.V	Lymph node	Spleen	Tongue	Skeletal Muscle	Lung	Kidney	Stomach	Adipose tissue	Skin
C1QA	0.36	0.48	0.45	0.62	0.61	1.06	0.49	0.64	0.82	0.73	0.52	0.39	0.44	7.08	3.31	0.72	0.37	0.55	0.45	0.65	0.58	0.70
C1QB	0.40	0.70	0.42	1.11	0.76	1.84	0.36	0.40	1.45	1.83	0.40	NA	0.27	4.92	2.95	0.47	0.26	0.67	0.29	0.43	0.62	0.44
C1QBP	0.68	0.85	0.93	1.03	1.09	0.88	0.88	2.47	0.98	0.67	0.88	0.69	1.08	0.79	0.89	0.93	0.75	0.75	1.21	0.93	1.75	0.87
C1QC	0.43	0.52	0.72	0.74	0.81	1.21	0.51	0.45	1.81	1.09	0.58	0.41	0.45	5.68	2.51	0.69	0.40	0.63	0.44	0.65	0.63	0.64
C1R	0.67	0.81	0.76	0.98	0.72	1.01	0.88	0.95	0.69	0.71	2.24	1.17	0.81	0.83	2.28	1.86	0.57	0.64	0.65	0.98	0.73	1.03
C1S	0.84	1.01	1.07	1.03	1.50	1.50	0.56	1.15	0.59	0.83	1.07	0.44	1.01	0.94	0.67	1.80	0.45	1.10	0.92	1.01	1.01	1.48
C3	0.50	0.67	0.70	0.69	1.05	1.11	0.66	0.70	1.92	0.90	3.83	0.51	0.50	2.64	0.69	0.65	0.48	1.12	0.50	0.70	0.88	0.61
C3AR1	0.83	1.02	1.12	0.95	1.19	0.95	1.08	0.90	1.14	0.97	0.99	1.07	0.94	0.96	0.83	0.84	1.07	1.23	0.92	1.03	0.89	1.10
C3P1	0.78	0.79	0.96	0.98	0.96	0.92	0.99	1.47	0.98	0.71	1.41	0.87	1.02	0.87	1.15	1.26	0.98	0.85	0.91	1.08	1.04	1.01
C4	0.86	0.92	1.17	0.99	1.11	0.73	0.58	0.98	1.04	0.90	1.29	1.17	1.23	1.68	1.05	0.70	0.60	0.88	0.97	1.02	1.45	0.70
C4BPA	0.38	0.43	0.97	0.71	2.35	1.62	0.53	0.45	1.54	1.77	0.51	0.27	0.31	3.16	1.81	0.47	0.30	0.70	0.33	0.34	1.76	1.29
C5	0.84	1.01	0.85	0.82	0.86	0.88	0.91	1.08	0.85	0.95	2.58	0.86	0.86	0.74	1.01	0.78	1.00	1.16	1.08	1.07	0.90	0.89
C5AR1	0.85	1.00	1.09	0.89	1.07	0.90	1.07	0.98	1.00	1.02	1.00	1.08	0.96	0.94	1.03	0.83	1.14	1.16	0.99	1.06	0.94	1.00
C6	0.92	1.01	0.94	1.12	1.03	0.97	1.01	0.88	0.89	1.16	1.09	0.97	0.99	1.03	0.97	NA	1.08	1.02	1.04	NA	0.93	0.94
C7	0.70	0.67	1.75	0.88	0.98	0.80	0.48	1.00	1.63	1.31	0.94	0.60	1.31	0.66	1.38	0.54	0.39	1.49	1.83	0.55	1.33	0.76
C8A	0.86	0.83	0.77	0.81	NA	0.91	NA	0.82	0.81	0.92	2.85	NA	0.79	NA	0.88	0.89	NA	NA	0.97	NA	0.90	NA
C8B	0.99	0.98	1.02	1.02	0.80	1.15	NA	0.97	1.05	1.14	1.21	0.93	0.96	1.14	1.03	1.04	0.79	0.96	1.15	0.85	1.05	0.78
C8G	0.89	0.84	1.08	0.96	1.03	0.95	1.05	1.07	1.08	0.69	1.92	1.01	0.90	1.03	0.87	1.01	0.86	0.87	0.88	0.96	0.93	1.11
C9	0.62	0.68	0.70	0.70	0.77	0.69	0.75	0.71	0.70	0.74	5.45	0.70	0.76	0.75	0.72	0.70	1.13	1.32	0.93	0.72	1.00	0.74
CD46	0.77	0.92	0.83	0.84	0.79	NA	0.97	0.75	NA	0.89	0.83	0.96	0.84	0.90	3.05	NA	1.08	0.85	0.78	0.96	NA	NA
CD55	1.21	1.03	0.99	1.06	1.00	1.26	NA	0.89	1.06	1.06	0.87	1.01	0.96	1.00	0.99	0.92	NA	0.99	0.98	0.97	0.91	0.83
CD59	1.06	1.56	1.20	1.67	2.19	1.84	0.46	0.84	0.52	1.16	0.28	0.28	0.35	0.60	0.32	0.87	0.39	0.45	0.71	1.97	2.78	0.50
CFAB	0.72	0.68	0.58	0.63	0.66	0.69	0.53	0.70	0.60	0.86	6.73	0.57	0.58	0.71	0.76	0.67	0.59	1.67	0.77	NA	0.72	0.59
CFD	0.91	NA	0.90	1.06	0.88	0.91	1.12	1.05	0.93	NA	1.04	1.06	0.88	1.27	1.00	1.00	0.87	0.91	0.91	1.39	0.86	1.03
CFH	0.70	0.91	1.09	0.95	0.84	0.92	0.72	1.10	1.05	1.06	1.07	0.68	1.11	1.01	0.93	0.83	NA	1.40	1.63	0.68	1.30	1.01
CFI	0.89	0.94	0.85	0.94	0.92	1.29	NA	0.78	0.85	1.92	2.27	0.76	0.67	0.94	1.01	0.70	0.65	1.12	1.20	0.94	0.80	0.59
CFP	1.15	0.69	2.25	1.16	1.90	0.63	0.30	0.67	2.81	0.83	0.33	0.73	0.85	1.24	0.89	0.34	0.35	0.96	1.76	0.51	1.05	0.60
CR1L	0.98	0.96	0.85	0.92	0.97	0.93	1.11	1.04	0.95	0.92	1.41	0.93	0.91	0.96	0.88	1.14	1.33	0.97	0.83	1.04	0.89	1.08
CR2	1.06	1.03	0.91	0.96	0.98	0.98	0.91	0.98	1.07	1.07	0.92	1.01	0.99	1.03	0.98	0.92	1.08	0.99	1.05	1.04	1.05	0.99
DF	0.21	0.37	1.23	0.83	1.94	1.56	0.33	0.42	0.57	0.82	0.24	0.95	0.71	4.58	0.49	0.60	0.33	0.63	0.22	0.24	3.78	0.94
ERCC2	NA	1.03	0.96	0.96	0.98	NA	0.99	0.90	1.34	0.97	0.95	0.97	1.00	0.83	0.88	NA	1.16	0.99	0.96	1.18	0.92	1.01
ITGB2	0.45	0.43	0.31	0.50	0.47	0.73	0.34	0.34	1.53	0.69	0.44	5.18	1.52	3.03	2.14	0.45	0.33	1.56	0.31	0.41	0.47	0.38

2.3. De Novo *Pathway Reconstruction: Topological Parameters*

Pathway analysis fails to consider many known genes and miRNAs that are not annotated in any pathway. To fill these gaps, we used *de novo* network reconstruction using both mRNA and miRNA profiles. Using a correlation measure with a permutation-based threshold of 0.9 of mutual information (0.9 was the maximum value of mutual information of the network generated by the permuted expression matrix), we generated a network with 7762 nodes (7647 genes and 115 miRNAs) and 44,092 edges (Figure 4). The global architecture of the network is characterized by two large clusters, which are shown as the blue and violet nodes in Figure 4. As expected (Figure 2A), these two clusters are composed of genes prevalently expressed in heart (the most different tissue) and in blood vessels (Figure S1). Thus, we separated these two clusters to create a vessel-specific and a heart-specific network.

To gain insight into the structure of complex networks of this type, various topological parameters were calculated (Table 2). The heart network is sparser and less connected than the vessel network. This is reflected by a larger number of connected components, a higher diameter and a smaller number of neighborhood genes of the heart network.

Figure 4. Regulatory network reconstructed using mutual information. The edges of the network are colored according to their prevalent expression. Heart-specific genes are shown in violet, vessel-specific genes are shown in blue, and genes without tissue-specific expression are shown in pink.

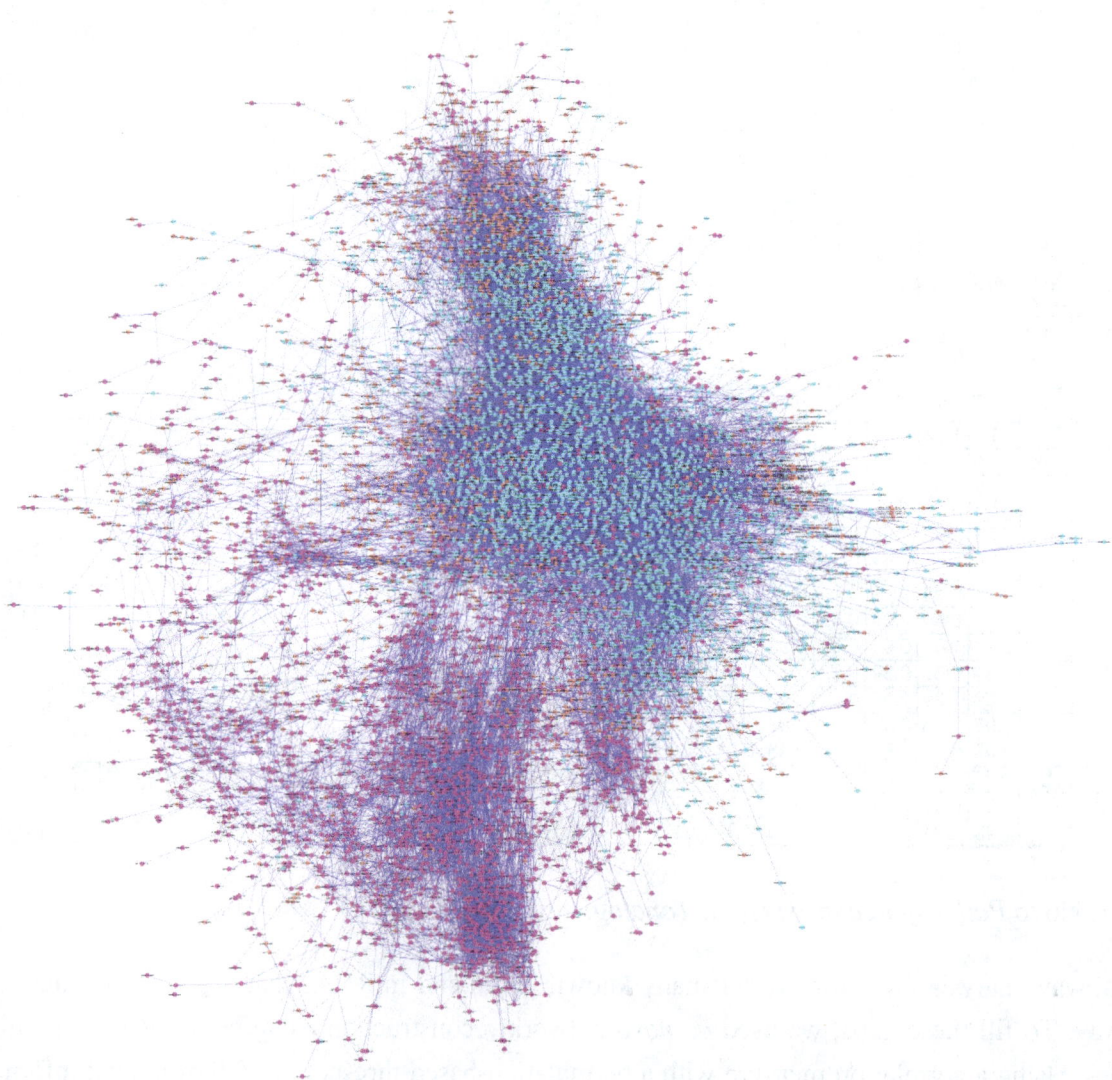

The degree of a node, also referred to as its connectivity, is the number of edges connected to the node. Based on this definition, the nodes with the highest connectivities are called hubs. In general, hub genes are master regulators and play important roles in the biology of the cell. In our networks, we define as hubs the top 5% of genes in the connectivity distribution. We found 162 and 128 hubs in the vessel and heart networks, respectively. The hub genes of the vessel network encode proteins that participate in two main processes: RNA processing and the regulation of apoptotic events (Table S2). During normal development as well as in pathology, the formation of new vessels and the regression of pre-existing ones depend on the balance between endothelial cell proliferation and endothelial cell apoptosis. In mature vessels, endothelial cell turnover is also under the control of these tightly regulated phenomena. Among the hubs of the heart network, we identified genes involved in cell membrane structure and signal transduction through MAPK activity as well as genes encoding various ion transporters (e.g., Na^{2+}, K^+) (Table S2). The members of the MAPK family are involved in the regulation of many cellular processes, including cell growth, differentiation, development, the cell cycle, death, and survival. Activation of genes in the MAPK family plays a key role in the pathogenesis of various processes in the heart, including myocardial hypertrophy and its transition to heart failure, ischemic and reperfusion injury, and cardioprotection conferred by ischemia- or drug-induced preconditioning [86].

Table 2. Summary of the principal topological parameters estimated for the *de novo* reconstructed network.

Topological parameters	Heart network	Vessels network
Average clustering coefficient	0.195	0.234
Connected components	237	86
Avg. number of neighbors	6.329	15.611
Network radius	1	1
Network diameter	36	16
Network centralization	0.020	0.036
Network density	0.002	0.005
Network heterogeneity	1.198	1.183

The *de novo* reconstructed network (Figure 4) is characterized by the presence of different miRNAs (Table S3) that are responsible for the regulation of vessel specificity. Figure 5 represents the sub-network of the neighboring genes of miRNAs. Interestingly, the central part of the network (the densely connected portion of the sub-network) is characterized by genes involved in smooth muscle contraction (Table S4) that show differential expression in arteries and veins (Figure 6). As discussed previously, a thicker ring of smooth muscle is present in arteries than in veins (see Section 2.2). Our results suggest that this difference may be regulated by specific miRNAs that display anti-correlated expression with their putative targets (Figure 6).

Specifically, the *α 2-actin* (ACTA2) smooth muscle gene in aorta (ENSSSCG00000010447) is regulated by a specific miRNA (prediction_15_14390446_14390503_-_3p) that is down-regulated in the aorta and up-regulated in venous tissue (Figure 6). Defects in ACTA2 are the cause of aortic aneurysm familial thoracic type 6 (AAT6) [MIM:611788]. AATs are characterized by permanent dilation of the thoracic aorta, usually due to degenerative changes in the aortic wall. RHOB (Ssc#S35170885), an important gene involved in vasoconstriction, is also regulated by miR-133a (Figure 6). RHO gene family is involved in vascular morphogenesis [87], and miR-133a contributes to the phenotypic state of smooth muscle cells both *in vitro* and *in vivo*, suggesting a potential for therapeutic application of this miRNA in vascular disease [88]. In fact, miR-133a, in association with miR143/145, is fundamental for the maintenance of the contractile smooth muscle cell phenotype [88]. The expression of miRNAs prediction_15_14390446_14390503_-_3p and miR-133 and their targets ACAT2 and RHOB was confirmed by qRT-PCR (Figure 6C).

Figure 5. Gene and miRNA interaction sub-network describing vessel specificity. Triangles represent miRNAs; circles represent mRNAs. Gene expression in the ascending aorta according to $\log_2$ (gene expression/average gene expression) is represented by color; green indicates down-regulation, red indicates up-regulation. Under each node, histograms representing $\log_2$ (gene expression/average gene expression) in the ascending aorta, descending aorta, inferior vena cava, and superior vena cava (reading from left to right) are shown. The area highlighted by the circle indicates the densely connected portion of the sub-network (an enlarged view of this area is available in Figure 6).

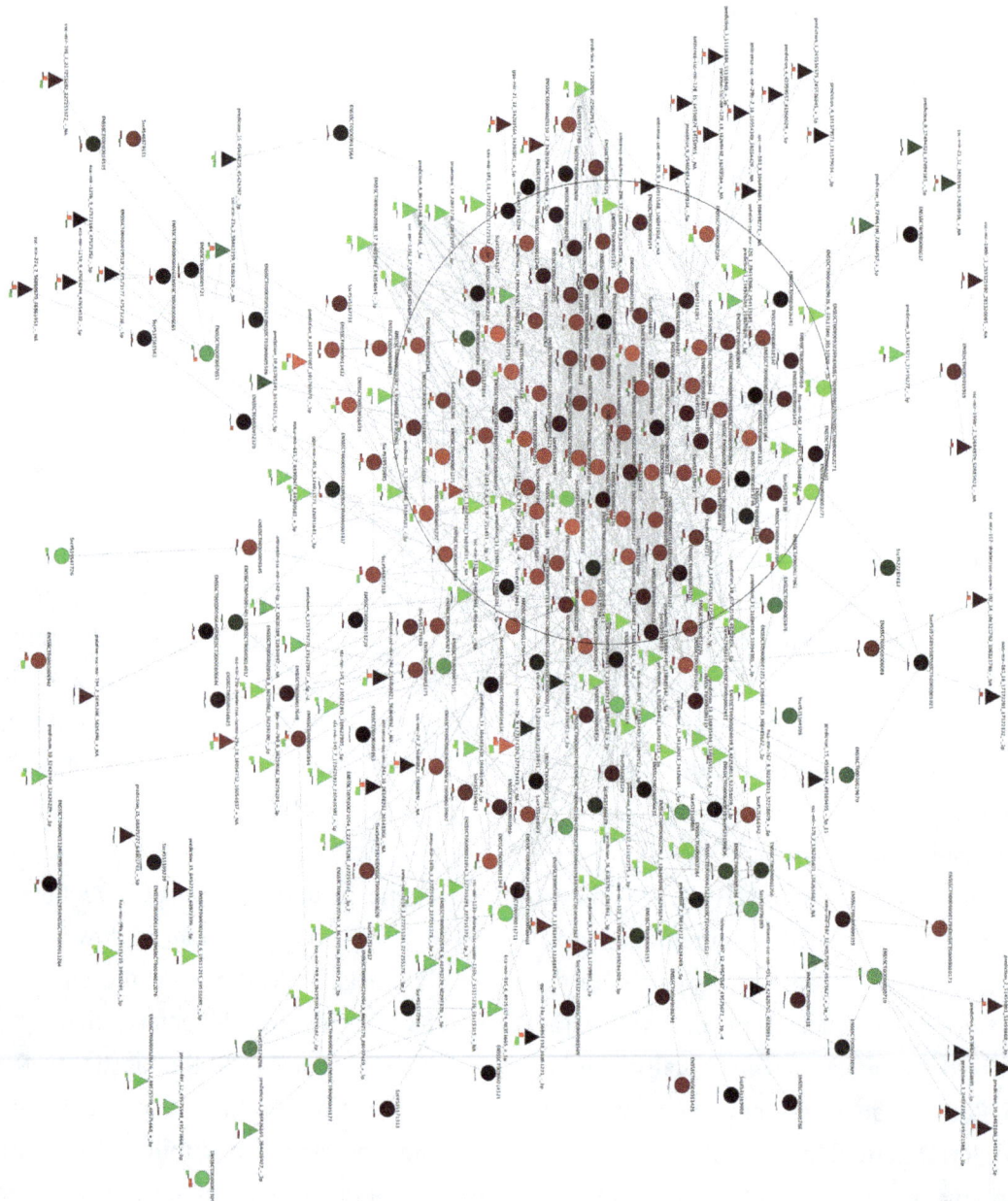

Figure 6. Enlarged view of the densely connected area of Figure 5. (**A**) The colors indicate expression in the aorta; (**B**) The colors indicate expression in veins. The triangles represent miRNAs; circles represent mRNAs. Up-regulated = red; down-regulated = green; * = nodes discussed in the text; (**C**) qRT-PCR results confirm that there is an inverse relationship between miRNAs and their targets. P_15 is for prediction_15_14390446_14390503_-_3p. In *Y* axis the original expression level related to H3. Bars are for standard deviation between three replicates.

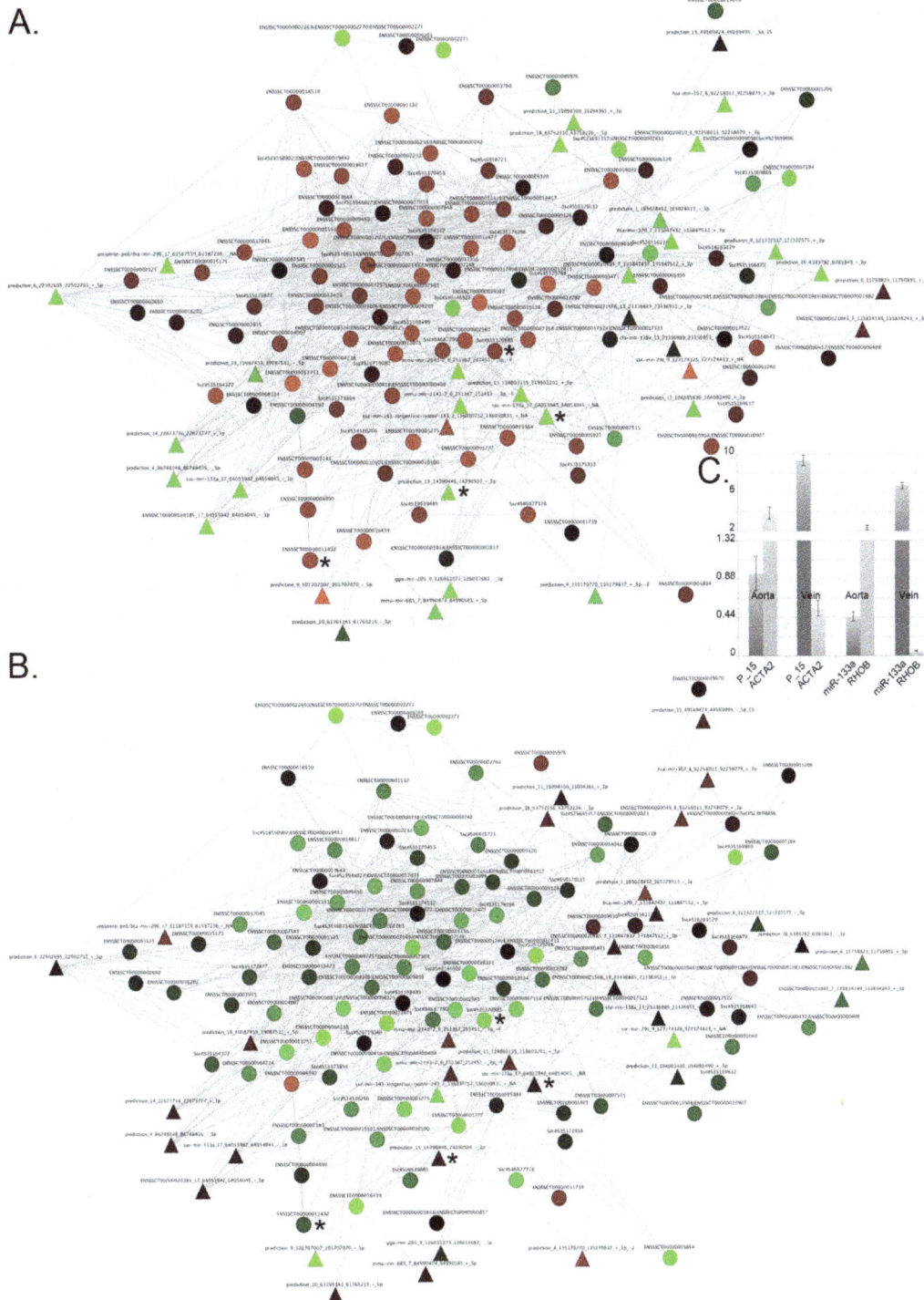

2.4. Integration of Supervised and Unsupervised Approaches

Supervised and unsupervised approaches gave similar results in terms of biological processes involved in tissue specificity. However, their complementary behavior might be better exploited through the use of an integrative approach. Specifically, our aim is to combine the topology of the discovered pathways with that of the *de novo* reconstructed network. The advantage of combining the topologies obtained in sections 2.2 and 2.3 is two-fold: (i) it allows the expansion of pathway definitions to include genes currently without pathway annotation; and (ii) it permits the inclusion of miRNAs. Using the topological structure of the pathway as a backbone, we include new genes in the pathway, following two rules: (i) a gene/miRNA is added only if it presents an edge in the *de novo* network with at least one gene in the pathway; and (ii) additional miRNAs are included if they share an edge with previously added non-annotated genes. Here, we will use this strategy to discuss one of the most interesting pathways significantly activated in arteries: the smooth muscle contraction pathway (Figure 7A). The genes used to expand this pathway (the γ isoform of the catalytic subunit of *protein phosphatase 1* (*PPP1CC*), *transgelin* (*TAGLN*), and smooth muscle and non-muscle *myosin light chain 6* (*Myl6*), among others) are primarily involved in membrane and actin filament organization, actomyosin function and responses to specific stimuli (NF-κB binding and response to unfolded protein) (Table S5), reflecting their functional congruence with the smooth muscle contraction pathway. Indeed, the membrane organization category includes the organismation of the sarcoplasmic reticulum, which is involved in the regulation of intracellular Ca^{2+} concentration (Figure 7). All of these genes are prevalently expressed in smooth muscle; in particular, TAGLN was purified from bovine aorta [89]. Moreover, we added 61 miRNAs that putatively regulate genes involved, directly or indirectly in the smooth muscle contraction pathway (Figure 7A). Interestingly, 23 miRNAs are involved in the regulation of the original genes of the pathway (core genes). Among these miRNAs, miR-542 (ENSSSCT00000021275), which was shown in a previous work to be involved in the epithelial-mesenchymal transition [90], was found to be associated with *vimentin* (*VIM*) regulation (Figure 7B). Finally, it is worth noting that many other miRNAs important for vascular remodeling and smooth muscle phenotypic control, such as miR-133 [88], miR-143 [91], miR-99b [92], miR-23a [93], miR-138 (ENSSSCT00000021566) [94], miR-29c [95], miR-125a (ENSSSCT00000020936) [95], and miR-24 [96]), are included in this network.

Figure 7. (A) Combination of pathway topology and *ab initio* reconstructed network. Nodes corresponding to the Reactome pathway (core nodes) are shown in red; additional genes in the first neighborhood of the core nodes obtained from the *ab initio* network are shown in light blue, and miRNAs are shown in grey; **(B)** Portion of **(A)** representing the miRNAs regulating the core nodes.

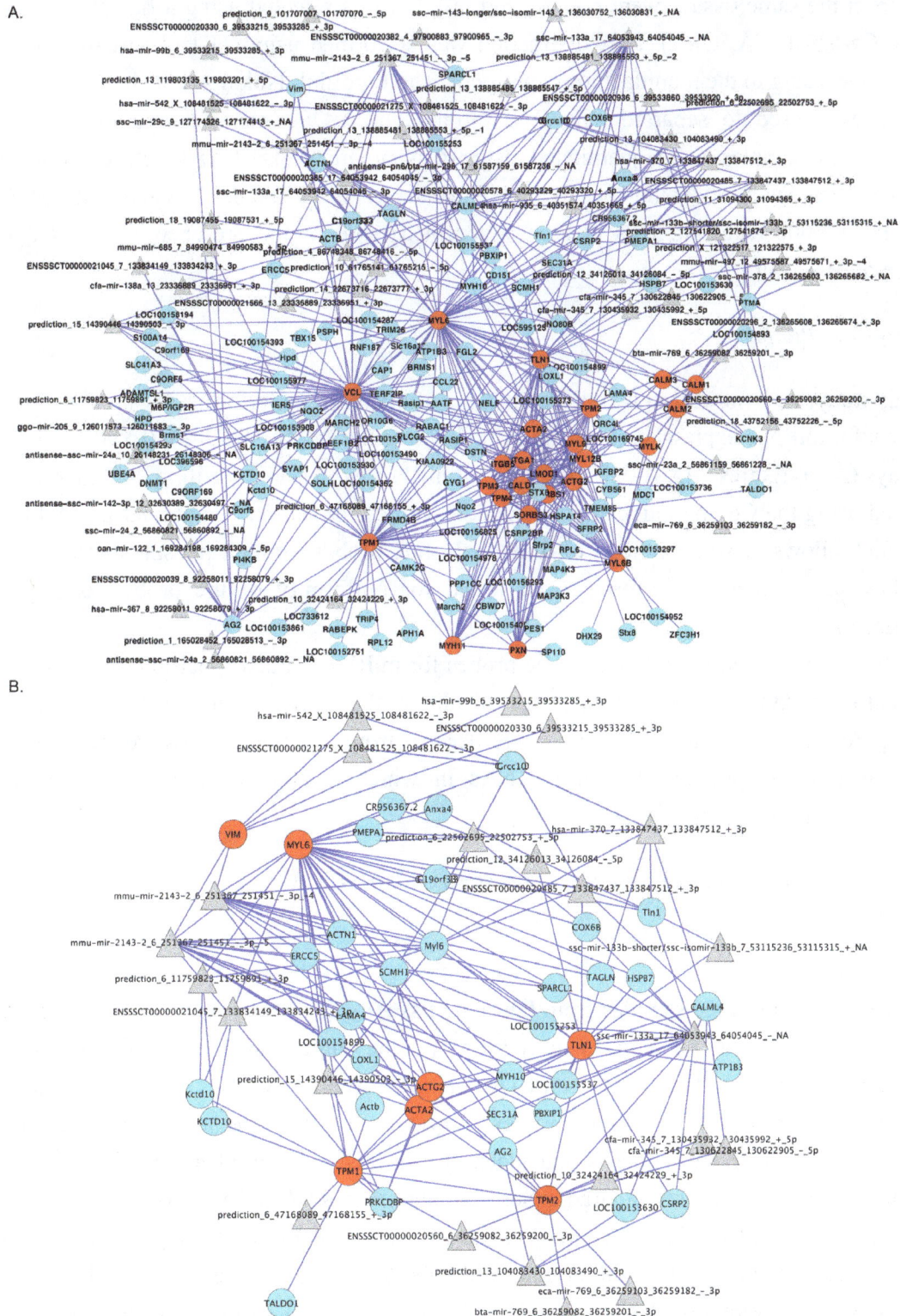

3. Experimental Section

3.1. Sample Preparation

RNA samples (total RNA and small RNAs) were extracted from the analyzed tissues of three non-inbred pigs and kept at80 °C until use. Before the experiments were performed, the three samples from the same tissues were pooled, and miRNA was selected using a flashPAGE instrument (Ambion, Carlsbad, CA, USA). RNA extraction was performed using TRIzol (Invitrogen, Carlsbad, CA, USA) according to the manufacturer's protocol. The PureLink Isolation Kit (Invitrogen, Carlsbad, CA, USA) was used to separate long RNA from short (<200 nt, after use in the flashPAGE instrument). All samples were quantitated using a NanoDrop ND-1000 spectrophotometer; RNA quality was then analyzed using the Agilent Bioanalyser 2100 (Agilent, Santa Clara, CA, USA) (Agilent RNA 6000 nano kit; RIN at least 7 accepted) and for the presence of miRNA using the Agilent small RNA kit.

3.2. Microarray Platforms

For this study, we synthesized two different types of microarray platforms: (a) 4 × 2 K Combimatrix microarrays for miRNA expression profiling (ArrayExpress ID: A-MEXP-2348); (b) 90 K Combimatrix microarrays (ArrayExpress ID: A-MEXP-2351) for mRNA expression profiling. All microarrays were synthesized using the Combimatrix oligonucleotide synthesizer station (Combimatrix, Mukilteo, WA, USA), which allows *in situ* synthesis of oligonucleotide probes through phosphoramidite chemistry. All synthesized microarray platforms were tested for uniformity of the probes as suggested by the manufacturer.

The 4 × 2 K microarrays contain specific probes for miRNAs. Each specific probe is flanked by a background probe that is used in the analysis to subtract the corresponding background fluorescence signal (Figure 1). The background probes were derived from a previous RAKE experiment aimed at the identification of specific ends of miRNAs in which a tiling microarray was used for the scope (Figure S2) [74].

3.3. Microarray mRNA and miRNA Gene Expression and qRT-PCR

3.3.1. mRNA

Pooled RNA (1 µg; three samples from the same tissue) was linearly amplified and labeled by the addition of biotinylated nucleotides according to the procedure described in the Ambion MessageAmp™ II aRNA Amplification kit (Ambion, Carlsbad, CA, USA). The procedure includes reverse transcription with an oligo-dT primer carrying a T7 promoter to produce the first-strand cDNA. After second-strand synthesis and clean-up, the cDNA is used as template in an *in vitro* transcription reaction to generate a large quantity of antisense RNA (aRNA). Biotinylated UTPs were incorporated into the aRNA during the *in vitro* transcription reaction. Following purification, 18 µg of aRNA was fragmented using the Ambion Fragmentation Kit (Ambion, Carlsbad, CA, USA). Intact and fragmented aRNAs were tested on an Agilent Bioanalyzer 2100 (Agilent, Santa Clara, CA, USA) using the RNA 6000 Nano LabChip (Agilent, Santa Clara, CA, USA). The size of intact aRNAs ranged from 300 to 4000 nucleotides, while that of fragmented aRNAs ranged from 50 to 250 nucleotides. Fragmented aRNA was hybridized to pre-hybridized 90 K Combimatrix microarrays. The pre-hybridization step was performed for 2 h at 42 °C in a solution containing 5× Denhardt's solution, 100 ng/μL salmon sperm DNA and 0.05% SDS in 1× Hyb solution prepared as suggested by Combimatrix. Hybridizations were carried out with 4.8 μg of fragmented aRNA in 25% DI formamide, 100 ng/μL salmon sperm DNA and 0.04% SDS in 1× Hybridization solution at 42 °C for 18 h with constant mixing. After hybridization, the microarray platforms were washed with the following:

- 6× SSPET (SSPE added with 0.05% of Tween-20) preheated at 42 °C for 5 min;
- 3× SSPET for 1 min at room temperature;
- 0.5× SSPET for 1 min at room temperature; and
- PBST for 1 min at room temperature.

The microarray chamber was then filled with biotin blocking solution (0.1% Tween-20 and 10 mg/mL BSA in 2× PBS) and incubated at room temperature for 1 h. Labeling was performed by incubating the microarray with dye labeling solution (0.1% Tween-20, 10 mg/mL BSA and 1.6 ng of Cy3-streptavidin (Amersham, Little Chalfont, UK) in 2× PBS) for 1 h at room temperature. After the washing steps (PBST for 1 min at room temperature two times; PBS for 1 min at room temperature), microarrays were scanned at 3 μm resolution with the VersArray ChiprRaderTM (BioRad, Hercules, CA, USA) (ArrayExpress ID: E-MTAB-1941).

3.3.2. miRNA

A sample of the miRNA pool (350 ng) was hybridized for 20 h at 37 °C in a static hybridization oven in hybridization buffer consisting of 6× SSPE, 8 mg/mL BSA, 700 ng of small RNAs and spike-in. After hybridization, the microarrays were washed with the following stringent procedure:

- 1 min at room temperature with 6× SSPET (SSPE containing 0.05% Tween-20);
- 1 min at room temperature with 3× SSPET;
- 1 min at room temperature with 2× PBS;
- 1 min at room temperature with 1× Buffer 2 (the buffer for the Klenow enzyme).

The RAKE reaction was performed at 36.5 °C by incubating the microarray for 1.5 h in 1× Buffer 2 containing 16 μM biotin-14-dATP (Invitrogen, Carlsbad, CA, USA) and 0.25 U/μL Klenow fragment (3'→5' exo−) (NEB, Ipswich, MA, USA). The microarrays were washed two times in 1× Buffer 2 and incubated in biotin blocking solution for 1 h at room temperature. Extended miRNAs (primers) were labeled by incubating the microarray in the dye labeling solution for 1 h at room temperature. The microarrays were rinsed in PBST (0.1% Tween-20 in 2× PBS) for 1 min at room temperature and in 2× PBS for 1 min at room temperature and scanned (ArrayExpress ID: E-MTAB-1938).

qRT-PCR was used to validate the expression of miRNAs and mRNAs. For mRNA, the SYBR green approach was used in association with the *Power* SYBR® Green PCR Master Mix (Applied Biosystems, Carlsbad, CA, USA); for miRNA, the NCode™ SYBR® Green miRNA qRT-PCR Kit (Life Technologies, Carlsbad, CA, USA) was used according to the manufacturer's specifications. The primers used were GCATGCAGAAGGAGATCACA (left) and GCTGGAAGGTGGACAGA

GAG (right) for ACTA2, TATGTGCTTCTCGGTGGACA (left) and CGAGGTAGTCGTA GGCTTGG (right) for RHOB, and GGTTCCCAGGCTAGGGGTCG (specific) for prediction_15_14390446_14390503_-_3p and CAGCTGGTTGAAGGGGACCA for miR-133a. The reference genes used were GAPDH for mRNA and snU6 for miRNA. The results shown are normalized to the expression of histone H3.

3.4. Data Analysis

Images of hybridized mRNA microarrays were quantitated using the Combimatrix imaging software. The raw data were normalized using the quantile method. The goal of the quantile method is to normalize the distribution of probe intensities across a set of microarrays. After normalization, the fluorescence intensities of probe spots presenting values lower than the average of the medians of all negative control probes were set as missing values (NA). The negative control probes were used to calculate the background value (filter). Probe spots presenting NA in more than six experiments were excluded from data analysis. Before performing the analysis, the intensity values of the replicated probes were averaged. Differentially expressed genes were identified using the MeV suite [97] and applying PCA (Principal Components Analysis) [98] and SAM (Significance Analysis of Microarrays) [2] analysis. COA (Correspondence Analysis) analysis [99] was used to determine the specificity of the *de novo* reconstructed network. Gene enrichment was performed using the DAVID web application [100]; pathway analysis was performed using GraphiteWeb [20].

miRNA data were pre-processed as previously described except that cyclic lowess normalization was applied [101]. After inter-array normalization, the fluorescence intensity of the specific miRNA probe was subtracted from the corresponding background fluorescence and used to extrapolate the miRNA concentration from the spike-in-derived curve. The spike-in curve was extrapolated using spline interpolation [102].

Pig gene symbols from Ensembl were converted to human gene symbols using the Ensembl orthologous database through the BioMart service. For UniGene clusters, we extracted the most similar protein or gene curated by NCBI (http://www.ncbi.nlm.nih.gov/ (accessed on 13 November 2013)) based on sequence similarity and then used the NCBI HomoloGene database to translate the protein or gene to its human homolog. This method is commonly used to map genes to pathways in non-model organisms or to map genes that are poorly annotated in model organisms [103]; it is also common to use the well-curated human pathways to extrapolate pathways for non-model organisms. GSEA [10] was then performed using the GraphiteWeb web tool [20].

Mutual information (MI) between all pairs of genes and miRNAs was estimated using the parmigene Bioconductor package [31]; miRNA-miRNA interactions have been removed. To assess MI significance, we estimated the null distribution using a permutational approach. The expression profiles of miRNAs and mRNAs were randomly shuffled, and MI was then estimated on the shuffled matrices. To generate the global network, we included only interactions with MI that were greater than the maximum MI value obtained from the null distribution, which was 0.9 (corresponding to quantile 0.999 in the empirical distribution).

The Cytoscape tool [104] with the Networkanalyser [105] plugin was used to estimate the topological properties of heart and vessel networks.

The topologies of the most interesting pathways derived from pathway annotation (graphite Bioconductor package) were integrated with the topology of the *de novo* reconstructed network. The combination was performed using the pathway topology as backbone; new genes/miRNAs were then added based on fulfilment of one of the following criteria: (i) if the new gene/miRNA shares an edge in the *de novo* network with at least one gene in the pathway; and (ii) if an miRNA shares an edge in the *de novo* network with at least one previously added gene.

4. Conclusions

Gene set analyses have been shown to provide better insights and more robust results in array experiments than classical gene-by-gene approaches. Here, we reviewed various strategies used in gene set analysis and showed how to address their integration. We combined genome and pathway information with expression data and applied this approach to a case study, the analysis of the pig cardiocirculatory system. Two new platforms for pig transcriptome analysis (mRNA and miRNA) were presented and applied to the study of tissue specificity. Different expression patterns were identified in heart and vessels; within these, arteries show distinct profiles from those of veins. These findings seem to be associated with the functional and structural composition of the vessels. In agreement with histochemical evidence, pathway analysis revealed the greater importance of smooth muscle in arteries than in veins. We showed that miRNAs participate in the definition of arterial and venous pathways; specifically, for smooth muscle, our data indicate the importance of miR-133a in regulating the *RHOB* gene. The use of a combination of supervised and unsupervised approaches allowed us to expand the compositions of known pathways to include new genes involved in membrane and actin filament organization, actomyosin function and response to stimuli and new miRNAs, most of which are known to be associated with vascular remodeling and control of the smooth muscle phenotype. These results demonstrate the feasibility and usefulness of combining these two approaches in identifying new candidate genes whose expression is associated with specific experimental conditions.

Acknowledgments

The authors acknowledge the CARIPARO Foundation (Project for Excellence 2012: "Role of coding and non-coding RNA in chronic myeloproliferative neoplasms: from bioinformatics to translational research") and the CRIBI Center for high-performance computing resources funded by the Regione Veneto (RISIB project SMUPR n. 4145). The authors wish to thank the University of Padova for support of this work (CPDR075919 and CPDA119031 to Chiara Romualdi; CPDR070805 to Gabriele Sales) and MicroCribi service for microarray synthesis support (http://microcribi.cribi.unipd.it).

References

1. Smyth, G.K. Linear models and empirical bayes methods for assessing differential expression in microarray experiments. *Stat. Appl. Genet. Mol. Biol.* **2004**, *3*, 1–28.

2. Tusher, V.G.; Tibshirani, R.; Chu, G. Significance analysis of microarrays applied to the ionizing radiation response. *Proc. Natl. Acad. Sci. USA* **2001**, *98*, 5116–5121.

3. Shen, K.; Tseng, G.C. Meta-analysis for pathway enrichment analysis when combining multiple genomic studies. *Bioinformatics* **2010**, *26*, 1316–1323.

4. Callegaro, A.; Basso, D.; Bicciato, S. A locally adaptive statistical procedure (LAP) to identify differentially expressed chromosomal regions. *Bioinformatics* **2006**, *22*, 2658–2666.

5. Toedling, J.; Schmeier, S.; Heinig, M.; Georgi, B.; Roepcke, S. MACAT—Microarray chromosome analysis tool. *Bioinformatics* **2005**, *21*, 2112–2113.

6. Turkheimer, F.E.; Roncaroli, F.; Hennuy, B.; Herens, C.; Nguyen, M.; Martin, D.; Evrard, A.; Bours, V.; Boniver, J.; Deprez, M. Chromosomal patterns of gene expression from microarray data: Methodology, validation and clinical relevance in gliomas. *BMC Bioinform.* **2006**, *7*, 526.

7. Barry, W.T.; Nobel, A.B.; Wright, F.A. Significance analysis of functional categories in gene expression studies: A structured permutation approach. *Bioinformatics* **2005**, *21*, 1943–1949.

8. Goeman, J.J.; van de Geer, S.A.; de Kort, F.; van Houwelingen, H.C. A global test for groups of genes: Testing association with a clinical outcome. *Bioinformatics* **2004**, *20*, 93–99.

9. Subramanian, A.; Tamayo, P.; Mootha, V.K.; Mukherjee, S.; Ebert, B.L.; Gillette, M.A.; Paulovich, A.; Pomeroy, S.L.; Golub, T.R.; Lander, E.S.; *et al.* Gene set enrichment analysis: A knowledge-based approach for interpreting genome-wide expression profiles. *Proc. Natl. Acad. Sci. USA* **2005**, *102*, 15545–15550.

10. Tian, L.; Greenberg, S.A.; Kong, S.W.; Altschuler, J.; Kohane, I.S.; Park, P.J. Discovering statistically significant pathways in expression profiling studies. *Proc. Natl. Acad. Sci. USA* **2005**, *102*, 13544–13549.

11. Levin, A.M.; Ghosh, D.; Cho, K.R.; Kardia, S.L. A model-based scan statistic for identifying extreme chromosomal regions of gene expression in human tumors. *Bioinformatics* **2005**, *21*, 2867–2874.

12. Efron, B.; Tibshirani, R. On testing the significance of sets of genes. *Ann. Appl. Stat.* **2007**, *1*, 107–129.

13. Tarca, A.L.; Draghici, S.; Khatri, P.; Hassan, S.S.; Mittal, P.; Kim, J.S.; Kim, C.J.; Kusanovic, J.P.; Romero, R. A novel signaling pathway impact analysis. *Bioinformatics* **2009**, *25*, 75–82.

14. Martini, P.; Sales, G.; Massa, M.S.; Chiogna, M.; Romualdi, C. Along signal paths: An empirical gene set approach exploiting pathway topology. *Nucleic Acids Res.* **2013**, *41*, e19.

15. Ackermann, M.; Strimmer, K. A general modular framework for gene set enrichment analysis. *BMC Bioinform.* **2009**, *10*, 47.

16. Nam, D.; Kim, S.Y. Gene-set approach for expression pattern analysis. *Brief. Bioinform.* **2008**, *9*, 189–197.

17. Sales, G.; Calura, E.; Cavalieri, D.; Romualdi, C. Graphite—A Bioconductor package to convert pathway topology to gene network. *BMC Bioinform.* **2012**, *13*, 20.

18. Martini, P.; Risso, D.; Sales, G.; Romualdi, C.; Lanfranchi, G.; Cagnin, S. Statistical Test of Expression Pattern (STEPath): A new strategy to integrate gene expression data with genomic information in individual and meta-analysis studies. *BMC Bioinform.* **2011**, *12*, 92.

19. Massa, M.S.; Chiogna, M.; Romualdi, C. Gene set analysis exploiting the topology of a pathway. *BMC Syst. Biol.* **2010**, *4*, 121.

20. Sales, G.; Calura, E.; Martini, P.; Romualdi, C. Graphite Web: Web tool for gene set analysis exploiting pathway topology. *Nucleic Acids Res.* **2013**, *41*, W89–W97.

21. Morgat, A.; Coissac, E.; Coudert, E.; Axelsen, K.B.; Keller, G.; Bairoch, A.; Bridge, A.; Bougueleret, L.; Xenarios, I.; Viari, A. UniPathway: A resource for the exploration and annotation of metabolic pathways. *Nucleic Acids Res.* **2012**, *40*, D761–D769.

22. Kanehisa, M.; Goto, S.; Furumichi, M.; Tanabe, M.; Hirakawa, M. KEGG for representation and analysis of molecular networks involving diseases and drugs. *Nucleic Acids Res.* **2010**, *38*, D355–D360.

23. Caspi, R.; Foerster, H.; Fulcher, C.A.; Kaipa, P.; Krummenacker, M.; Latendresse, M.; Paley, S.; Rhee, S.Y.; Shearer, A.G.; Tissier, C.; *et al.* The MetaCyc Database of metabolic pathways and enzymes and the BioCyc collection of Pathway/Genome Databases. *Nucleic Acids Res.* **2008**, *36*, D623–D631.

24. Joshi-Tope, G.; Gillespie, M.; Vastrik, I.; D'Eustachio, P.; Schmidt, E.; de Bono, B.; Jassal, B.; Gopinath, G.R.; Wu, G.R.; Matthews, L.; *et al.* Reactome: A knowledgebase of biological pathways. *Nucleic Acids Res.* **2005**, *33*, D428–D432.

25. Friedman, N. Inferring cellular networks using probabilistic graphical models. *Science* **2004**, *303*, 799–805.

26. Butte, A.J.; Tamayo, P.; Slonim, D.; Golub, T.R.; Kohane, I.S. Discovering functional relationships between RNA expression and chemotherapeutic susceptibility using relevance networks. *Proc. Natl. Acad. Sci. USA* **2000**, *97*, 12182–12186.

27. Schafer, J.; Strimmer, K. An empirical Bayes approach to inferring large-scale gene association networks. *Bioinformatics* **2005**, *21*, 754–764.

28. Markowetz, F.; Spang, R. Inferring cellular networks—A review. *BMC Bioinform.* **2007**, *8*, S5.

29. Basso, K.; Margolin, A.A.; Stolovitzky, G.; Klein, U.; Dalla-Favera, R.; Califano, A. Reverse engineering of regulatory networks in human B cells. *Nat. Genet.* **2005**, *37*, 382–390.

30. Margolin, A.A.; Nemenman, I.; Basso, K.; Wiggins, C.; Stolovitzky, G.; Dalla Favera, R.; Califano, A. ARACNE: An algorithm for the reconstruction of gene regulatory networks in a mammalian cellular context. *BMC Bioinform.* **2006**, *7*, S7.

31. Sales, G.; Romualdi, C. Parmigene—A parallel R package for mutual information estimation and gene network reconstruction. *Bioinformatics* **2011**, *27*, 1876–1877.

32. Marbach, D.; Costello, J.C.; Kuffner, R.; Vega, N.M.; Prill, R.J.; Camacho, D.M.; Allison, K.R.; Consortium, D.; Kellis, M.; Collins, J.J.; *et al.* Wisdom of crowds for robust gene network inference. *Nat. Methods* **2012**, *9*, 796–804.

33. Macias, S.; Plass, M.; Stajuda, A.; Michlewski, G.; Eyras, E.; Caceres, J.F. DGCR8 HITS-CLIP reveals novel functions for the Microprocessor. *Nat. Struct. Mol. Biol.* **2012**, *19*, 760–766.

34. Thomson, D.W.; Bracken, C.P.; Goodall, G.J. Experimental strategies for microRNA target identification. *Nucleic Acids Res.* **2011**, *39*, 6845–6853.

35. Hafner, M.; Landthaler, M.; Burger, L.; Khorshid, M.; Hausser, J.; Berninger, P.; Rothballer, A.; Ascano, M., Jr.; Jungkamp, A.C.; Munschauer, M.; *et al.* Transcriptome-wide identification of RNA-binding protein and microRNA target sites by PAR-CLIP. *Cell* **2010**, *141*, 129–141.

36. Chi, S.W.; Zang, J.B.; Mele, A.; Darnell, R.B. Argonaute HITS-CLIP decodes microRNA-mRNA interaction maps. *Nature* **2009**, *460*, 479–486.

37. Yousef, M.; Showe, L.; Showe, M. A study of microRNAs *in silico* and *in vivo*: Bioinformatics approaches to microRNA discovery and target identification. *FEBS J.* **2009**, *276*, 2150–2156.

38. Witkos, T.M.; Koscianska, E.; Krzyzosiak, W.J. Practical aspects of microRNA target prediction. *Curr. Mol. Med.* **2011**, *11*, 93–109.

39. Sales, G.; Coppe, A.; Bisognin, A.; Biasiolo, M.; Bortoluzzi, S.; Romualdi, C. MAGIA, a web-based tool for miRNA and Genes Integrated Analysis. *Nucleic Acids Res.* **2010**, *38*, W352–W359.

40. Bisognin, A.; Sales, G.; Coppe, A.; Bortoluzzi, S.; Romualdi, C. MAGIA^2: From miRNA and genes expression data integrative analysis to microRNA-transcription factor mixed regulatory circuits (2012 update). *Nucleic Acids Res.* **2012**, *40*, W13–W21.

41. Nam, S.; Li, M.; Choi, K.; Balch, C.; Kim, S.; Nephew, K.P. MicroRNA and mRNA integrated analysis (MMIA): A web tool for examining biological functions of microRNA expression. *Nucleic Acids Res.* **2009**, *37*, W356–W362.

42. Ross, J.W.; Fernandez de Castro, J.P.; Zhao, J.; Samuel, M.; Walters, E.; Rios, C.; Bray-Ward, P.; Jones, B.W.; Marc, R.E.; Wang, W.; *et al.* Generation of an inbred miniature pig model of retinitis pigmentosa. *Investig. Ophthalmol. Vis. Sci.* **2012**, *53*, 501–507.

43. Maxmen, A. Model pigs face messy path. *Nature* **2012**, *486*, 453.

44. Sandrin, M.S.; Loveland, B.E.; McKenzie, I.F. Genetic engineering for xenotransplantation. *J. Card. Surg.* **2001**, *16*, 448–457.

45. Ekser, B.; Rigotti, P.; Gridelli, B.; Cooper, D.K. Xenotransplantation of solid organs in the pig-to-primate model. *Transpl. Immunol.* **2009**, *21*, 87–92.

46. Zhang, Q.; Widmer, G.; Tzipori, S. A pig model of the human gastrointestinal tract. *Gut Microbes* **2013**, *4*, 193–200.

47. Kragh, P.M.; Nielsen, A.L.; Li, J.; Du, Y.; Lin, L.; Schmidt, M.; Bogh, I.B.; Holm, I.E.; Jakobsen, J.E.; Johansen, M.G.; *et al.* Hemizygous minipigs produced by random gene insertion and handmade cloning express the Alzheimer's disease-causing dominant mutation APPsw. *Transgenic Res.* **2009**, *18*, 545–558.

48. Granada, J.F.; Kaluza, G.L.; Wilensky, R.L.; Biedermann, B.C.; Schwartz, R.S.; Falk, E. Porcine models of coronary atherosclerosis and vulnerable plaque for imaging and interventional research. *EuroIntervention* **2009**, *5*, 140–148.

49. Groenen, M.A.; Archibald, A.L.; Uenishi, H.; Tuggle, C.K.; Takeuchi, Y.; Rothschild, M.F.; Rogel-Gaillard, C.; Park, C.; Milan, D.; Megens, H.J.; *et al.* Analyses of pig genomes provide insight into porcine demography and evolution. *Nature* **2012**, *491*, 393–398.

50. Li, M.; Wu, H.; Luo, Z.; Xia, Y.; Guan, J.; Wang, T.; Gu, Y.; Chen, L.; Zhang, K.; Ma, J.; *et al.* An atlas of DNA methylomes in porcine adipose and muscle tissues. *Nat. Commun.* **2012**, *3*, 850.

51. Fairbairn, L.; Kapetanovic, R.; Beraldi, D.; Sester, D.P.; Tuggle, C.K.; Archibald, A.L.; Hume, D.A. Comparative analysis of monocyte subsets in the pig. *J. Immunol.* **2013**, *190*, 6389–6396.

52. Martins, R.P.; Lorenzi, V.; Arce, C.; Lucena, C.; Carvajal, A.; Garrido, J.J. Innate and adaptive immune mechanisms are effectively induced in ileal Peyer's patches of Salmonella typhimurium infected pigs. *Dev. Comp. Immunol.* **2013**, *41*, 100–104.

53. Hulst, M.; Smits, M.; Vastenhouw, S.; de Wit, A.; Niewold, T.; van der Meulen, J. Transcription networks responsible for early regulation of Salmonella-induced inflammation in the jejunum of pigs. *J. Inflamm.* **2013**, *10*, 18.

54. Adler, M.; Murani, E.; Brunner, R.; Ponsuksili, S.; Wimmers, K. Transcriptomic response of porcine PBMCs to vaccination with tetanus toxoid as a model antigen. *PLoS One* **2013**, *8*, e58306.

55. Freeman, T.C.; Ivens, A.; Baillie, J.K.; Beraldi, D.; Barnett, M.W.; Dorward, D.; Downing, A.; Fairbairn, L.; Kapetanovic, R.; Raza, S.; *et al.* A gene expression atlas of the domestic pig. *BMC Biol.* **2012**, *10*, 90.

56. McDaneld, T.G.; Smith, T.P.; Harhay, G.P.; Wiedmann, R.T. Next-generation sequencing of the porcine skeletal muscle transcriptome for computational prediction of microRNA gene targets. *PLoS One* **2012**, *7*, e42039.

57. Zhou, B.; Liu, H.L.; Shi, F.X.; Wang, J.Y. MicroRNA expression profiles of porcine skeletal muscle. *Anim. Genet.* **2010**, *41*, 499–508.

58. Liu, Y.; Li, M.; Ma, J.; Zhang, J.; Zhou, C.; Wang, T.; Gao, X.; Li, X. Identification of differences in microRNA transcriptomes between porcine oxidative and glycolytic skeletal muscles. *BMC Mol. Biol.* **2013**, *14*, 7.

59. Siengdee, P.; Trakooljul, N.; Murani, E.; Schwerin, M.; Wimmers, K.; Ponsuksili, S. Transcriptional profiling and miRNA-dependent regulatory network analysis of longissimus dorsi muscle during prenatal and adult stages in two distinct pig breeds. *Anim. Genet.* **2013**, *44*, 398–407.

60. McDaneld, T.G.; Smith, T.P.; Doumit, M.E.; Miles, J.R.; Coutinho, L.L.; Sonstegard, T.S.; Matukumalli, L.K.; Nonneman, D.J.; Wiedmann, R.T. MicroRNA transcriptome profiles during swine skeletal muscle development. *BMC Genomics* **2009**, *10*, 77.

61. Huang, T.H.; Zhu, M.J.; Li, X.Y.; Zhao, S.H. Discovery of porcine microRNAs and profiling from skeletal muscle tissues during development. *PLoS One* **2008**, *3*, e3225.

62. Shen, H.; Liu, T.; Fu, L.; Zhao, S.; Fan, B.; Cao, J.; Li, X. Identification of microRNAs involved in dexamethasone-induced muscle atrophy. *Mol. Cell. Biochem.* **2013**, *381*, 105–113.

63. Timoneda, O.; Balcells, I.; Nunez, J.I.; Egea, R.; Vera, G.; Castello, A.; Tomas, A.; Sanchez, A. miRNA expression profile analysis in kidney of different porcine breeds. *PLoS One* **2013**, *8*, e55402.

64. Li, A.; Song, T.; Wang, F.; Liu, D.; Fan, Z.; Zhang, C.; He, J.; Wang, S. MicroRNAome and expression profile of developing tooth germ in miniature pigs. *PLoS One* **2012**, *7*, e52256.

65. Sharbati, S.; Friedlander, M.R.; Sharbati, J.; Hoeke, L.; Chen, W.; Keller, A.; Stahler, P.F.; Rajewsky, N.; Einspanier, R. Deciphering the porcine intestinal microRNA transcriptome. *BMC Genomics* **2010**, *11*, 275.

66. Podolska, A.; Kaczkowski, B.; Kamp Busk, P.; Sokilde, R.; Litman, T.; Fredholm, M.; Cirera, S. MicroRNA expression profiling of the porcine developing brain. *PLoS One* **2011**, *6*, e14494.

67. Zhou, Y.; Tang, X.; Song, Q.; Ji, Y.; Wang, H.; Jiao, H.; Ouyang, H.; Pang, D. Identification and characterization of pig embryo microRNAs by Solexa sequencing. *Reprod. Domest. Anim.* **2013**, *48*, 112–120.

68. Lian, C.; Sun, B.; Niu, S.; Yang, R.; Liu, B.; Lu, C.; Meng, J.; Qiu, Z.; Zhang, L.; Zhao, Z. A comparative profile of the microRNA transcriptome in immature and mature porcine testes using Solexa deep sequencing. *FEBS J.* **2012**, *279*, 964–975.

69. Li, M.; Liu, Y.; Wang, T.; Guan, J.; Luo, Z.; Chen, H.; Wang, X.; Chen, L.; Ma, J.; Mu, Z.; *et al.* Repertoire of porcine microRNAs in adult ovary and testis by deep sequencing. *Int. J. Biol. Sci.* **2011**, *7*, 1045–1055.

70. Curry, E.; Safranski, T.J.; Pratt, S.L. Differential expression of porcine sperm microRNAs and their association with sperm morphology and motility. *Theriogenology* **2011**, *76*, 1532–1539.

71. Luo, L.; Ye, L.; Liu, G.; Shao, G.; Zheng, R.; Ren, Z.; Zuo, B.; Xu, D.; Lei, M.; Jiang, S.; *et al.* Microarray-based approach identifies differentially expressed microRNAs in porcine sexually immature and mature testes. *PLoS One* **2010**, *5*, e11744.

72. Li, H.; Xi, Q.; Xiong, Y.; Cheng, X.; Qi, Q.; Yang, L.; Shu, G.; Wang, S.; Wang, L.; Gao, P.; *et al.* A comprehensive expression profile of microRNAs in porcine pituitary. *PLoS One* **2011**, *6*, e24883.

73. Li, H.Y.; Xi, Q.Y.; Xiong, Y.Y.; Liu, X.L.; Cheng, X.; Shu, G.; Wang, S.B.; Wang, L.N.; Gao, P.; Zhu, X.T.; *et al.* Identification and comparison of microRNAs from skeletal muscle and adipose tissues from two porcine breeds. *Anim. Genet.* **2012**, *43*, 704–713.

74. Martini, P.; Sales, G.; Brugiolo, M.; Gandaglia, A.; Naso, F.; De Pitta', C.; Spina, M.; Gerosa, G.; Romualdi, C.; Cagnin, S.; *et al.* Tissue-specific expression and regulatory networks of pig microRNAome. *PLoS One* **2013**, unpublished work.

75. Nelson, P.T.; Baldwin, D.A.; Kloosterman, W.P.; Kauppinen, S.; Plasterk, R.H.; Mourelatos, Z. RAKE and LNA-ISH reveal microRNA expression and localization in archival human brain. *RNA* **2006**, *12*, 187–191.

76. Nelson, P.T.; Baldwin, D.A.; Scearce, L.M.; Oberholtzer, J.C.; Tobias, J.W.; Mourelatos, Z. Microarray-based, high-throughput gene expression profiling of microRNAs. *Nat. Methods* **2004**, *1*, 155–161.

77. Kronick, M.N. Creation of the whole human genome microarray. *Expert Rev. Proteomics* **2004**, *1*, 19–28.

78. Wong, A.P.; Nili, N.; Strauss, B.H. *In vitro* differences between venous and arterial-derived smooth muscle cells: Potential modulatory role of decorin. *Cardiovasc. Res.* **2005**, *65*, 702–710.

79. Ross, J.M.; McIntire, L.V.; Moake, J.L.; Rand, J.H. Platelet adhesion and aggregation on human type VI collagen surfaces under physiological flow conditions. *Blood* **1995**, *85*, 1826–1835.

80. Smedegard, G.; Hedqvist, P.; Dahlen, S.E.; Revenas, B.; Hammarstrom, S.; Samuelsson, B. Leukotriene C4 affects pulmonary and cardiovascular dynamics in monkey. *Nature* **1982**, *295*, 327–329.

81. Pawloski, J.R.; Chapnick, B.M. Antagonism of LTD4-evoked relaxation in canine renal artery and vein. *Am. J. Physiol.* **1993**, *265*, H980–H985.

82. Brink, C.; Dahlen, S.E.; Drazen, J.; Evans, J.F.; Hay, D.W.; Nicosia, S.; Serhan, C.N.; Shimizu, T.; Yokomizo, T. International Union of Pharmacology XXXVII. Nomenclature for leukotriene and lipoxin receptors. *Pharmacol. Rev.* **2003**, *55*, 195–227.

83. Inoue, H.; Taba, Y.; Miwa, Y.; Yokota, C.; Miyagi, M.; Sasaguri, T. Transcriptional and posttranscriptional regulation of cyclooxygenase-2 expression by fluid shear stress in vascular endothelial cells. *Arterioscler. Thromb. Vasc. Biol.* **2002**, *22*, 1415–1420.

84. Dahboul, F.; Leroy, P.; Maguin Gate, K.; Boudier, A.; Gaucher, C.; Liminana, P.; Lartaud, I.; Pompella, A.; Perrin-Sarrado, C. Endothelial γ-glutamyltransferase contributes to the vasorelaxant effect of *S*-nitrosoglutathione in rat aorta. *PLoS One* **2012**, *7*, e43190.

85. Yousaf, N.; Howard, J.C.; Williams, B.D. Studies in the rat of antibody-coated and *N*-ethylmaleimide-treated erythrocyte clearance by the spleen. I. Effects of *in vivo* complement activation. *Immunology* **1986**, *59*, 75–79.

86. Ravingerova, T.; Barancik, M.; Strniskova, M. Mitogen-activated protein kinases: A new therapeutic target in cardiac pathology. *Mol. Cell. Biochem.* **2003**, *247*, 127–138.

87. Howe, G.A.; Addison, C.L. RhoB controls endothelial cell morphogenesis in part via negative regulation of RhoA. *Vasc. Cell* **2012**, *4*, 1.

88. Torella, D.; Iaconetti, C.; Catalucci, D.; Ellison, G.M.; Leone, A.; Waring, C.D.; Bochicchio, A.; Vicinanza, C.; Aquila, I.; Curcio, A.; *et al.* MicroRNA-133 controls vascular smooth muscle cell phenotypic switch *in vitro* and vascular remodeling *in vivo*. *Circ. Res.* **2011**, *109*, 880–893.

89. Kobayashi, R.; Kubota, T.; Hidaka, H. Purification, characterization, and partial sequence analysis of a new 25-kDa actin-binding protein from bovine aorta: A SM22 homolog. *Biochem. Biophys. Res. Commun.* **1994**, *198*, 1275–1280.

90. Rhodes, L.V.; Tilghman, S.L.; Boue, S.M.; Wang, S.; Khalili, H.; Muir, S.E.; Bratton, M.R.; Zhang, Q.; Wang, G.; Burow, M.E.; *et al.* Glyceollins as novel targeted therapeutic for the treatment of triple-negative breast cancer. *Oncol. Lett.* **2012**, *3*, 163–171.

91. Cordes, K.R.; Sheehy, N.T.; White, M.P.; Berry, E.C.; Morton, S.U.; Muth, A.N.; Lee, T.H.; Miano, J.M.; Ivey, K.N.; Srivastava, D. miR-145 and miR-143 regulate smooth muscle cell fate and plasticity. *Nature* **2009**, *460*, 705–710.

92. Ikeda, S.; Pu, W.T. Expression and function of microRNAs in heart disease. *Curr. Drug Targets* **2010**, *11*, 913–925.

93. Chhabra, R.; Dubey, R.; Saini, N. Cooperative and individualistic functions of the microRNAs in the miR-23a~27a~24-2 cluster and its implication in human diseases. *Mol. Cancer* **2010**, *9*, 232.

94. Li, S.; Ran, Y.; Zhang, D.; Chen, J.; Zhu, D. MicroRNA-138 plays a role in hypoxic pulmonary vascular remodelling by targeting Mst1. *Biochem. J.* **2013**, *452*, 281–291.

95. Park, C.; Yan, W.; Ward, S.M.; Hwang, S.J.; Wu, Q.; Hatton, W.J.; Park, J.K.; Sanders, K.M.; Ro, S. MicroRNAs dynamically remodel gastrointestinal smooth muscle cells. *PLoS One* **2011**, *6*, e18628.

96. Talasila, A.; Yu, H.; Ackers-Johnson, M.; Bot, M.; van Berkel, T.; Bennett, M.; Bot, I.; Sinha, S. Myocardin regulates vascular response to injury through miR-24/-29a and platelet-derived growth factor recepto-β. *Arterioscler. Thromb. Vasc. Biol.* **2013**, *33*, 2355–2365.

97. Saeed, A.I.; Sharov, V.; White, J.; Li, J.; Liang, W.; Bhagabati, N.; Braisted, J.; Klapa, M.; Currier, T.; Thiagarajan, M.; *et al.* TM4: A free, open-source system for microarray data management and analysis. *Biotechniques* **2003**, *34*, 374–378.

98. Raychaudhuri, S.; Stuart, J.M.; Altman, R.B. Principal components analysis to summarize microarray experiments: Application to sporulation time series. *Pac. Symp. Biocomput.* **2000**, *5*, 455–466.

99. Fellenberg, K.; Hauser, N.C.; Brors, B.; Neutzner, A.; Hoheisel, J.D.; Vingron, M. Correspondence analysis applied to microarray data. *Proc. Natl. Acad. Sci. USA* **2001**, *98*, 10781–10786.

100. Huang da, W.; Sherman, B.T.; Lempicki, R.A. Systematic and integrative analysis of large gene lists using DAVID bioinformatics resources. *Nat. Protoc.* **2009**, *4*, 44–57.

101. Risso, D.; Massa, M.S.; Chiogna, M.; Romualdi, C. A modified LOESS normalization applied to microRNA arrays: A comparative evaluation. *Bioinformatics* **2009**, *25*, 2685–2691.

102. Spath, H. *Two Dimensional Spline Interpolation Algorithms*; A K Peters/CRC Press: Wellesley, MA, USA, 1995.

103. Kanehisa, M.; Goto, S.; Kawashima, S.; Okuno, Y.; Hattori, M. The KEGG resource for deciphering the genome. *Nucleic Acids Res.* **2004**, *32*, D277–D280.

104. Saito, R.; Smoot, M.E.; Ono, K.; Ruscheinski, J.; Wang, P.L.; Lotia, S.; Pico, A.R.; Bader, G.D.; Ideker, T. A travel guide to Cytoscape plugins. *Nat. Methods* **2012**, *9*, 1069–1076.

105. Doncheva, N.T.; Assenov, Y.; Domingues, F.S.; Albrecht, M. Topological analysis and interactive visualization of biological networks and protein structures. *Nat. Protoc.* **2012**, *7*, 670–685.

5

Long Non-Coding RNA in Cancer

Nina Hauptman [†] **and Damjan Glavač** [†,*]

Department of Molecular Genetics, Institute of Pathology, University of Ljubljana, SI-1000 Ljubljana, Slovenia; E-Mail: nina.hauptman@mf.uni-lj.si

[†] These authors contributed equally to this work.

[*] Author to whom correspondence should be addressed; E-Mail: damjan.glavac@mf.uni-lj.si;

Abstract: Long non-coding RNAs (lncRNAs) are pervasively transcribed in the genome and are emerging as new players in tumorigenesis due to their various functions in transcriptional, posttranscriptional and epigenetic mechanisms of gene regulation. LncRNAs are deregulated in a number of cancers, demonstrating both oncogenic and tumor suppressive roles, thus suggesting their aberrant expression may be a substantial contributor in cancer development. In this review, we will summarize their emerging role in human cancer and discuss their perspectives in diagnostics as potential biomarkers.

Keywords: long non-coding RNA; cancer; oncogenic lncRNA; tumor suppressor lncRNA

1. Introduction

The central dogma of molecular biology postulates gene-coding through storage of genetic information and proteins as the main molecules of cellular functions, while RNA has the role of an intermediary between DNA sequence and encoded protein. The findings of the human genome project thus came as a surprise, since only 1.5% of the human genome encodes protein-coding genes [1–5]. Development of new techniques revolutionized the molecular world with evidence that at least 90% of the human genome is actively transcribed [6,7]. The human transcriptome has shown more complexity than previously assumed since the protein-coding transcripts are being a minority, compared to a more complex group of non-coding RNAs (ncRNAs), such as microRNAs (miRNAs), long non-coding RNAs (lncRNAs), small nucleolar RNAs (snoRNAs), small interfering RNAS (siRNAs), small nuclear (snRNAs), and piwi-interacting RNAs (piRNAs) [8–15]. Although initially thought to be transcriptional noise, ncRNA may play a crucial role in cellular development, physiology and pathologies.

Depending on their size, ncRNAs are divided into two major groups. Transcripts shorter than 200 nucleotides are referred to as small ncRNAs, which include miRNA, siRNA, piRNA, *etc.* The other group is composed of lncRNA, where the transcripts lack a significant open reading frame, and have length of 200 nt up to 100 kilobases. A lncRNA can be placed into one or more of five broad categories: (1) sense, or (2) antisense, when overlapping one or more exons of another transcript on the same, or opposite, strand, respectively; (3) bidirectional, when the sequence is located on the opposite strand from a neighboring coding transcript whose transcription is initiated less than 1000 base pairs away, (4) intronic, when it is derived wholly from within an intron of a second transcript, or (5) intergenic, when it lies within the genomic interval between two genes [16] (Figure 1). There are some lncRNAs that are transcribed by RNA polymerase III while the majority of lncRNAs are transcribed by RNA polymerase II, spliced and polyadenylated [17]. Most of the lncRNAs are located in the cytoplasm, although there are some found in both cytoplasm and nucleus [18].

Figure 1. Categories of long non-coding RNA (lncRNA).

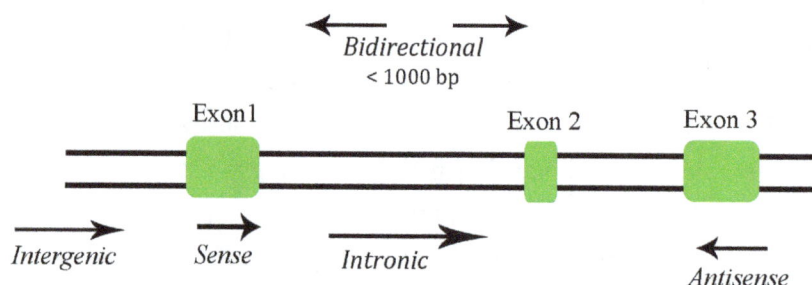

2. Long Non-Coding RNA Functions

LncRNAs are involved in almost every step of a life cycle of genes and regulate diverse functions. Several lncRNAs can regulate gene expression at various levels, including chromatin modification, transcription, and posttranscriptional processing [19].

So far, their role was extensively studied in epigenetic regulation, such as imprinting. Diploid organisms carry two alleles of each of the parents' autosomal genes. In most cases, both of the alleles are expressed equally, except when a subset of genes shows imprinting in which expression is restricted by epigenetic mechanism to either maternal or paternal allele [17]. X-inactivation (XCI) is a process that equalizes gene expression between males and females by inactivating one X in female cells [17]. Some lncRNAs participate in global cellular behavior by controlling apoptosis, cell death and cell growth [15,20]. LncRNA can also mediate epigenetic modification by recruiting chromatin remodeling complex to specific chromatin loci, e.g., HOTAIR by polycomb repression complex 2 (PCR2) and/or lysine-specific demethylase 1 (LSD1), CCND1 by protein termed translocated in liposarcoma (TLS), and ANRIL by polycomb repression complex 1 and 2 (PCR1 and PCR2) [5,21–25]. The mode of action of some lncRNAs is interaction with their intracellular steroid receptors. Other lncRNAs function by regulating transcription through a variety of mechanisms that include interacting with RNA-binding proteins, acting as a coactivator of transcription factors, or repressing a major promoter of their target gene [22]. In addition to chromatin modification and transcriptional regulation, lncRNAs can regulate gene expression at the posttranscriptional level.

3. Oncogenic lncRNA

SRA—Steroid Receptor RNA Activator is a coactivator for steroid receptors and acts as an ncRNA found in the nucleus and cytoplasm. SRA regulates gene expression mediated by steroid receptors through complexing with proteins also containing steroid receptor coactivator 1 (SRC-1) [26]. The SRA1 gene can also encode a protein that acts as a coactivator and corepressor [27]. SRA levels have been found to be upregulated in breast tumors where it is assumed that increased SRA levels change the steroid receptors' actions, contributing to breast tumorigenesis. While the expression of SRA in normal tissues is low, it is highly up-regulated in various tumors of the human breast, uterus and ovary. This evidence supports that SRA is a potential biomarker of steroid-dependent tumors [26].

HOTAIR—HOX Antisense Intergenic RNA with a length of 2.2 kb was found in the *HOXC* locus and is transcribed in antisense manner [28]. It is the first lncRNA discovered to be involved in tumorigenesis. In breast cancer, both primary and metastatic, the expression is up regulated; in the latter case up to 2000-fold increase was shown [23]. The high expression level of HOTAIR in primary breast cancer is also correlated to metastasis, and poor survival rate [23]. The level of HOTAIR expression is higher in patients with lymph node metastasis in hepatocellular cancer [29].

Polycomb group proteins mediate repression of transcription of thousands of genes that control differentiation pathways during development, and have roles in stem cell pluripotency and human cancer [23,30–34]. The target of PRC2 is the *HOXD* locus on chromosome 2 where the PRC2 in association with HOTAIR causes the transcriptional silencing of several metastasis suppressor genes resulting in breast epithelial cells having the expression of embryonic fibroblast. Alternating the level of HOTAIR results in enhanced PRC2 repressive activity [23]. HOTAIR acts as a molecular scaffold having two known chromatin modification complexes. The 5′ region of lncRNA binds to the PRC2 complex responsible for H3K27 methylation and the 3′ region binds to LSD1, which mediates enzymatic demethylation of H3K4 [24,30,35]. This result suggests the possible function of HOTAIR as a scaffold binding to selected histone modification enzymes and therefore causing histone modification on target genes [30]. Although the precise mechanism is still not known, it is clear that HOTAIR remodels chromatin to promote cancer invasiveness.

HOTAIR as an epigenetic regulator in gene expression is deregulated in different cancers [23,36–38]. In hepatocellular carcinoma (HCC) and HCC patients with liver transplantation, the levels of HOTAIR compared with normal liver tissue are elevated. Expression levels of HOTAIR can also be used as an independent prognostic marker for HCC recurrence and lower survival rate [31]. HOTAIR can be a potential biomarker for the existence of lymph node metastasis in HCC [29].

ANRIL—Antisense ncRNA in the INK4 locus

Many transcripts coding for proteins have anti-sense partners, whose perturbation can alter the expression of the sense transcripts [39]. Some of these genes are tumor suppressors, which can be epigenetically silenced by antisense ncRNA [40].

ANRIL activates two polycomb repressor complexes, PRC1 and PRC2 [21,25], resulting in chromatin reorganization which silences the INK4b-ARF-INK4a locus encoding tumor suppressors $p15^{INK4b}$, $p14^{ARF}$, $p16^{INK4a}$, which are active in cell cycle inhibition, senescence and stress-induced apoptosis. Overexpression of ANRIL in prostate cancer has shown silencing of INK4b-ARF-INK4a and p15/CDKN2B by heterochromatin reformation [25,41]. The repression is mediated by direct binding to combox 7 (CBX 7) and SUZ12, members of PRC1 and PRC2, respectively [21,25].

MALAT 1—Metastasis-Associated Lung Adenocarcinoma Transcript 1

This lncRNA was first associated with high metastatic potential and poor patient prognosis during a comparative screen of non-small cell lung cancer patients with or without metastatic tumors [42]. MALAT1 is widely expressed in normal human tissues [42,43] and is found to be upregulated in a variety of human cancers of the breast, prostate, colon, liver and uterus [44–47]. The MALAT1 locus at 11q13.1 has been identified to harbor chromosomal translocation break points associated with cancer [48–50]. MALAT1 is localized in nuclear speckles and widely expressed in normal tissues [42,43], but was found to be upregulated in hepatocellular carcinoma, breast, pancreas, osteosarcoma, colon and prostate cancers [44–47,51]. It has been shown that increased expression of MALAT1 can be used as a prognostic marker for HCC patients following liver transplantation [52].

A number of studies have implicated MALAT1 in the regulation of cell mobility, due to its high levels of expression in cancers. For example, RNA interference-mediated silencing of MALAT1 reduced the *in vitro* migration of lung adenocarcinoma cells by influencing the expression of motility-related genes [53]. Recent studies on knockout MALAT1 mice have not displayed any cellular phenotype. Future studies will be needed where mice will be exposed to different stresses, such as induction of cancer, which will potentially unveil its function. It is known that MALAT1 as well as HOTAIR play vital roles in human cells but it is possible that they have no significant role in living animals under normal physiological conditions [54–56].

4. Oncogenic and Tumor Suppressor lncRNA

H19 is expressed from the maternal allele and has a pivotal role in genomic imprinting during cell growth and development [57]. The locus contains *H19* and insulin-like growing factor 2 (*IGF2*), which are imprinted. This leads to differential expression of both genes, *H19* from maternal and *IGF2* from paternal allele [57,58]. The loss of imprinting results in misexpression of *H19* and was observed in many tumors including hepatocellular and bladder cancer [59,60]. This lncRNA has been linked to oncogenic and tumor suppressor properties [57]. cMYC induces the expression of *H19* in different cell types where *H19* potentiates tumorigenesis [58]. In addition c-MYC also down-regulates expression of *IGF2* imprinted gene. *H19* transcripts are precursors for miR-675 which functionally down-regulates the tumor suppressor gene for retinoblastoma in human colorectal cancer [61]. Data support *H19* deregulation causing either oncogenic or tumor suppressor properties, although the exact mechanism is still elusive.

5. Tumor Suppressor lncRNA

MEG3—Maternally Expressed Gene 3

LncRNA MEG3 is a transcript of the maternally imprinted gene. In normal pituitary cells MEG3 is expressed, the loss of expression is observed in pituitary adenomas and the majority of meningiomas and meningioma cell lines [62,63]. MEG3 activates regulation of tumor suppressor protein p53. Normally, p53 protein levels are extremely low due to its rapid degradation via the ubiquitin-proteasome pathway. The ubiquitination of p53 is mainly mediated by MDM2, an E3 ubiquitin ligase. MEG3 down-regulates MDM2 expression, which suggests that MDM2 down-regulation is one of the mechanisms whereby MEG3 activates p53 [64]. MEG3 significantly increases p53 protein level and stimulates p53-dependent transcription [65]. MEG3 enhances p53 binding to target promoters such as GDF15 but not p21 and is also able to inhibit cell proliferation in the absence p53, suggesting that MEG3 is a p53 dependent and independent tumor suppressor [62–65].

GAS5—Growth Arrest-Specific 5 is widely expressed in embryonic and adult tissues. Expression is almost undetectable in growing leukemia cells and abundant in saturation density-arrested cells [66,67]. GAS5 functions as a starvation or growth arrest-linked riborepressor for the glucocorticoid receptors by binding to their DNA binding domain inhibiting the association of these receptors with their DNA recognition sequence. This suppresses the induction of several responsive genes including the gene encoding cellular inhibitor of apoptosis 2 (cIAP2), reducing cell metabolism and synthesizes cells to apoptosis [4,67]. GAS5 can induce apoptosis directly or indirectly in the prostate and breast cancer cell lines, where it was shown that GAS5 has a significantly lower expression in breast cancers compared to normal breast epithelial tissues [68].

CCND1/Cyclin D1 is a heterogenous lncRNA transcribed from the promoter region of the Cyclin D1 gene. Cyclin D1 is a cell cycle regulator that is frequently mutated, amplified and over expressed in a variety of cancers [69]. LncRNA recruits the RNA-binding proteinTLS, which is a key transcriptional regulatory sensor of DNA damage signals. Upon binding TLS undergoes allosteric modification, modulating activities of CREB-binding protein (CBP) and p300, resulting in inhibition of the cyclin D1 gene expression [22].

LincRNA-p21 expression is directly induced by the p53 signaling pathway. It is required for global repression of genes that interfere with p53 function to regulate cellular apoptosis. Lincrna-p21 mediated gene repression occurs through physical interaction with RNA-binding protein hnRNP K leading to the promoters of genes being repressed in a p53 dependent manner [70].

6. Diagnostic Benefits of lncRNA

So far, the majority of cancer biomarkers are protein-coding genes, their transcripts or the proteins. The non-coding regions are evolving as a biomarker hotspots only recently. By the advent of high-throughput sequencing, we are now able to identify deregulated expression of transcriptome at much higher resolution, what allow us to decipher smaller changes in the expression level. In the case of lncRNAs, where their main function is regulation of other genes expression, the importance of lncRNAs maintained expression is evident. Since cancer is a complicated disease, which involves many factors, molecular biomarkers are valuable diagnostic and prognostic tools that could ease the disease management. Compared to protein-coding RNAs, using lncRNA as markers is of advantage since their own expression is a better indicator of the tumor status. Many lncRNAs are now connected to cancer due to new technologies and are emerging into the field of molecular biology as new regulatory players. Several lncRNA were found to be deregulated in a wide variety of cancers (Table 1).

Table 1. Cancer associated lncRNAs (adapted from [15,71]).

Name	Cytoband (Size)	Cancer Types	References
AK023948	8q24.22 (2807 nt)	Papillary thyroid carcinoma (down regulated)	[72]
ANRIL	9p21.3 (~3.9kb)	Prostate, leukemia	[36,41]
BC200	2p21 (200 nt)	Breast, cervix, esophagus, lung, ovary, parotid, tongue	[73,74]
PRNCR1	8q24.2 (13 kb)	Prostate	[75]
H19	11p15.5 (2.3 kb)	Bladder, lung, liver, breast, esophagus, choriocarcinoma, colon	[57,58,76–80]
HOTAIR	12q13.13 (2.2 kb)	Breast, hepatocellular	[23,29,30,36]
HULC	6p24.3 (~500 nt)	Hepatocellular	[4,81,82]
LincRNA-p21	~3.1 kb	Represses p53 pathway; induces apoptosis	[70]
Loc285194	3q13.31 (2105 nt)	Osteosarcoma	[83]
Malat1	11q13.1 (7.5 kb)	breast, prostate, colon, liver, uterus	[44–47]
MEG3	14q32.2 (1.6 kb)	Brain (down-regulated)	[62,65]
PTNEP1	9p13.3 (3.9 kb)	Prostate	[84]
Spry4-it1	5q31.3 (~700 nt)	Melanoma	[85]
SRA	5q31.3 (1965 nt)	Breast, uterus, ovary (down-regulated)	[26,27]
UCA1/CUDR	19p13.12 (1.4, 2.2, 2.7 kb)	Bladder, colon, cervix, lung, thyroid, liver, breast, esophagus, stomach	[86,87]
Wt1-as	11p13 (isoforms)	acute myeloid leukemia	[88]
PCA3	9q21.22 (0.6–4 kb)	Prostate	[89]
GAS5	1q25.1 (isoforms)	Breast (down-regulated)	[68]

In breast cancer research higher expressions of SRA and SRAP, compared to normal tissue were observed. Possibly SRAP expression contributes to higher survival for patient undergoing Tamoxifen treatment [90].

The expression of MALAT 1 is elevated in osteosarcoma patients with poor response to chemotherapy, which suggests that this transcript plays a crucial role in the pathology of tumors [53]. Additionally MALAT 1 serves as an independent prognostic marker for patient survival in early stage non-small cell lung cancer [42].

In hepatocellular carcinoma, (HCC) definitive diagnosis of lymph node metastasis is difficult without histological evidence. It has been demonstrated that a significant correlation between *HOTAIR* gene expression and lymph node metastasis exists, suggesting that measuring HOTAIR lncRNA is a potential biomarker for predicting lymph node metastasis [29]. Upregulation of HOTAIR is closely associated with gastrointestinal stromal tumor (GIST) aggressiveness and metastasis and it can be used as a potential biomarker [38,91].

MALAT1 is a powerful biomarker for HCC recurrence prediction following liver transplantation. Moreover, silencing MALAT1 activity in HCC would be a potential anticancer therapy to prevent tumor recurrence after orthotopic liver transplantation [52].

SPRY4-IT1 expression is substantially increased in patient melanoma cell samples compared to melanocytes. The elevated expression of SPRY4-IT1 in melanoma cells, its accumulation in the cell cytoplasm, and its effects on cell dynamics suggest that the misexpression of SPRY4-IT1 may have an important role in melanoma development, and could be an early biomarker and a key regulator for melanoma pathogenesis in humans [85].

The novel potential biomarkers can be discovered through certain types of highly expressed cancer-associated lncRNAs [92]. Therapeutic benefit can be obtained through pathways mediating transcriptional gene silencing, especially those of tumor suppressors and oncogenes [93]. For patients' comfort, biomarkers should be detected in samples obtained in a non-invasive way. Desirable samples are body fluids, such as serum or urine, where circulating nuclear acids (CNAs), both DNA and RNA species, are found. CNAs are found in plasma, cell-free serum, sputum and urine [29,94–97].

PRNCR1 (prostate cancer non-coding RNA1) expression was upregulated in some of the prostate cancer cells as well as precursor lesion prostatic intraepithelial neoplasia and considered as a tumor marker [75].

Suggestions that lncRNA can be used as biomarkers and/or drug targets have arisen from numerous studies observing the expression patterns of tumor tissues comparing to normal ones [14]. The possible therapies arising from this knowledge would be beneficial in cases where protein target drugs have not been effective. A recent study has shown that reduced expression of ncRAN enhanced the chemotherapeutic drug *in vitro* [98]. This opens another possibility of cancer treatment, where a combination of drugs would have much higher effect.

Often lncRNAs exhibit tissue specific patterns that distinguish them from miRNAs and protein-coding mRNAs that are expressed from multiple tissue types. Their specificity makes them precise biomarkers for cancer diagnostics [99]. PCA3 is a prostate-specific lncRNA overexpressed in prostate cancer. Although its functions are not understood, it was still utilized as a biomarker in a clinical test. Expression of the PCA3 transcript is determined from prostate cells in urine samples of patients [100,101]. Another lncRNA detected in body fluids is HULC, expression of which is disrupted in hepatocellular carcinomas and can be monitored in patients' blood sera [102].

To understand the biology of cancer it will be essential to identify, annotate lncRNAs and study their expression profiles in human tissues and diseases [103,104]. With this, the potential of lncRNAs on biology and medicine will be revealed. Long non-coding RNAs have recently arisen as new discoveries in the field of molecular biology. Since only a few individual lncRNAs have been functionally studied, still a lot of questions remain to be addressed [4]. At the moment the full potential of cancer therapy is not yet developed. The future of it lies in specific targeting of cancer cells and specific delivery of the drugs. LncRNAs are a possible resource for developing diagnostics and therapies, although a better understanding of their function and precise mechanism through which they function are needed first [4]. Another possibility for cancer treatment lies in combination of drugs, where one would change the expression of lncRNA in a way for chemotherapeutic drug to have a higher effect. Since probably the lncRNA function through their secondary structure special molecules could be developed to disrupt their secondary structure or bind to them to form complexes through which an inactivation of lncRNA would occur. These molecules should be highly specific in order not to disrupt other molecules and mechanisms. To discover the right molecules more studies of the complex mechanisms involving lncRNA are needed.

7. Conclusions

RNA used to be just a messenger between coding genes and proteins encoded by them. However, "transcriptional noise" is turning out to be a very important part of regulation processes. With the discovery of LncRNA and their functions, the new world of molecular biology is emerging. There is much research still on the way towards a deeper understanding of regulation processes in which lncRNA is one of the important players. LncRNA deregulation in human disease is unveiling the complexity of cellular processes. Studying the mechanisms of lncRNA involvement in oncogenic and tumor suppressive pathways will lead to new cancer diagnostic markers and will pave the way to novel therapeutic targets.

Acknowledgments

This work was supported by program P3-0054 of the Slovenian Research Agency.

References

1.   Stein, L.D. Human genome: End of the beginning. *Nature* **2004**, *431*, 915–916.

2.   Ponting, C.P.; Belgard, T.G. Transcribed dark matter: meaning or myth? *Hum. Mol. Genet.* **2010**, *19*, R162–R168.

3.   Lander, E.S.; Linton, L.M.; Birren, B.; Nusbaum, C.; Zody, M.C.; Baldwin, J.; Devon, K.; Dewar, K.; Doyle, M.; FitzHugh, W.; *et al.* Initial sequencing and analysis of the human genome. *Nature* **2001**, *49*, 860–921.

4.   Gutschner, T.; Diederichs, S. The Hallmarks of Cancer: A long non-coding RNA point of view. *RNA Biol.* **2012**, *9*, 703–719.

5.   Nie, L.; Wu, H.-J.; Hsu, J.-M.; Chang, S.-S.; LaBaff, A.; Li, C.-W.; Wang, Y.; Hsu, J.L.; Hung, M.-C. Long non-coding RNAs: Versatile master regulators of gene expression and crucial players in cancer. *Am. J. Transl. Res.* **2012**, *4*, 127–150.

6.   Birney, E.; Stamatoyannopoulos, J.A.; Dutta, A.; Guigó, R.; Gingeras, T.R.; Margulies, E.H.; Weng, Z.; Snyder, M.; Dermitzakis, E.T.; Thurman, R.E.; *et al.* Identification and analysis of functional elements in 1% of the human genome by the ENCODE pilot project. *Bioessays* **2007**, *32*, 599–608.

7.   Costa, F.F. Non-coding RNAs: Meet thy masters. *Bioessays* **2010**, *32*, 599–608.

8.   Kapranov, P.; Willingham, A.T.; Gingeras, T.R. Genome-wide transcription and the implications for genomic organization. *Nat. Rev. Genet.* **2007**, *8*, 413–423.

9.   Frith, M.C.; Pheasant, M.; Mattick, J.S. The amazing complexity of the human transcriptome. *Eur. J. Hum. Genet.* **2005**, *13*, 894–897.

10.   Khachane, A.N.; Harrison, P.M. Mining mammalian transcript data for functional long non-coding RNAs. *PLoS One* **2010**, *5*, doi:10.1371/journal.pone.0010316.

11.   Mattick, J.S.; Makunin, I.V. Non-coding RNA. *Hum. Mol. Genet.* **2006**, *15*, R17–R29.

12. Guttman, M.; Amit, I.; Garber, M.; French, C.; Lin, M.F.; Feldser, D.; Huarte, M.; Zuk, O.; Carey, B.W.; Cassady, J.P.; *et al.* Chromatin signature reveals over a thousand highly conserved large non-coding RNAs in mammals. *Nature* **2009**, *458*, 223–227.

13. Washietl, S.; Hofacker, I.L.; Lukasser, M.; Huttenhofer, A.; Stadler, P.F. Mapping of conserved RNA secondary structures predicts thousands of functional noncoding RNAs in the human genome. *Nat. Biotechnol.* **2005**, *23*, 1383–1390.

14. Taft, R.J.; Pang, K.C.; Mercer, T.R.; Dinger, M.; Mattick, J.S. Non-coding RNAs: Regulators of disease. *J. Pathol.* **2010**, *220*, 126–139.

15. Sana, J.; Faltejskova, P.; Svoboda, M.; Slaby, O. Novel classes of non-coding RNAs and cancer. *J. Transl. Med.* **2012**, *10*, doi:10.1186/1479-5876-10-103.

16. Ponting, C.P.; Oliver, P.L.; Reik, W. Evolution and functions of long noncoding RNAs. *Cell* **2009**, *136*, 629–641.

17. Wang, K.C.; Chang, H.Y. Molecular mechanisms of long noncoding RNAs. *Mol. Cell* **2011**, *43*, 904–914.

18. Banfai, B.; Jia, H.; Khatun, J.; Wood, E.; Risk, B.; Gundling, W.E.; Kundaje, A.; Gunawardena, H.P.; Yu, Y.; Xie, L.; *et al.* Long noncoding RNAs are rarely translated in two human cell lines. *Genome Res.* **2012**, *22*, 1646–1657.

19. Wilusz, J.E.; Sunwoo, H.; Spector, D.L. Long noncoding RNAs: Functional surprises from the RNA world. *Genes Dev.* **2009**, *23*, 1494–1504.

20. Wapinski, O.; Chang, H.Y. Long noncoding RNAs and human disease. *Trends. Cell Biol.* **2011**, *21*, 354–361.

21. Kotake, Y.; Nakagawa, T.; Kitagawa, K.; Suzuki, S.; Liu, N.; Kitagawa, M.; Xiong, Y. Long non-coding RNA ANRIL is required for the PRC2 recruitment to and silencing of p15(INK4B) tumor suppressor gene. *Oncogene* **2011**, *30*, 1956–1962.

22. Wang, X.; Arai, S.; Song, X.; Reichart, D.; Du, K.; Pascual, G.; Tempst, P.; Rosenfeld, M.G.; Glass, C.K.; Kurokawa, R. Induced ncRNAs allosterically modify RNA-binding proteins in cis to inhibit transcription. *Nature* **2008**, *454*, 126–130.

23. Gupta, R.A.; Shah, N.; Wang, K.C.; Kim, J.; Horlings, H.M.; Wong, D.J.; Tsai, M.C.; Hung, T.; Argani, P.; Rinn, J.L.; *et al.* Long non-coding RNA HOTAIR reprograms chromatin state to promote cancer metastasis. *Nature* **2010**, *464*, 1071–1076.

24. Hayami, S.; Kelly, J.D.; Cho, H.S.; Yoshimatsu, M.; Unoki, M.; Tsunoda, T.; Field, H.I.; Neal, D.E.; Yamaue, H.; Ponder, B.A.; *et al.* Overexpression of LSD1 contributes to human carcinogenesis through chromatin regulation in various cancers. *Int. J. Cancer* **2011**, *128*, 574–586.

25. Yap, K.L.; Li, S.; Munoz-Cabello, A.M.; Raguz, S.; Zeng, L.; Mujtaba, S.; Gil, J.; Walsh, M.J.; Zhou, M.M. Molecular interplay of the noncoding RNA ANRIL and methylated histone H3 lysine 27 by polycomb CBX7 in transcriptional silencing of INK4a. *Mol. Cell* **2010**, *38*, 662–674.

26. Lanz, R.B.; Chua, S.S.; Barron, N.; Soder, B.M.; DeMayo, F.; O'Malley, B.W. Steroid receptor RNA activator stimulates proliferation as well as apoptosis *in vivo*. *Mol. Cell. Biol.* **2003**, *23*, 7163–7176.

27. Chooniedass-Kothari, S.; Vincett, D.; Yan, Y.; Cooper, C.; Hamedani, M.K.; Myal, Y.; Leygue, E. The protein encoded by the functional steroid receptor RNA activator is a new modulator of ER alpha transcriptional activity. *FEBS Lett.* **2010**, *584*, 1174–1180.

28. Rinn, J.L.; Kertesz, M.; Wang, J.K.; Squazzo, S.L.; Xu, X.; Brugmann, S.A.; Goodnough, L.H.; Helms, J.A.; Farnham, P.J.; Segal, E.; *et al.* Functional Demarcation of Active and Silent Chromatin Domains in Human HOX Loci by Noncoding RNAs. *Cell* **2007**, *129*, 1311–1323.

29. Geng, Y.J.; Xie, S.L.; Li, Q.; Ma, J.; Wang, G.Y. Large intervening non-coding RNA HOTAIR is associated with hepatocellular carcinoma progression. *J. Int. Med. Res.* **2011**, *39*, 2119–2128.

30. Tsai, M.C.; Manor, O.; Wan, Y.; Mosammaparast, N.; Wang, J.K.; Lan, F.; Shi, Y.; Segal, E.; Chang, H.Y. Long noncoding RNA as modular scaffold of histone modification complexes. *Science* **2010**, *329*, 689–693.

31. Morey, L.; Helin, K. Polycomb group protein-mediated repression of transcription. *Trends. Biochem. Sci.* **2010**, *35*, 323–332.

32. Zhao, J.; Ohsumi, T.K.; Kung, J.T.; Ogawa, Y.; Grau, D.J.; Sarma, K.; Song, J.J.; Kingston, R.E.; Borowsky, M.; Lee, J.T. Genome-wide identification of polycomb-associated RNAs by RIP-seq. *Mol. Cell* **2010**, *40*, 939–953.

33. Zhang, Z.; Jones, A.; Sun, C.W.; Li, C.; Chang, C.W.; Joo, H.Y.; Dai, Q.; Mysliwiec, M.R.; Wu, L.C.; Guo, Y.; *et al.* PRC2 complexes with JARID2, MTF2, and esPRC2p48 in ES cells to modulate ES cell pluripotency and somatic cell reprogramming. *Stem Cells* **2011**, *29*, 229–240.

34. Simon, J.A.; Lange, C.A. Roles of the EZH2 histone methyltransferase in cancer epigenetics. *Mutat. Res.* **2008**, *647*, 21–29.

35. Sirchia, S.M.; Tabano, S.; Monti, L.; Recalcati, M.P.; Gariboldi, M.; Grati, F.R.; Porta, G.; Finelli, P.; Radice, P.; Miozzo, M. Misbehaviour of XIST RNA in breast cancer cells. *PLoS One* **2009**, *4*, doi:10.1371/journal.pone.0005559.

36. Yang, Z.; Zhou, L.; Wu, L.M.; Lai, M.C.; Xie, H.Y.; Zhang, F.; Zheng, S.S. Overexpression of long non-coding RNA HOTAIR predicts tumor recurrence in hepatocellular carcinoma patients following liver transplantation. *Ann. Surg. Oncol.* **2011**, *18*, 1243–1250.

37. Kogo, R.; Shimamura, T.; Mimori, K.; Kawahara, K.; Imoto, S.; Sudo, T.; Tanaka, F.; Shibata, K.; Suzuki, A.; Komune, S.; *et al.* Long noncoding RNA HOTAIR regulates polycomb-dependent chromatin modification and is associated with poor prognosis in colorectal cancers. *Cancer Res.* **2011**, *71*, 6320–6326.

38. Niinuma, T.; Suzuki, H.; Nojima, M.; Nosho, K.; Yamamoto, H.; Takamaru, H.; Yamamoto, E.; Maruyama, R.; Nobuoka, T.; Miyazaki, Y.; *et al.* Upregulation of miR-196a and HOTAIR drive malignant character in gastrointestinal stromal tumors. *Cancer Res.* **2012**, *72*, 1126–1136.

39. Katayama, S.; Tomaru, Y.; Kasukawa, T.; Waki, K.; Nakanishi, M.; Nakamura, M.; Nishida, H.; Yap, C.C.; Suzuki, M.; Kawai, J.; *et al.* Antisense transcription in the mammalian transcriptome. *Science* **2005**, *309*, 1564–1566.

40. Kim, W.Y.; Sharpless, N.E. The regulation of INK4/ARF in cancer and aging. *Cell* **2006**, *127*, 265–275.

41. Yu, W.; Gius, D.; Onyango, P.; Muldoon-Jacobs, K.; Karp, J.; Feinberg, A.P.; Cui, H. Epigenetic silencing of tumour suppressor gene p15 by its antisense RNA. *Nature* **2008**, *451*, 202–206.

42. Ji, P.; Diederichs, S.; Wang, W.; Boing, S.; Metzger, R.; Schneider, P.M.; Tidow, N.; Brandt, B.; Buerger, H.; Bulk, E.; *et al.* MALAT-1, a novel noncoding RNA, and thymosin beta4 predict metastasis and survival in early-stage non-small cell lung cancer. *Oncogene* **2003**, *22*, 8031–8041.

43. Hutchinson, J.N.; Ensminger, A.W.; Clemson, C.M.; Lynch, C.R.; Lawrence, J.B.; Chess, A. A screen for nuclear transcripts identifies two linked noncoding RNAs associated with SC35 splicing domains. *BMC Genomics* **2007**, *8*, doi:10.1186/1471-2164-8-39.

44. Guffanti, A.; Iacono, M.; Pelucchi, P.; Kim, N.; Solda, G.; Croft, L.J.; Taft, R.J.; Rizzi, E.; Askarian-Amiri, M.; Bonnal, R.J.; *et al.* A transcriptional sketch of a primary human breast cancer by 454 deep sequencing. *BMC Genomics* **2009**, *10*, doi:10.1186/1471-2164-10-163.

45. Yamada, K.; Kano, J.; Tsunoda, H.; Yoshikawa, H.; Okubo, C.; Ishiyama, T.; Noguchi, M. Phenotypic characterization of endometrial stromal sarcoma of the uterus. *Cancer Sci.* **2006**, *97*, 106–112.

46. Lin, R.; Maeda, S.; Liu, C.; Karin, M.; Edgington, T.S. A large noncoding RNA is a marker for murine hepatocellular carcinomas and a spectrum of human carcinomas. *Oncogene* **2007**, *26*, 851–858.

47. Luo, J.H.; Ren, B.; Keryanov, S.; Tsang, G.C.; Reo, U.N.M.; Monga, S.P.; Storm, A.; Demetris, A.J.; Nalesnik, M.; Yu, Y.P.; *et al.* Transcriptomic and genomic analysis of human hepatocellular carcinomas and hepatoblastomas. *Hepatology* **2006**, *44*, 1012–1024.

48. Davis, I.J.; Hsi, B.L.; Arroyo, J.D.; Vargas, S.O.; Yeh, Y.A.; Motyckova, G.; Valencia, P.; Perez-Atayde, A.R.; Argani, P.; Ladanyi, M.; *et al.* Cloning of an Alpha-TFEB fusion in renal tumors harboring the t(6;11)(p21;q13) chromosome translocation. *Proc. Natl. Acad Sci USA* **2003**, *100*, 6051–6056.

49. Kuiper, R.P.; Schepens, M.; Thijssen, J.; van Asseldonk, M.; van den Berg, E.; Bridge, J.; Schuuring, E.; Schoenmakers, E.F.; van Kessel, A.G. Upregulation of the transcription factor TFEB in t(6;11)(p21;q13)-positive renal cell carcinomas due to promoter substitution. *Hum. Mol. Genet.* **2003**, *12*, 1661–1669.

50. Rajaram, V.; Knezevich, S.; Bove, K.E.; Perry, A.; Pfeifer, J.D. DNA sequence of the translocation breakpoints in undifferentiated embryonal sarcoma arising in mesenchymal hamartoma of the liver harboring the t(11;19)(q11;q13.4) translocation. *Genes Chromosomes Cancer* **2007**, *46*, 508–513.

51. Fellenberg, J.; Bernd, L.; Delling, G.; Witte, D.; Zahlten-Hinguranage, A. Prognostic significance of drug-regulated genes in high-grade osteosarcoma. *Mod. Pathol.* **2007**, *20*, 1085–1094.

52. Lai, M.C.; Yang, Z.; Zhou, L.; Zhu, Q.Q.; Xie, H.Y.; Zhang, F.; Wu, L.M.; Chen, L.M.; Zheng, S.S. Long non-coding RNA MALAT-1 overexpression predicts tumor recurrence of hepatocellular carcinoma after liver transplantation. *Med. Oncol.* **2012**, *29*, 1810–1816.

53. Tano, K.; Mizuno, R.; Okada, T.; Rakwal, R.; Shibato, J.; Masuo, Y.; Ijiri, K.; Akimitsu, N. MALAT-1 enhances cell motility of lung adenocarcinoma cells by influencing the expression of motility-related genes. *FEBS Lett.* **2010**, *584*, 4575–4580.

54. Nakagawa, S.; Ip, J.Y.; Shioi, G.; Tripathi, V.; Zong, X.; Hirose, T.; Prasanth, K.V. Malat1 is not an essential component of nuclear speckles in mice. *RNA* **2012**, *18*, 1487–1499.

55. Bickmore, W.A.; Schorderet, P.; Duboule, D. Structural and Functional Differences in the Long Non-Coding RNA Hotair in Mouse and Human. *PLoS Genet.* **2011**, *7*, doi:10.1371/journal.pgen.1002071.

56. Eißmann, M.; Gutschner, T.; ämmerle, M.; Günther, S.; Caudron -Herger, M.; Groß, M.;
     Schirmacher, P.; Rippe, K.; Braun, T.; Zörnig, M.; Diederichs, S. Loss of the abundant nuclear
     non-coding RNA MALAT1 is compatible with life and development. *RNA Biol.* **2012**, *9*, 1076–
     1087.

57. Gabory, A.; Jammes, H.; Dandolo, L. The *H19* locus: role of an imprinted non-coding RNA in
     growth and development. *Bioessays* **2010**, *32*, 473–480.

58. Barsyte-Lovejoy, D.; Lau, S.K.; Boutros, P.C.; Khosravi, F.; Jurisica, I.; Andrulis, I.L.;
     Tsao, M.S.; Penn, L.Z. The c-Myc oncogene directly induces the *H19* noncoding RNA by
     allele-specific binding to potentiate tumorigenesis. *Cancer Res.* **2006**, *66*, 5330–5337.

59. van Bakel, H.; Nislow, C.; Blencowe, B.J.; Hughes, T.R. Most "dark matter" transcripts are
     associated with known genes. *PLoS Biol.* **2010**, *8*, doi:10.1371/journal.pbio.1000371.

60. Oosumi, T.; Belknap, W.R.; Garlick, B. Mariner transposons in humans. *Nature* **1995**, *378*,
     672–672.

61. Tsang, W.P.; Ng, E.K.; Ng, S.S.; Jin, H.; Yu, J.; Sung, J.J.; Kwok, T.T. Oncofetal *H19*-derived
     miR-675 regulates tumor suppressor RB in human colorectal cancer. *Carcinogenesis* **2010**, *31*,
     350–358.

62. Gejman, R.; Batista, D.L.; Zhong, Y.; Zhou, Y.; Zhang, X.; Swearingen, B.; Stratakis, C.A.;
     Hedley-Whyte, E.T.; Klibanski, A. Selective loss of MEG3 expression and intergenic
     differentially methylated region hypermethylation in the MEG3/DLK1 locus in human clinically
     nonfunctioning pituitary adenomas. *J. Clin. Endocrinol. Metab.* **2008**, *93*, 4119–4125.

63. Zhang, X.; Gejman, R.; Mahta, A.; Zhong, Y.; Rice, K.A.; Zhou, Y.; Cheunsuchon, P.; Louis,
     D.N.; Klibanski, A. Maternally expressed gene 3, an imprinted noncoding RNA gene, is
     associated with meningioma pathogenesis and progression. *Cancer Res.* **2010**, *70*, 2350–2358.

64. Zhou, Y.; Zhang, X.; Klibanski, A. MEG3 noncoding RNA: A tumor suppressor.
     J. Mol. Endocrinol. **2012**, *48*, R45–R53.

65. Benetatos, L.; Vartholomatos, G.; Hatzimichael, E. MEG3 imprinted gene contribution in
     tumorigenesis. *Int. J. Cancer* **2011**, *129*, 773–779.

66. Coccia, E.M.; Cicala, C.; Charlesworth, A.; Ciccarelli, C.; Rossi, G.B.; Philipson, L.; Sorrentino, V.
     Regulation and expression of a growth arrest-specific gene (gas5) during growth,differentiation,
     and development. *Mol. Cell. Biol.* **1992**, *12*, 3514–3521.

67. Kino, T.; Hurt, D.E.; Ichijo, T.; Nader, N.; Chrousos, G.P. Noncoding RNA gas5 is a growth
     arrest- and starvation-associated repressor of the glucocorticoid receptor. *Sci Signal* **2010**, *3*, ra8.

68. Mourtada-Maarabouni, M.; Pickard, M.R.; Hedge, V.L.; Farzaneh, F.; Williams, G.T. GAS5, a
     non-protein-coding RNA, controls apoptosis and is downregulated in breast cancer. *Oncogene*
     2009, *28*, 195–208.

69. Diehl, J.A. Cycling to Cancer with Cyclin D1. *Cancer Biol. Ther.* **2002**, *1*, 226–231.

70. Huarte, M.; Guttman, M.; Feldser, D.; Garber, M.; Koziol, M.J.; Kenzelmann-Broz, D.; Khalil,
     A.M.; Zuk, O.; Amit, I.; Rabani, M.; *et al.* A large intergenic noncoding RNA induced by p53
     mediates global gene repression in the p53 response. *Cell* **2010**, *142*, 409–419.

71. Amaral, P.P.; Clark, M.B.; Gascoigne, D.K.; Dinger, M.E.; Mattick, J.S. lncRNAdb: A reference
     database for long noncoding RNAs. *Nucleic Acids Res.* **2011**, *39*, D146–D151.

72. He, H.; Nagy, R.; Liyanarachchi, S.; Jiao, H.; Li, W.; Suster, S.; Kere, J.; de la Chapelle, A. A susceptibility locus for papillary thyroid carcinoma on chromosome 8q24. *Cancer Res.* **2009**, *69*, 625–631.

73. Chen, W.; Böcker, W.; Brosius, J.; Tiedge, H. Expression of neural BC200 RNA in human tumours. *J. Pathol.* **1997**, *183*, 345–351.

74. Iacoangeli, A.; Lin, Y.; Morley, E.J.; Muslimov, I.A.; Bianchi, R.; Reilly, J.; Weedon, J.; Diallo, R.; Bocker, W.; Tiedge, H. BC200 RNA in invasive and preinvasive breast cancer. *Carcinogenesis* **2004**, *25*, 2125–2133.

75. Chung, S.; Nakagawa, H.; Uemura, M.; Piao, L.; Ashikawa, K.; Hosono, N.; Takata, R.; Akamatsu, S.; Kawaguchi, T.; Morizono, T.; *et al.* Association of a novel long non-coding RNA in 8q24 with prostate cancer susceptibility. *Cancer Sci.* **2011**, *102*, 245–252.

76. Hibi, K.; Nakamura, H.; Hirai, A.; Fujikake, Y.; Kasai, Y.; Akiyama, S.; Ito, K.; Takagi, H. Loss of *H19* imprinting in esophageal cancer. *Cancer Res.* **1996**, *56*, 480–482.

77. Fellig, Y.; Ariel, I.; Ohana, P.; Schachter, P.; Sinelnikov, I.; Birman, T.; Ayesh, S.; Schneider, T.; de Groot, N.; Czerniak, A.; *et al. H19* expression in hepatic metastases from a range of human carcinomas. *J. Clin. Pathol.* **2005**, *58*, 1064–1068.

78. Matouk, I.J.; de Groot, N.; Mezan, S.; Ayesh, S.; Abu-lail, R.; Hochberg, A.; Galun, E. The *H19* non-coding RNA is essential for human tumor growth. *PLoS One* **2007**, *2*, doi:10.1371/journal.pone.0000845.

79. Arima, T.; Matsuda, T.; Takagi, N.; Wake, N. Association of IGF2 and *H19* imprinting with choriocarcinoma development. *Cancer Genet. Cytogenet.* **1997**, *93*, 39–47.

80. Berteaux, N.; Lottin, S.; Monte, D.; Pinte, S.; Quatannens, B.; Coll, J.; Hondermarck, H.; Curgy, J.J.; Dugimont, T.; Adriaenssens, E. *H19* mRNA-like noncoding RNA promotes breast cancer cell proliferation through positive control by E2F1. *J. Biol. Chem.* **2005**, *280*, 29625–29636.

81. Matouk, I.J.; Abbasi, I.; Hochberg, A.; Galun, E.; Dweik, H.; Akkawi, M. Highly upregulated in liver cancer noncoding RNA is overexpressed in hepatic colorectal metastasis. *Eur. J. Gastroenterol. Hepatol.* **2009**, *21*, 688–692.

82. Panzitt, K.; Tschernatsch, M.M.; Guelly, C.; Moustafa, T.; Stradner, M.; Strohmaier, H.M.; Buck, C.R.; Denk, H.; Schroeder, R.; Trauner, M.; *et al.* Characterization of HULC, a novel gene with striking up-regulation in hepatocellular carcinoma, as noncoding RNA. *Gastroenterology* **2007**, *132*, 330–342.

83. Pasic, I.; Shlien, A.; Durbin, A.D.; Stavropoulos, D.J.; Baskin, B.; Ray, P.N.; Novokmet, A.; Malkin, D. Recurrent focal copy-number changes and loss of heterozygosity implicate two noncoding RNAs and one tumor suppressor gene at chromosome 3q13.31 in osteosarcoma. *Cancer Res.* **2010**, *70*, 160–171.

84. Poliseno, L.; Salmena, L.; Zhang, J.; Carver, B.; Haveman, W.J.; Pandolfi, P.P. A coding-independent function of gene and pseudogene mRNAs regulates tumour biology. *Nature* **2010**, *465*, 1033–1038.

85. Khaitan, D.; Dinger, M.E.; Mazar, J.; Crawford, J.; Smith, M.A.; Mattick, J.S.; Perera, R.J. The melanoma-upregulated long noncoding RNA SPRY4-IT1 modulates apoptosis and invasion. *Cancer Res.* **2011**, *71*, 3852–3862.

86. Wang, F.; Li, X.; Xie, X.; Zhao, L.; Chen, W. UCA1, a non-protein-coding RNA up-regulated in bladder carcinoma and embryo, influencing cell growth and promoting invasion. *FEBS Lett.* **2008**, *582*, 1919–1927.

87. Wang, X.S.; Zhang, Z.; Wang, H.C.; Cai, J.L.; Xu, Q.W.; Li, M.Q.; Chen, Y.C.; Qian, X.P.; Lu, T.J.; Yu, L.Z.; *et al.* Rapid identification of UCA1 as a very sensitive and specific unique marker for human bladder carcinoma. *Clin. Cancer Res.* **2006**, *12*, 4851–4858.

88. Dallosso, A.R.; Hancock, A.L.; Malik, S.; Salpekar, A.; King-Underwood, L.; Pritchard-Jones, K.; Peters, J.; Moorwood, K.; Ward, A.; Malik, K.T.; *et al.* Alternately spliced WT1 antisense transcripts interact with WT1 sense RNA and show epigenetic and splicing defects in cancer. *RNA* **2007**, *13*, 2287–2299.

89. de Kok, J.B.; Verhaegh, G.W.; Roelofs, R.W.; Hessels, D.; Kiemeney, L.A.; Aalders, T.W.; Swinkels, D.W.; Schalken, J.A. DD3(PCA3), a very sensitive and specific marker to detect prostate tumors. *Cancer Res.* **2002**, *62*, 2695–2689.

90. Leygue, E. Steroid receptor RNA activator (SRA1): Unusual bifaceted gene products with suspected relevance to breast cancer. *Nucl. Recept. Signal.* **2007**, *4*, doi:10.1621/nrs.05006.

91. Qi, P.; Du, X. The long non-coding RNAs, a new cancer diagnostic and therapeutic gold mine. *Mod. Pathol.* **2012**, doi:10.1038/modpathol.2012.160.

92. Gibb, E.A.; Brown, C.J.; Lam, W.L. The functional role of long non-coding RNA in human carcinomas. *Mol. Cancer* **2011**, *10*, doi:10.1186/1476-4598-10-38.

93. Morris, K.V. RNA-directed transcriptional gene silencing and activation in human cells. *Oligonucleotides* **2009**, *19*, 299–306.

94. Schöler, N.; Langer, C.; Döhner, H.; Buske, C.; Kuchenbauer, F. Serum microRNAs as a novel class of biomarkers: A comprehensive review of the literature. *Exp. Hematol.* **2010**, *38*, 1126–1130.

95. Xie, Y.; Todd, N.W.; Liu, Z.; Zhan, M.; Fang, H.; Peng, H.; Alattar, M.; Deepak, J.; Stass, S.A.; Jiang, F. Altered miRNA expression in sputum for diagnosis of non-small cell lung cancer. *Lung Cancer* **2010**, *67*, 170–176.

96. Xing, L.; Todd, N.W.; Yu, L.; Fang, H.; Jiang, F. Early detection of squamous cell lung cancer in sputum by a panel of microRNA markers. *Mod. Pathol.* **2010**, *23*, 1157–1164.

97. Kosaka, N.; Iguchi, H.; Ochiya, T. Circulating microRNA in body fluid: a new potential biomarker for cancer diagnosis and prognosis. *Cancer Sci.* **2010**, *101*, 2087–2092.

98. Zhu, Y.; Yu, M.; Li, Z.; Kong, C.; Bi, J.; Li, J.; Gao, Z. ncRAN, a newly identified long noncoding RNA, enhances human bladder tumor growth, invasion, and survival. *Urology* **2011**, *77*, 510.e1–510.e5.

99. Prensner, J.R.; Iyer, M.K.; Balbin, O.A.; Dhanasekaran, S.M.; Cao, Q.; Brenner, J.C.; Laxman, B.; Asangani, I.A.; Grasso, C.S.; Kominsky, H.D.; *et al.* Transcriptome sequencing across a prostate cancer cohort identifies PCAT-1, an unannotated lincRNA implicated in disease progression. *Nat. Biotechnol.* **2011**, *29*, 742–749.

100. Hessels, D.; Klein Gunnewiek, J.M.T.; van Oort, I.; Karthaus, H.F.M.; van Leenders, G.J.L.; van Balken, B.; Kiemeney, L.A.; Witjes, J.A.; Schalken, J.A. DD3PCA3-based Molecular Urine Analysis for the Diagnosis of Prostate Cancer. *Eur. Urol.* **2003**, *44*, 8–16.

101. Tinzl, M.; Marberger, M.; Horvath, S.; Chypre, C. DD3PCA3 RNA Analysis in Urine – A New Perspective for Detecting Prostate Cancer. *Eur. Urol.* **2004**, *46*, 182–187.

102. Muro, E.M.; Andrade-Navarro, M.A. Pseudogenes as an alternative source of natural antisense transcripts. *BMC Evol. Biol.* **2010**, *10*, doi:10.1186/1471-2148-10-338.

103. Morris, K.V.; Santoso, S.; Turner, A.M.; Pastori, C.; Hawkins, P.G. Bidirectional transcription directs both transcriptional gene activation and suppression in human cells. *PLoS Genet.* **2008**, *4*, doi:10.1371/journal.pgen.1000258.

104. Lyle, R.; Watanabe, D.; te Vruchte, D.; Lerchner, W.; Smrzka, O.W.; Wutz, A.; Schageman, J.; Hahner, L.; Davies, C.; Barlow, D.P. The imprinted antisense RNA at the Igf2r locus overlaps but does not imprint Mas1. *Nat. Genet.* **2000**, *25*, 19–21.

6

SnoRNA U50 Levels are Regulated by Cell Proliferation and rRNA Transcription

Annalisa Pacilli, Claudio Ceccarelli, Davide Treré and Lorenzo Montanaro *

Department of Experimental, Diagnostic and Specialty Medicine, University of Bologna, Sant'Orsola-Malpighi University Hospital, via Massarenti, 9, Bologna 40138, Italy;
E-Mails: annalisa.pacilli@gmail.com (A.P.); claudio.ceccarelli@unibo.it (C.C.);
davide.trere@unibo.it (D.T.)

* Author to whom correspondence should be addressed; E-Mail: lorenzo.montanaro@unibo.it;

Abstract: rRNA post transcriptional modifications play a role in cancer development by affecting ribosomal function. In particular, the snoRNA U50, mediating the methylation of C2848 in 28S rRNA, has been suggested as a potential tumor suppressor-like gene playing a role in breast and prostate cancers and B-cell lymphoma. Indeed, we observed the downregulation of U50 in colon cancer cell lines as well as tumors. We then investigated the relationship between U50 and proliferation in lymphocytes stimulated by phytohemagglutinin (PHA) and observed a strong decrease in U50 levels associated with a reduced C2848 methylation. This reduction was due to an alteration of U50 stability and to an increase of its consumption. Indeed, the blockade of ribosome biogenesis induced only an early decrease in U50 followed by a stabilization of U50 levels when ribosome biogenesis was almost completely blocked. Similar results were found with other snoRNAs. Lastly, we observed that U50 modulation affects ribosome efficiency in IRES-mediated translation, demonstrating that changes in the methylation levels of a single specific site on 28S rRNA may alter ribosome function. In conclusion, our results link U50 to the cellular proliferation rate and ribosome biogenesis and these findings may explain why its levels are often greatly reduced in cancers.

Keywords: snoRNA U50; ribosome methylation; IRES-mediated translation; colon cancer

1. Introduction

Ribosome biogenesis is a highly coordinated process occurring in the nucleolus, where a polycistronic pre-ribosomal RNA (pre-rRNA) transcript is processed to generate the mature 18S, 5.8S, and 28S rRNA. During this processing, the rRNA sequences undergo extensive covalent nucleotide modification, largely directed by small nucleolar RNA (snoRNA)-protein complexes (snoRNP) [1,2]. SnoRNA may be divided into two classes: The H/ACA and the C/D box, mediating pseudouridylation and 2'O-methylation of specific sites, respectively [3,4]. In particular, the methylation reaction is guided by an extensive region (10–21 nt) of complementarity between the C/D box snoRNA and rRNA sequences flanking the modification site [5–8]. In mammals, snoRNAs are transcribed by the RNA Polymerase II being localized within the introns of snoRNA host genes. These host genes are also transcribed for either protein coding or noncoding mRNAs [9] which often contain a 5' terminal oligopyrimidine (5' TOP) sequence responsible for their translational upregulation in response to growth factors or other conditions requiring increased protein synthesis (reviewed by Meyuhas *et al.* [10]); however, the precise function of 5' TOP motif with respect to snoRNA synthesis is unknown [11]. Furthermore, the development of three-dimensional maps of the modified nucleotides in the ribosomes of Escherichia coli and yeast has revealed that rRNA modifications occur in conserved and functionally important regions for subunit–subunit and nascent protein interactions, for tRNA and mRNA binding, but not in those interacting with proteins (see [12,13]). This correlation indicates that modifications influence both the structure and the function of the ribosome [14]. Indeed, there is evidence that post-transcriptional rRNA modifications, including pseudouridylation and methylation, affect ribosomal function [15–17] and that alterations in this modification pattern might be involved in human diseases, such as ribosomopathies and tumorigenesis [18,19]. Recent reports have demonstrated that somatic rearrangements, mutations, or the reduction in the expression of the C/D snoRNA U50 have been found in breast carcinomas, prostate cancer, and B-cell lymphomas [20–23]. snoRNA U50 is known to mediate the methylation of ribose residues corresponding to the cytosines in positions 2848 and 2863 in 28S ribosomal RNA [5,23]. In breast cancer cell lines, the reintroduction of U50 is able to induce cell death, suggesting a tumor-suppressor-like behavior for this snoRNA [23]. The human snoRNA U50 sequence is localized in the 5th intron of the non-coding host gene named small nucleolar RNA host gene 5 (SNHG5) [20,24], which is a member of the 5' TOP gene family.

In this paper we investigated the relationship between snoU50 and cancer in colon cancer cell lines and tumors with particular regard to proliferation.

2. Results and Discussion

2.1. Evaluation of U50 Levels in Colorectal Cancer Tissues and Cell Lines

The evaluation of snoRNA U50 levels on colon cancer tissues was performed on both tumor and normal tissues in a series of 34 patients. We found that U50 was downregulated in tumor tissues if compared to the normal counterpart and this reduction was statistically significant in a subgroup of low-stage tumors ($p = 0.047$) (Figure 1A, left). The decrease in U50 levels in tumors was in line with previous reports, demonstrating the same behavior in prostate and breast cancers [21,23]. In order to find a key element linking U50 and tumorigenesis, we performed a correlation analysis with the available clinical and bio-pathological features of tumors (see Table S1) and we found a significant association with the tumor grade. Indeed, high-grade tumors displayed lower U50 levels in comparison to those observed in low-grade tumors ($p = 0.049$) (Figure 1A, right). We then evaluated U50 expression in a panel of eleven colon cancer cell lines. We found that U50 expression is highly variable between lines, but always lower than normal colon tissues (NT) (Figure 1B, left), while the overall comparison between NT and colon cancer cell lines showed a statistically significant difference for U50 expression ($p = 0.0004$) (Figure 1B, right-top). Furthermore, we found that the U50 expression in lines derived from primary tumors (HCT, SW480, RKO, HCA7, CaCo-2, La174T, HT29, SW48) was significantly different from that in those derived from metastatic tumors (Colo205, SW620, LoVo-$p = 0.0121$) (Figure 1B, right-bottom).

Figure 1. Downregulation of U50 levels in colon cancer tumors and cell lines. (**A**) Evaluation of U50 levels in both tumoral and adjacent untransformed colon tissues on all samples ($n = 34$) and in a subgroup of low-stage tumors ($n = 20$) (left); comparison of U50 levels between low- and high-grade tumors ($n = 27$ and $n = 7$, respectively); (**B**) Evaluation of U50 levels in a cohort of colon cancer cell lines (left); comparison between U50 levels in normal colon tissues and cell lines (right-top) and between cell lines derived from primary or metastatic tumors (right-bottom). The results correspond to means ± S.E.M. of three different experiments. * $p \leq 0.05$; *** $p \leq 0.001$; NT, normal tissues; TT, tumoral tissues; T, tumors.

A

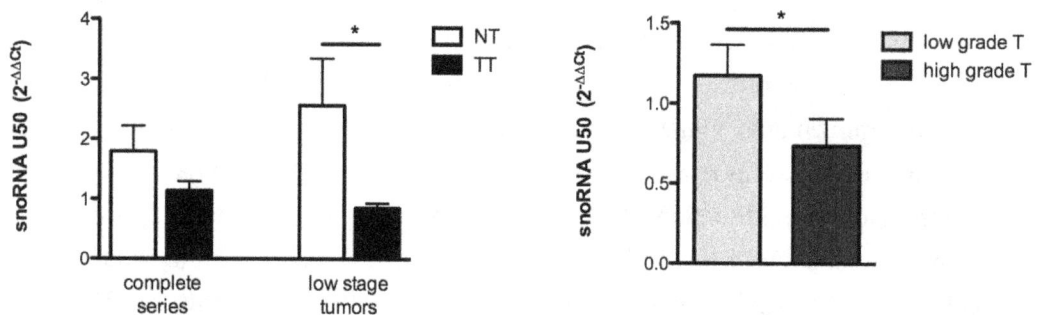

B

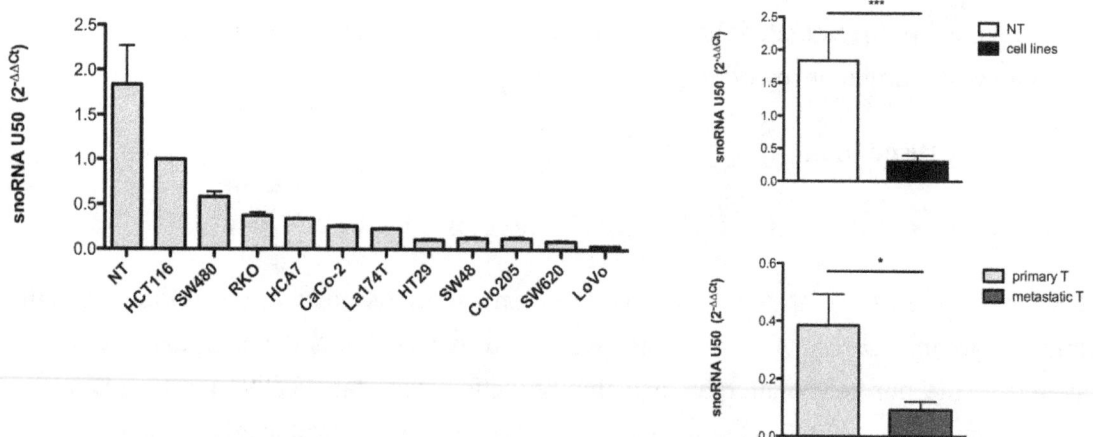

2.2. Relationship between U50 and Proliferation

Compared to untransformed tissues, U50 was downregulated in tumors. Being that one of the major distinguishing characteristics between these tissues is their proliferation activity, we investigated whether there is a relationship between U50 and proliferation.

In order to study changes in U50 synthesis depending on cellular proliferation, we used a well-established model of investigation, *i.e.*, the comparison of resting and stimulated primary human lymphocytes. Thus we isolated lymphocytes from healthy donors and performed lymphocyte proliferation assays by using phytohemoagglutinin (PHA). After PHA stimulation, we evaluated the expressions of both U50 and its host gene SNHG5 and observed a strong reduction of U50 levels in PHA stimulated cells, compared to controls (Figure 2A). Conversely, the SNHG5 host gene was upregulated (Figure 2B). Since snoU50 and SNHG5 are simultaneously transcribed, this discrepancy might be explained by a different stability of the two RNAs. The massive decrease in U50 led us to hypothesize that during cellular proliferation U50 stability might be extremely reduced. To prove this hypothesis, we performed mRNA stability assay on lymphocytes—Both stimulated and not stimulated by PHA—By treating cells with the transcriptional inhibitor Actinomycin D (Act-D) at an 80 nM concentration. At this concentration, Act-D is known to abolish the activity of RNA polymerases I and II. We observed that the Act-D treatment did not reduce U50 stability in non-proliferative control cells, but rather induced a progressive accumulation of this snoRNA depending on the length of Act-D treatment (Figure 3). On the contrary, in PHA stimulated cells, we found an early decrease in U50 levels up to 3 h of treatment, followed by a general stabilization of U50 levels (Figure 3). The fluctuation of U50 levels observed at 3 h may be the result of the combined effect of time needed to obtain the full inhibition of ribosome processing by Act-D treatment and the regulation of SNHG5 transcription as documented in Figure S1. Furthermore, PHA treatment induced a notable decrease in U50 levels but an increase in SNHG5 levels. The different levels of our transcripts in the two conditions before Act-D treatment and the behavior of mature SNHG5 mRNA levels observed after 80 nM Act-D treatment excludes the possibility that the increase of U50 levels after treatment was due to an increase of its transcription.

Figure 2. Regulation of U50 levels in response to cellular proliferation. Evaluation of U50 (**A**); and its host gene SNHG5 (**B**) levels in human lymphocytes stimulated or not to proliferate with PHA for 72 h. The results correspond to means ± S.E.M. of four different experiments. * $p \leq 0.05$; ** $p \leq 0.01$. CTR, controls; PHA, phytohemoagglutinin.

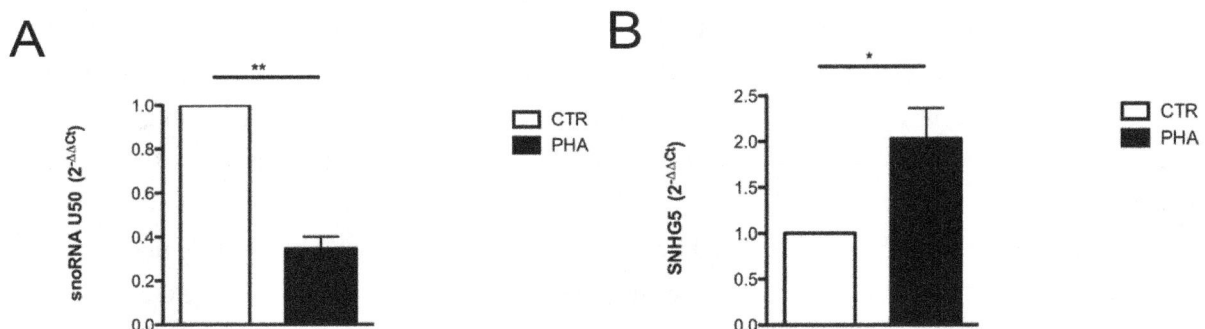

We hypothesized that the observed stabilization of U50 was a consequence of the concomitant transcriptional inhibition of rRNA genes, which implied a reduced post-transcriptional modification of rRNA, and thus a lower U50 consumption. This would be in line with the variation of 45S rRNA levels found in both control and PHA-treated lymphocytes (Figure 3). The different regulation of 45S levels over time observed in control and PHA treated cells can be explained by the different activity of rRNA processing in the two conditions. Taken together, these results showed that U50 and 45S had opposite behaviors. This is particularly clear in control cells, since the ribosome biogenesis is not as intensive and U50 did not have to be massively consumed.

Figure 3. Relationship between U50 and rRNA transcription. Evaluation of U50 and 45S levels in human lymphocytes controls or PHA-stimulated in the presence or absence of high doses of the transcriptional inhibitor Act-D (80 nM) in order to block both polymerases I and II. The analyses were performed 20 min, 1 h, 3 h, 8 h, and 24 h after Act-D treatment. The results correspond to means ± S.E.M. of three different experiments. CTR, controls; PHA, phytohemoagglutinin; Act-D, actinomycin-D.

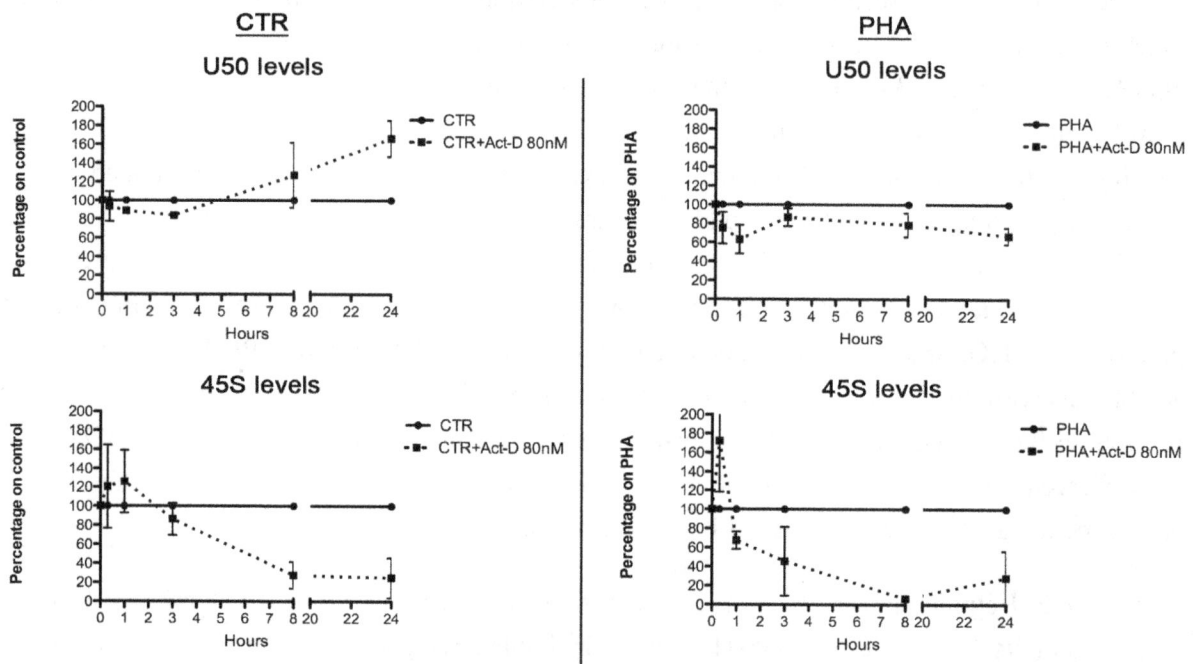

2.3. Relationship between U50 and rRNA Transcription

The accumulation of U50 induced by the block of total transcription and the inverse relationship between U50, proliferation, and 45S levels led us to suppose that U50 synthesis could be regulated by the ribosome biogenesis rate, and so we selectively reduced rRNA transcription in SW620 colon cancer cells. We chose this line since the levels of U50 are quite low in presence of a very active ribosome biogenesis activity.

A tenfold lower Act-D concentration was used to selectively inhibit polymerase I only. This selectivity was proved by measuring the transcription levels of a housekeeping mRNA (β-glucuronidase-GUS) and SNHG5, whose genes were both transcribed by polymerase II. Results demonstrated that none were affected by Act-D 8nM treatment (Figure S2). The SNHG5 transcript was also measured to exclude the possibility that U50 levels could be regulated by its transcription. After 45S synthesis inhibition, U50 levels increased (Figure 4). This result obtained after the blockade of rRNA transcription confirmed our hypothesis that U50 and cellular proliferation are inversely associated because of its massive consumption and not because of its transcription. Indeed, U50 is implicated in post-transcriptional methylation of 28S rRNA, and it is well known that during proliferation there is an increase in ribosome biogenesis requiring modification of rRNA molecules. The levels of U50 and 45S transcript displayed opposite behaviors, thus confirming that the strict dependence of U50 on the rRNA transcription rate might be due to its biological function. Taken together, these data indicated a putative role of U50 and snoRNAs in growth, tumorigenicity, and metastasis. We then investigated whether the relationship between ribosome biogenesis and U50 is shared by other snoRNAs and evaluated the cellular levels of 3 additional C/D box snoRNAs such as U33, U34, and U56 housed in genes coding for the ribosomal protein L13A (U33 and U34) and for the nucleolar protein 5A, which is a component of snoRNPs complex (Nop56). We found that these selected snoRNAs displayed the same regulation and that, in all cases, the accumulation started after 3 h and progressively increased with the length of treatment (Figure S3).

Figure 4. Relationship between U50 and *de novo* rRNA transcription in SW620 cell line. Evaluation of U50 stability performed by measuring U50 and 45S levels in SW620 cells after blockade of Polymerase I with low doses of Act-D. The analyses were performed 20 min, 1 h, 3 h, 8 h, and 24 h after Act-D treatment. The results correspond to means ± S.E.M. of three different experiments. CTR, controls; Act-D, actinomycin-D.

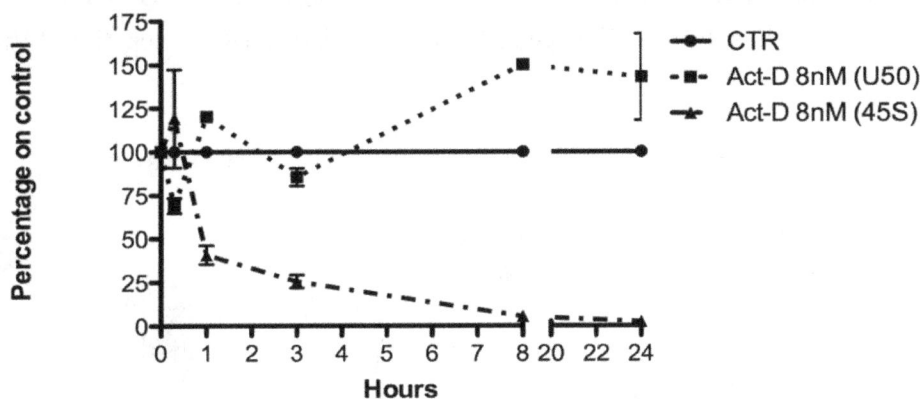

2.4. Effect of U50 Regulation on rRNA Site-Specific Methylation and Ribosome Activity

Prompted by these observations, we investigated the possible effect of U50 modulation on mRNA translation. Indeed, U50 is responsible for the site-specific C2848 methylation on 28s rRNA, and It has recently been demonstrated that incorrect methylation of rRNA is associated with an impaired capacity to initiate translation through Internal Ribosome Entry Site (IRES) [19,25] by influencing the mechanism of 80S complex formation on IRES elements [25]. Actually there are studies reporting that changes in snoRNAs are crucial for ribosome biogenesis or function [26,27], and showing an increase in non-coding host genes either after inhibition of translation and elongation [11,28] or in growth arrest conditions [29]. However, it is not clear whether the modulation of a single snoRNA might also contribute to altering ribosome efficiency. In this paper, we focused both on the role of U50 modulation in C2848 site-specific methylation, and on IRES-mediated translation. Therefore, we semi-quantitatively evaluated the changes in C2848 methylation in PHA stimulated cells and observed a decrease in site-specific C2848 methylation in response to proliferation (Figure S4). To better understand the role of U50 and its influence on methylation and IRES-mediated translation, we regulated U50 levels by specific knockdown (KD) and overexpression. We then evaluated the effect on C2848 of 28S methylation, on the translation of known viral IRESes types (Cricket Paralysis Virus-CrPV, Encephalomyocarditis virus-EMCV and Hepatitis C Virus-HCV), and on a cellular IRES (c-Myc) known to be affected by rRNA methylation [25]. We performed U50 KD on HCT116 cell lines and observed that, 72 h after LNA transfection, we obtained a rather strong U50 KD efficiency associated with a decrease in C2848 methylation (Figure 5A,C). Thus we decided to evaluate IRES-mediated translation after 72 h of U50 KD. We found that U50 KD did not significantly alter ribosome activity (Figure S3, top), although we observed a slight increase in CrPV- and HCV IRESes-mediated translation (Figure 5E). It is likely that we could not appreciate notable differences in ribosome translation activity in the U50 KD model because the experiments were performed in a tumor cell line with already low U50 levels in comparision to untransformed cells.

Figure 5. Effect of U50 modulation on C2848 methylation and IRES-mediated translation. Evaluation of U50 levels (**A** and **B**); C2848 methylation status (**C** and **D**); and IRES-mediated translation (**E** and **F**) in HCT116 cells after 72 h of U50 knockdown (left panel) and U50 overexpression (right panel). The results correspond to means ± S.E.M. of three different experiments. CTR, controls; KD, knockdown; pSrQ, pSIREN-RetroQ vector (empty vector); pSrQ U50, pSIREN-RetroQ-U50 vector; CrPV, CRicket Paralysis Virus; EMCV, EncephaloMyoCarditis Virus; HCV, Hepatitis C Virus.

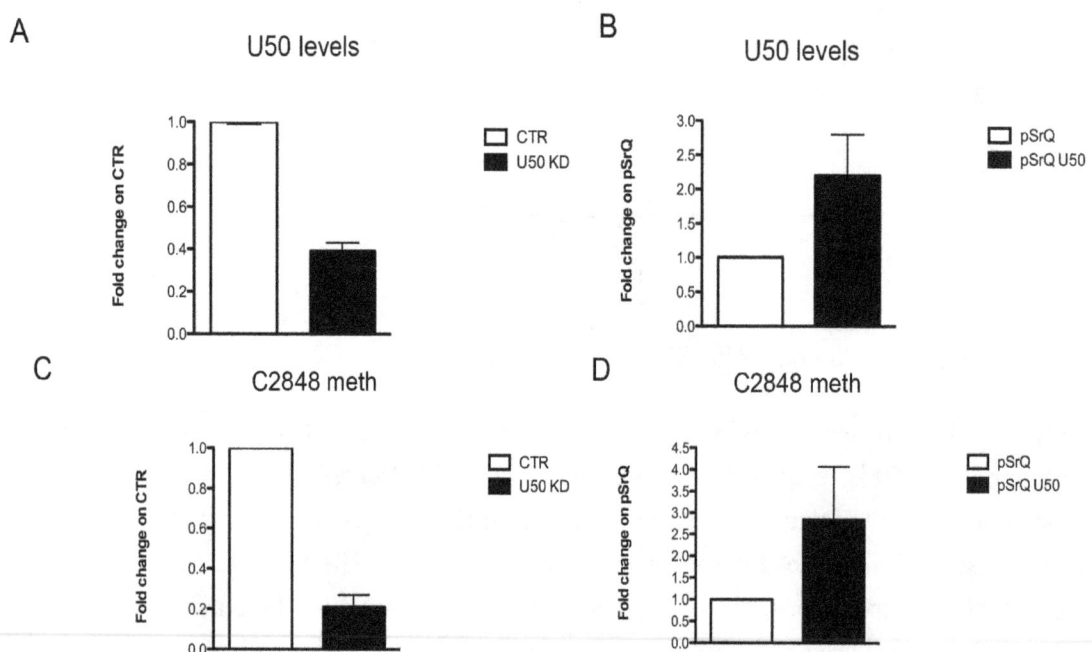

Figure 5. *Cont.*

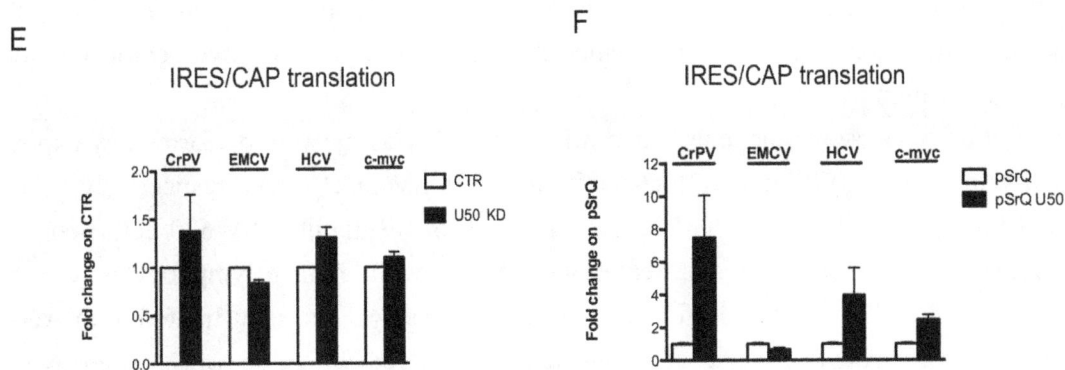

Furthermore, in order to investigate the effect of U50 overexpression, we used the pSrQ-U50 vector to stably transfect HCT116, SW620, and LoVo cell lines. In addition to HCT116, we chose SW620 and LoVo cells since they displayed the lowest U50 basal level in our cohort of colon cancer lines, as previously reported. We successfully obtained all U50 overexpressing lines, but an increase in C2848 methylation was achieved for the HCT116 line only (Figures 5B,D and S6). In contrast with results after U50 KD, in the presence of higher C2848 methylation levels, we found a general decrease in CAP-mediated translation and an increase in IRES-mediated translation (Figure S5, bottom). These observations were consistent for CrPV, HCV, and c-Myc IRESes, while EMCV IRES translation was not affected by changes in the methylation of this specific site (Figure 5F). These results might be explained by the augmented ribosome affinity for IRESes secondary structures, which was probably caused by changes in the ribosome structure mediated by C2848 methylation status.

3. Experimental Section

3.1. Patient Materials

Both tumor and adjacent non-tumor (hereinafter referred to as "normal tissue") colon tissues selected from a series of thirty-four consecutive patients who had undergone surgical resection for primary carcinoma at the Surgical Department of the University of Bologna, on the sole basis of frozen tissue availability. For each patient, clinical information was recorded and the corresponding tissue was histologically characterized and processed following standard procedures to define bio-pathological features. Specimen collection and tissue analyses were approved by the Bologna University Ethical Committee on human tissue research. Tissues were preserved at $-80\ °C$ until use. A 60 mg piece for each sample was minced in liquid nitrogen and then lysed for total RNA extraction using Tri-reagent solution (Ambion, Life Technologies Corporation, Monza, Italy) according to the manufacturer's instructions.

3.2. Cell Cultures and Treatments

All colon cancer cell lines were grown in their optimal culture medium and conditions at $37\ °C$ in $5\%\ CO_2$ according to the American Type Culture Collection instructions.

Lymphocytes were recovered from healthy donors' buffy coats collected by the transfusion center of Sant'Orsola-Malpighi Hospital in Bologna after ficoll gradient and monocyte depletion. Human lymphocytes were grown in RPMI-1640 medium (Sigma, Milan, Italy), supplemented with 10% FBS, 1% of penicillin and streptomycin, 1% of glutamine, and 0.5% of non-essential amino acids (all Sigma). Lymphocyte proliferation assays were performed by adding 10 μg/mL of phytohemoagglutinin (PHA) (Sigma) to the medium.

The inhibition of ribosome biogenesis and mRNA stability assay were performed by exposing cells to Actinomycin D (Act-D) (Sigma) at a concentration of 8 nM and 80 nM, respectively, and assessed at different time points (20 min, 1 h, 3 h, 8 h, and 24 h of treatments). SW620 cells were seeded at 100,000 cells/well in 6-well plates and treated with Act-D, while human lymphocytes were seeded at 1,000,000 cells/mL, PHA stimulated for 72 h (where necessary), and then treated with Act-D. PHA stimulation was evaluated by cell counting and considered effective only when lymphocytes proliferation was increased more than 5 fold compared to control cells.

3.3. Gene Expression Assays

Total RNA was extracted from cell lines using Tri-reagent (Ambion) and 1 μg was reverse-transcribed using the High-Capacity cDNA Archive Kit (Applied Biosystems, Carlsbad, CA, USA). Real-time PCR analysis was carried out in a Gene Amp 7000 Sequence Detection System (Applied Biosystems) using the TaqMan approach for β-glucuronidase (GUS) (Applied Biosystems) and SYBR green approach for all other genes, using specific couples of primers. The housekeeping GUS gene was used to calculate the relative amounts of the studied target genes in all experiments except for those in which Act-D treatment was performed. In the latter case, for each sample we calculated the 2^{-Ct}, assumed the values of untreated cells as reference (100%), and then calculated the relative percentage of investigated genes. For each sample, three replicates were analyzed. Analyses in cell lines and tumors were carried out by using aliquots of a single stock of cDNA obtained from HCT116 as internal calibrator. The analysis of U50 levels was performed by qPCR using a primer with a linker sequence attached to a U50-specific sequence for cDNA synthesis and reverse linker specific primer, and U50 forward specific primer for real-time PCR as described in Dong, 2008 [21]. All primers used were reported in Table S2.

3.4. SnoRNA U50 Knockdown and Upregulation

Since it has been shown that RNAi is an unsuitable tool for snoRNA KD [30], we used an oligomediated RNaseH cleavage strategy. A custom-made hybrid DNA-RNA Locked Nucleic Acid (LNA)-antisense specific for U50 (Exiqon, Vedbaek, Denmark) was transfected into cells with Lipofectamine RNAiMAX reagent (Invitrogen, Life Technologies, Carlsbad, CA, USA), while the efficiency of KD was evaluated after 24 h and 72 h. The same amount of a control oligo sequence (TCGAGCGGCCGCCCGGGC) was transfected with lipofectamine in control cells.

For U50 upregulation, we used a retroviral transduction system. We briefly transfected the Phoenix A cell line with 10 μg of pSIREN-RetroQ-U50 plasmid or the pSIRENRetroQ vector control (gifts of Prof. Dong) [21] using Lipofectamine 2000 reagent, and collected the supernatant (containing viruses) after 24 h and 48 h. After that we infected HCT116, SW620, and LoVo cell lines previously seeded in 6-well plates with viruses using standard spinoculation protocols. Seventy-two hours after infection, cell lines were selected for at least 10 days, while adding puromycin into the media at a final concentration of 0.7 µg/mL. The upregulation of U50 was verified by qPCR as described above.

3.5. Semi-Quantitative Site-Specific Methylation Assay

To evaluate site-specific (C2848) methylation levels, we followed a modified previously described method [19]. This modified method consists of the reverse transcription (RT) of total RNA with Moloney Murine Leukemia Virus Reverse Transcriptase (M-MLV) by using a 28S specific oligonucleotide (targeting the methylation site downstream, at either low or high dNTPs concentrations, either affecting cDNA synthesis or not, respectively, in the presence of modified sites. The cDNAs resulting from both RT are evalutated by qPCR; the ratio between high and low dNTPs-derived products semi-quantitatively indicates the modification of the site downstream of the oligonucleotide used. An endogenous housekeeping RNA (GUS) was used to evaluate the efficiency of reverse transcription.

3.6. mRNA Transfection and Internal Ribosome Entry Site (IRES) Translation

Capped mRNA was transcribed from linearized pR-CrPV-IRES-F (gift of Dr. D. Ruggero) [31], pF-EMCV-IRES-R (gift of Prof. A.C. Palmenberg) [32], pR-HCV-IRES-F (gift of Prof R.E. Lloyd) [33], and pR-c-MYC-IRES-F [34], by using the mMessage mMachine T7 or T3 kits (Ambion). Cells were transfected with 0.4 µg RNA/sample using Lipofectamine 2000 (Invitrogen) following the manufacturer's instructions. After 4 h transfection, media were changed and 2 h later cells were harvested and analyzed with a dual-luciferase assay kit (Promega, Madison, WI, USA) according to the manufacturer's instructions.

3.7. Statistical Analysis

All data were analyzed using Prism software, version 5.0a. A paired t-test was used for patient data set when comparing normal and correspondent tumoral tissues. A two-sided Student's t-test was used for the comparisons between two groups. p values <0.05 were regarded as statistically significant.

4. Conclusions

Recent reports have demonstrated that U50 was downregulated in prostate and breast cancers and in B-cell lymphomas [20,21,23]. In this study we have demonstrated for the first time that snoRNA U50 expression: (i) is also downregulated in colon cancer; (ii) is reduced during cell proliferation; and (iii) its levels are inversely associated with ribosome biogenesis. Furthermore, since U50 mediates site-specific methylation on C2848 of 28S rRNA, we investigated the effect of its modulation on ribosome activity and showed that proliferating cells with low levels of both U50 and C2848 methylation only moderately changed their translational activity, while the U50 and C2848 methylation increments brought about an IRES-mediated translation propensity. At the present time the biological significance of these observations in cancer progression is unclear, and further investigations will aim to investigate which genes are differently expressed and what their roles might be in human cancers and other pathologies.

Acknowledgements

We thank the "Centro Interdipartimentale di ricerche sul cancro-Giorgio Prodi (CIRC)" for supporting AP. We are grateful to Dong for the pSIREN-RetroQ-U50 plasmid. This work was supported by PRIN grant from the Italian Ministry of Education, University and Research n. 20104AE23N_002 to LM and by the Pallotti Legacy for Cancer Research.

References

1. Filipowicz, W.; Pogacić, V. Biogenesis of small nucleolar ribonucleoproteins. *Curr. Opin. Cell Biol.* **2002**, *14*, 319–327.

2. Smith, C.M.; Steitz, J.A. Sno storm in the nucleolus: New roles for myriad small RNPs. *Cell* **1997**, *89*, 669–672.

3. Ni, J.; Tien, A.L.; Fournier, M.J. Small nucleolar RNAs direct site-specific synthesis of pseudouridine in ribosomal RNA. *Cell* **1997**, *89*, 565–573.

4. Kiss, T. Small nucleolar RNAs: An abundant group of noncoding RNAs with diverse cellular functions. *Cell* **2002**, *109*, 145–148.

5. Kiss-Laszlo, Z.; Henry, Y.; Bachellerie, J.P.; Caizergues-Ferrer, M.; Kiss, T. Site-specific ribose methylation of preribosomal RNA: A novel function for small nucleolar RNAs. *Cell* **1996**, *85*, 1077–1088.

6. Cavaille, J.; Bachellerie, J.P. SnoRNA-guided ribose methylation of rRNA: Structural features of the guide RNA duplex influencing the extent of the reaction. *Nucleic Acids Res.* **1998**, *26*, 1576–1587.

7. Bachellerie, J.P.; Michot, B.; Nicoloso, M.; Balakin, A.; Ni, J.; Fournier, M.J. Antisense snoRNAs: A family of nucleolar RNAs with long complementarities to rRNA. *Trends Biochem. Sci.* **1995**, *20*, 261–264.

8. Kiss, T. Small nucleolar RNA-guided posttranscriptional modification of cellular RNAs. *EMBO J.* **2001**, *20*, 3617–3622.

9. Weinstein, L.B.; Steitz, J.A. Guided tours: From precursor snoRNA to functional snoRNP. *Curr. Opin. Cell Biol.* **1999**, *11*, 378–384.

10. Meyuhas, O.; Avni, D.; Shama, S. Translational Control of Ribosomal Protein mRNAs in Eukaryotes. In *Translational Control*; Hershey, J.W.B., Mathews, M.B., Sonenberg, N., Eds.; Cold Spring Harbor Laboratory Press: New York, NY, USA, 1996; Volume 30, pp. 363–388.

11. Smith, C.M.; Steitz, J.A. Classification of *gas5* as a multi-small-nucleolar-RNA (snoRNA) host gene and a member of the 5'-terminal oligopyrimidine gene family reveals common features of snoRNA host genes. *Mol. Cell Biol.* **1998**, *18*, 6897–6909.

12. Decatur, W.A.; Fournier, M.J. rRNA modifications and ribosome function. *Trends Biochem. Sci.* **2002**, *27*, 344–351.

13. Yusupov, M.M.; Yusupova, G.Z.; Baucom, A.; Lieberman, K.; Earnest, T.N.; Cate, J.H.; Noller, H.F. Crystal structure of the ribosome at 5.5 Å resolution. *Science* **2001**, *292*, 883–896.

14. Lane, B.G.; Ofengand, J.; Gray, M.W. Pseudouridine and *O*2'-methylated nucleosides. Significance of their selective occurrence in rRNA domains that function in ribosome-catalyzed synthesis of the peptide bonds in proteins. *Biochimie* **1995**, *77*, 7–15.

15. Green, R.; Noller, H.F. Reconstitution of functional 50S ribosomes from *in vitro* transcripts of *Bacillus stearothermophilus* 23S rRNA. *Biochemistry* **1999**, *38*, 1772–1779.

16. Khaitovich, P.; Tenson, T.; Kloss, P.; Mankin, A.S. Reconstitution of functionally active *Thermus aquaticus* large ribosomal subunits with *in vitro*-transcribed rRNA. *Biochemistry* **1999**, *38*, 1780–1788.

17. Liu, B.; Liang, X.H.; Piekna-Przybylska, D.; Liu, Q.; Fournier, M.J. Mis-targeted methylation in rRNA can severely impair ribosome synthesis and activity. *RNA Biol.* **2008**, *5*, 249–254.

18. Montanaro, L.; Calienni, M.; Bertoni, S.; Rocchi, L.; Sansone, P.; Storci, G.; Santini, D.; Ceccarelli, C.; Taffurelli, M.; Carnicelli, D.; *et al.* Novel dyskerin-mediated mechanism of p53 inactivation through defective mRNA translation. *Cancer Res.* **2010**, *70*, 4767–4777.

19. Belin, S.; Beghin, A.; Solano-Gonzàlez, E.; Bezin, L.; Brunet-Manquat, S.; Textoris, J.; Prats, A.C.; Mertani, H.C.; Dumontet, C.; Diaz, J.J. Dysregulation of ribosome biogenesis and translational capacity is associated with tumor progression of human breast cancer cells. *PLoS One* **2009**, *4*, doi:10.1371/journal.pone.0007147.

20. Tanaka, R.; Satoh, H.; Moriyama, M.; Satoh, K.; Morishita, Y.; Yoshida, S.; Watanabe, T.; Nakamura, Y.; Mori, S. Intronic U50 small-nucleolar-RNA (snoRNA) host gene of no protein-coding potential is mapped at the chromosome breakpoint t(3;6)(q27; q15) of human B-cell lymphoma. *Genes Cells* **2000**, *5*, 277–287.

21. Dong, X.Y.; Rodriguez, C.; Guo, P.; Sun, X.; Talbot, J.T.; Zhou, W.; Petros, J.; Li, Q.; Vessella, R.L.; Kibel, A.S.; *et al.* SnoRNA U50 is a candidate tumor-suppressor gene at 6q14.3 with a mutation associated with clinically significant prostate cancer. *Hum. Mol. Genet.* **2008**, *17*, 1031–1042.

22. Mourtada-Maarabouni, M.; Pickard, M.R.; Hedge, V.L.; Farzaneh, F.; Williams, G.T. GAS5, a non-protein-coding RNA, controls apoptosis and is downregulated in breast cancer. *Oncogene* **2009**, *28*, 195–208.

23. Dong, X.Y.; Guo, P.; Boyd, J.; Sun, X.; Li, Q.; Zhou, W.; Dong, J.T. Implication of snoRNA U50 in human breast cancer. *J. Genet .Genomics.* **2009**, *36*, 447–454.

24. Nakamura, Y.; Takahashi, N.; Kakegawa, E.; Yoshida, K.; Ito, Y.; Kayano, H.; Niitsu, N.; Jinnai, I.; Bessho, M. The GAS5 (growth arrest-specific transcript 5) gene fuses to BCL6 as a result of t(1;3)(q25;q27) in a patient with B-cell lymphoma. *Cancer Genet. Cytogenet.* **2008**, *182*, 144–149.

25. Basu, A.; Das, P.; Chaudhuri, S.; Bevilacqua, E.; Andrews, J.; Barik, S.; Hatzoglou, M.; Komar, A.A.; Mazumder, B. Requirement of rRNA methylation for 80S ribosome assembly on a cohort of cellular internal ribosome Entry Sites. *Mol. Cell Biol.* **2011**, *31*, 4482–4499.

26. Nicoloso, M.; Qu, L.H.; Michot, B.; Bachellerie, J.P. Intron-encoded, antisense small nucleolar RNAs: The characterization of nine novel species points to their direct role as guides for the 2'-*O*-ribose methylation of rRNAs. *J. Mol. Biol.* **1996**, *260*, 178–195.

27. Maxwell, E.S.; Fournier, M.J. The small nucleolar RNAs. *Annu. Rev. Biochem.* **1995**, *64*, 897–934.

28. Maquat, L.E. When cells stop making sense: Effects of nonsense codons on RNA metabolism in vertebrate cells. *RNA* **1995**, *1*, 453–465.

29. Kino, T.; Hurt, D.E.; Ichijo, T.; Nader, N.; Chrousos, G.P. Noncoding RNA Gas5 is a growth arrest and starvation-associated repressor of the glucocorticoid receptor. *Sci. Signal.* **2010**, *3*, doi:10.1126/scisignal.2000568.

30. Ploner, A.; Ploner, C.; Lukasser, M.; Niederegger, H.; Hüttenhofer, A. Methodological obstacles in knocking down small noncoding RNAs. *RNA* **2009**, *15*, 1797–1804.

31. Yoon, A.; Peng, G.; Brandenburger, Y.; Zollo, O.; Xu, W.; Rego, E.; Ruggero, D. Impaired control of IRES-mediated translation in X-linked dyskeratosis congenita. *Science* **2006**, *312*, 902–906.

32. Bochkov, Y.A.; Palmenberg, A.C. Translational efficiency of EMCV IRES in bicistronic vectors is dependent upon IRES sequence and gene location. *Biotechniques* **2006**, *41*, 283–284.

33. Van Eden, M.E.; Byrd, M.P.; Sherrill, K.W.; Lloyd, R.E. Demonstrating internal ribosome entry sites in eukaryotic mRNAs using stringent RNA test procedures. *RNA* **2004**, *10*, 720–730.

34. Thoma, C.; Fraterman, S.; Gentzel, M.; Wilm, M.; Hentze, M.W. Translation initiation by the c-Myc mRNA internal ribosome entry sequence and the poly(A) tail. *RNA* **2008**, *14*, 1579–1589.

7

Micromanaging Abdominal Aortic Aneurysms

Lars Maegdefessel [1,†], **Joshua M. Spin** [2,†], **Matti Adam** [2], **Uwe Raaz** [2], **Ryuji Toh** [2], **Futoshi Nakagami** [2] **and Philip S. Tsao** [2,*]

[1] Department of Medicine, Karolinska Institute, Stockholm SE-17176, Sweden;
E-Mail: lars.maegdefessel@ki.se

[2] Division of Cardiovascular Medicine, Stanford University, Stanford, CA 94305-5406, USA;
E-Mails: josh.spin@gmail.com (J.M.S.); matti.adam@stanford.edu (M.A.);
uraaz@stanford.edu (U.R.); rtoh1214@hotmail.com (R.T.); saruwaku@gmail.com (F.N.)

[†] These authors contributed equally to this work.

[*] Author to whom correspondence should be addressed; E-Mail: ptsao@stanford.edu;

Abstract: The contribution of abdominal aortic aneurysm (AAA) disease to human morbidity and mortality has increased in the aging, industrialized world. In response, extraordinary efforts have been launched to determine the molecular and pathophysiological characteristics of the diseased aorta. This work aims to develop novel diagnostic and therapeutic strategies to limit AAA expansion and, ultimately, rupture. Contributions from multiple research groups have uncovered a complex transcriptional and post-transcriptional regulatory milieu, which is believed to be essential for maintaining aortic vascular homeostasis. Recently, novel small noncoding RNAs, called microRNAs, have been identified as important transcriptional and post-transcriptional inhibitors of gene expression. MicroRNAs are thought to "fine tune" the translational output of their target messenger RNAs (mRNAs) by promoting mRNA degradation or inhibiting translation. With the discovery that microRNAs act as powerful regulators in the context of a wide variety of diseases, it is only logical that microRNAs be thoroughly explored as potential therapeutic entities. This current review summarizes interesting findings regarding the intriguing roles and benefits of microRNA expression modulation during AAA initiation and propagation. These studies utilize disease-relevant murine models, as well as human tissue from patients undergoing surgical aortic aneurysm repair. Furthermore, we critically examine future therapeutic strategies with regard to their clinical and translational feasibility.

Keywords: microRNA; aortic aneurysm; fibrosis; vascular smooth muscle cells; inflammation; biomarker

---

1. Abdominal Aortic Aneurysm Disease

Abdominal aortic aneurysms (AAAs) are defined as permanent dilations of the abdominal aorta. The diagnosis of AAA is commonly an accidental finding, although an increasing number of screening programs target particularly high-risk populations [1]. Screening demonstrates that disease prevalence ranges from 1.3% (45–54 years of age) to 12.5% in men (75–84 years of age), and in women from 0% in the youngest to 5.2% in the oldest age groups [2]. Some recently performed analyses, however, suggest lower prevalence in certain subpopulations [3]. The most feared clinical consequence of AAA progression is acute rupture, which carries a mortality of ~80% [4]. The number of deaths attributed to AAA rupture is around 15,000 annually in the United States [5]. However, this incidence is likely underestimated, since AAA rupture is often not recognized as the cause of death. As many as 60% of patients with AAAs die of other cardiovascular causes, such as myocardial infarction or stroke, thereby suggesting a relationship between AAAs and atherosclerosis [6].

Known predictors of AAA growth include diameter of the aorta at diagnosis and active smoking [7]. Some studies have demonstrated that the incidence and progression of AAA are also related to hypertension and age [8]. However, smoking is considered to be the major modifiable risk factor for development of AAA. Indeed, AAA is more closely associated with cigarette smoking than any other tobacco-related disease, excepting lung cancer. The vast majority of AAA patients (>90%) have a history of smoking [9]. As mentioned above, the prevalence of AAAs is greater in men than in women. However, there is emerging evidence that women present with an increased risk of AAA rupture at smaller aortic diameters than men [10,11].

To date, the only available treatment option for AAA has been surgical repair [1]. The classic approach includes the insertion of an intraluminal graft via open access to the aneurysmal aorta. This has now largely been replaced by endovascular stenting approaches. Besides their lack of indication in early stages of the disease, the current interventional methods carry significant operative risk, and thus appear effective only in preventing aortic rupture [4]. Until now, no pharmacological approach has been identified which effectively limits AAA progression or the risk of rupture in humans. What has been lacking is a detailed understanding of the mechanisms of AAA initiation and expansion.

2. Pathology and Cellular Mechanisms

Others have previously discussed the multiple potential cellular and molecular mechanisms associated with AAA development [12,13]. In this article, we will primarily focus on recognized crucial molecular and cellular patho-mechanisms in aneurysm development that are subject to microRNA (miR) regulatory control. Modulation of these miRs could evolve into new therapeutic strategies on the molecular level to combat the burden of aortic aneurysms.

The complex pathology of AAAs is characterized by progressive aortic dilation, promoted by an imbalance between vascular smooth muscle cell (VSMC) proliferation and apoptosis, as well as impairment of extracellular matrix (ECM) synthesis and degradation. These effects are due (at least in part) to transmural aortic inflammation and its disruptive effects on vessel wall homeostasis [8,14,15].

2.1. Impaired Homeostasis of Vascular Smooth Muscle Cells and Extracellular Matrix

Inherited syndromes associated with aneurysm formation suggest the importance of disruption of VSMC and ECM homeostasis in aortic dilation [16], although these familial conditions are more typically associated with ascending thoracic aortic aneurysms (TAAs). The aortic pathology of TAA is characterized by elastic fiber fragmentation and loss, proteoglycan accumulation, as well as focal or diffuse regional VSMC degradation and loss [17]. The role of TGF-β signaling dysregulation in this process is complex. Marfan Syndrome (MFS) and Loeys-Dietz Syndrome (LDS), caused respectively by mutations in fibrillin-1 and TGF-β receptors I and II, predispose to ascending thoracic aortic aneurysms (TAAs), but are much less often associated with AAA [18]. The same is true of familial SMAD3 mutations [19,20].

While considerable evidence points to excessive TGF-β signaling in the various familial TAA-associated conditions, animal models have connected AAA to decreased TGF-β activity [21]. While TGF-β receptor 2 is down-regulated in human AAA tissues [22], no association has been found between genetic polymorphisms in TGF-β receptors and serum TGF-β1 concentration in humans with AAA [23]. Systemic blockade of TGF-β activity augments AngII-induced AAAs in C57BL/6 mice as well as hypercholesterolemic mice, and appears associated with VSMC apoptosis, elastin degradation, and increased inflammatory activity in the aortic wall. In a rat model with chimeric aneurysms located in the infrarenal aorta, TGF-β1 overexpression via endovascular delivery of an adenoviral construct stabilizes pre-existing aortic aneurysms [21].

As a side note, TAAs also occur in families (without syndromic features) due to mutations in SMC contractile protein genes, including SMC-specific isoforms of α-actin (ACTA2) and myosin heavy chain (MYH11), along with the kinase that controls SMC contraction (MYLK) [17,24,25].

Multiple matrix metalloproteases (MMPs), which degrade ECM and are important regulators of aortic vessel wall integrity and morphology, have been extensively studied in human AAA, MMP-9 in particular. A recent meta-analysis included eight case-control studies comparing blood MMP-9 concentration between patients with AAAs and control subjects. Despite wide heterogeneity in circulating levels (30–750 ng/L), significantly higher MMP-9 concentrations were found in AAA patients [26].

2.2. Inflammation

Various inflammatory cell types are enriched in AAA tissues, especially macrophages. In a rabbit model of AAA induced by periaortic application of calcium chloride, there is striking macrophage accumulation within the adventitia [27]. This feature is also observed in porcine pancreatic elastase (PPE)-infusion induced AAAs in rats [28]. In ApoE$^{-/-}$ mice infused with angiotensin II (AngII), macrophage infiltration within the medial layers of the aorta is accompanied by medial rupture as an early characteristic [29], while profound accumulation of macrophages in the adventitia is seen throughout AAA progression [30].

Furthermore, macrophages appear to actively contribute to AAA development. CCR2 and monocyte chemoattractant protein-1 (MCP-1) interactions are important for macrophage-mediated inflammatory responses, including monocyte chemotaxis. Deficiency of CCR2 in mice limits the formation of AngII- and calcium chloride-induced AAAs [31,32]. Myeloid differentiation factor 88 (MyD88), an adaptor protein central to toll-like receptor signaling, also seems to play a pivotal role in macrophage-mediated vascular inflammation as deficiency of this molecule in macrophages diminishes murine AngII-induced AAAs [33].

T- and B-lymphocytes are frequently observed in AAAs [29,34]. A functional deficiency of CD4+ CD25+ T-regulatory cells was reported in patients with AAAs, and disruption of the balance of T-helper type 1 and type 2 cell function induces AAA in mice with allografted aortas [35].

Neutrophils are also present in human, and animal model AAAs [36]. The adhesion molecule L-selectin was found to be an important mediator for neutrophil recruitment in PPE-induced AAA formation in mice [37]. Neutrophil depletion in mice with aortic perfusion of PPE leads to attenuation of AAAs [36].

Further, many cytokines and chemokines play roles in AAA development [38]. Tumor necrosis factor (TNF)-α, a landmark cytokine in many inflammatory responses, is increased in plasma from patients with AAA and in human AAA tissues [39–41]. TNF-α-converting enzyme (TACE/ADAM17), and osteoprotegerin (a secreted glycoprotein member of the TNF receptor superfamily) are enhanced in human AAAs [39,40,42]. Genetic deficiency or pharmacological inhibition of TNF-α by administration of infliximab attenuates calcium chloride-induced AAAs in mice [43].

3. MicroRNA Biogenesis and Function

MicroRNA (miRNAs) are a class of well-conserved, short, non-coding RNAs that have emerged as key post-transcriptional regulators of gene expression in animals and plants. miRNAs have been described to play major roles in most, if not all, biological processes by influencing stability and translation of messenger RNAs [44]. miRNA genes are transcribed by RNA polymerase II as capped and polyadenylated primary miRNA transcripts (pri-miRNA) [45]. Pri-miRNA processing occurs in two steps, catalyzed by the enzymes Drosha and Dicer in cooperation with a dsRNA binding protein, "DiGeorge syndrome critical region gene 8" (DGCR8) [46]. In the first step, the Drosha-DGCR8 complex processes pri-miRNA into a ~70-nucleotide precursor hairpin (pre-miRNA), which is then exported to the cytoplasm. Some pre-miRNAs are produced from very short introns (mirtrons) as a result of splicing and debranching, bypassing the Drosha-DGCR8 step [47]. Nuclear export of pre-miRNAs is mediated by the transport receptor exportin 5 (XPO5) [48].

In the cytoplasm Dicer matures pre-miRNA into an imperfect RNA duplex. The strand with the weakest base pairing at the 5' terminus is loaded into the miRNA-induced silencing complex (miRISC), and is therefore considered to be biologically active [49]. While both strands of the duplex are produced in equal amounts by transcription, their accumulation into the miRISC is asymmetric [50]. Initially, the non-miRISC strand was assumed to be an inactive passenger designated the *(star)-strand.

However, systemic computational analysis has demonstrated that star-strands may contain well-conserved target recognition sites, indicating functional relevance [51]. Indeed, several recent publications have reported star-strands to be biologically active, widening the potential regulatory potency of miRNA-duplexes [52–54].

After the selected strand is loaded into the miRISC, the miRNA guides the miRISC to bind to the 3'UTR of its target sequence. The seed sequence (the first two to eight nucleotides) is considered the most important for target recognition and silencing of the mRNA [55,56]. Translation of the mRNA is inhibited after association of the miRISC with its target sequence. Efficient mRNA targeting requires continuous base pairing of the seed region to the target mRNA. Furthermore, Ago(argonaute)-proteins and the glycine-tryptophan protein of 182 kDa (GW182), core components of the miRISC, are directly associated with miRNAs, and are needed for effective translational repression, mRNA destabilization, and degradation. The exact mechanisms of translational arrest by the miRNA:mRNA complex are still a matter of debate, although both initiation and elongation steps of translation are thought to be affected [57,58].

4. miRs in AAA Disease

In recent years, several miRs have been found to regulate vascular pathologies, in general, and aortic aneurysm (thoracic and abdominal) disease, in particular (Table 1). We performed a systematic published literature search on articles investigating miRNA expression and function in aortic aneurysm disease. This current review focuses mainly on miRNAs that have not only been detected as being potentially dys-regulated in human aneurysmal tissue, but have also been thoroughly studied in functional experiments, thus accessing a therapeutic strategy of beneficially altering miRNA expression to limit AAA progression (Figure 1).

Table 1. Regulatory role of microRNAs in murine abdominal aortic aneurysm (AAA) disease models (AFB = adventitial fibroblasts; AngII = angiotensin II; ASMC = aortic smooth muscle cells; PPE = porcine pancreatic elastase).

microRNA	Model of AAA induction	Effect on AAA progression
miR-21	PPE-infusion in C57BL/6 mice and AngII-infusion in $ApoE^{-/-}$ mice [59]	Regulates proliferation and apoptosis in ASMCs via PTEN/PI3K/AKT; induction of miR-21 through NFκB
miR-26a	PPE-infusion in C57BL/6 mice and AngII-infusion in $ApoE^{-/-}$ mice [60]	Inhibition of ASMC-differentiation via SMAD-1 and SMAD-4 depression
miR-29b	AngII in 1.5-year-old C57BL/6 [61]; PPE-infusion in C57BL/6 mice and AngII in $ApoE^{-/-}$ mice [62]	Modulating the fibrotic response in aortic wall through several collagen isoforms; repression of miR-29b in AFBs through TGF-β
miR-143/145	miR-143/145 knockout and $ApoE^{-/-}$ mice [63]	Regulation of ASMC homeostasis and differentiation

Figure 1. Association between microRNAs and murine abdominal aortic aneurysm formation. microRNAs (miRs) in bold and underlined have been established as regulators of aneurysm disease, utilizing gain- and loss-of function studies. All other miRs are suspected and potential disease-related modulators.

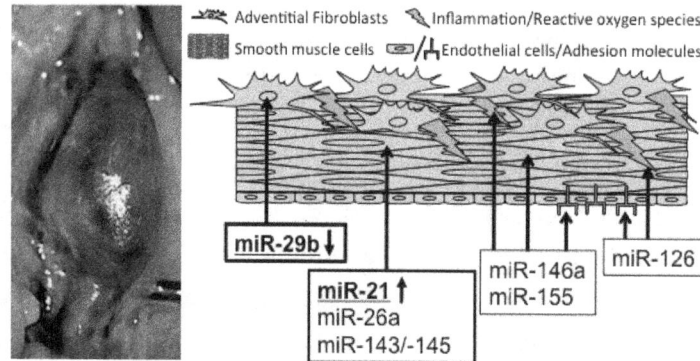

4.1. miR-21

miR-21 is considered an onco-miRNA, with increased expression in many solid tumors, where it promotes cell proliferation, migration and anti-apoptosis [64]. Data indicate that miR-21 is also highly expressed in VSMCs, and implicate it in the regulation of SMC phenotype in vascular disorders, such as post-injury neointimal lesions [65,66].

Interestingly, miR-21 stimulation induces up-regulation of smooth-muscle restricted contractile proteins through silencing of "programmed cell death protein" (PDCD)-4 expression, a known tumor suppressor protein. These findings suggest that miR-21 could regulate both VSMC contractile function [67] and proliferation [68]. miR-21 also targets multiple members of the dedicator of cytokinesis (DOCK) superfamily and modulates the activity of ras-related C3 botulinum toxin substrate 1 (Rac1) small GTPase to regulate VSMC phenotype [69].

miR-21 regulates growth and survival of VSMCs by decreasing the expression of "phosphatase and tensin homolog" (PTEN) and inducing expression of Bcl-2, resulting in pro-proliferative and anti-apoptotic effects in a carotid injury model in rats [68]. Regarding homeostasis, miR-21 promotes VSMC differentiation in response to TGF-β1 and BMP-4 [67]. These factors were shown to stimulate the processing of miR-21 in human pulmonary artery smooth muscle cells from the pri-miR to the mature miR via SMAD proteins. Additionally, miR-21 has been shown to regulate hypoxia-induced pulmonary VSMC proliferation and migration by regulating PDCD4, Sprouty 2 (SPRY2), and peroxisome proliferator-activated receptor-α (PPARα), known for their anti-proliferative and anti-migratory effects on VSMCs [70].

Interestingly, a recent report indicates that miR-21 is induced in tissue of arteriosclerosis obliterans of the lower extremities, even with <10% stenosis, and also is induced in VSMCs in response to platelet-derived growth factor (PDGF)-BB and/or hypoxia. In this report, tropomyosin 1 (TPM1) was identified as a target gene for miR-21. TPM1 reduction leads to a reduction in cytoskeletal stability, promoting VSMC proliferation and migration [71].

Furthermore, cyclic stretch has been shown to modulate miR-21 expression at the transcriptional level via FBJ murine osteosarcoma viral oncogene homolog (c-fos/AP-1) in cultured human aortic

SMCs [72]. While moderate stretch is essential for maintaining vessel wall structure and vascular homeostasis [73], exacerbated stretch, as in hypertension, could promote pathological vascular remodeling by stimulating SMC proliferation, apoptosis, and migration and abnormal extracellular matrix deposition [74,75].

In endothelial cells (ECs), prolonged shear stress up-regulates the expression of miR-21 through modulation of the phosphatidylinositol-4,5-bisphosphate 3-kinase (PI3K)/v-akt murine thymoma viral oncogene (Akt) pathway, which leads to an increase of nitric oxide (NO) production while reducing apoptosis [76]. miR-21 is also expressed in endothelial progenitor cells (EPCs), where it suppresses high mobility group AT-hook 2 (Hmga2) expression, a chromatin-associated protein that modulates transcription through altering chromatin structure. Thus, inducing overexpression of miR-21 decreases proliferation and limits EPC angiogenesis *in vitro* and *in vivo* [77].

In regards to aortic dilatation, we discovered that miR-21 was significantly up-regulated in two established murine models of AAA disease, the PPE-infusion model in C57B/L6 mice and the AngII-infusion in ApoE$^{-/-}$ mice [59]. Out of the aforementioned VSMC-specific miR-21 target genes that alter proliferation and apoptosis, PTEN was the only target gene to be significantly down-regulated at three different time points during aneurysm development and progression. PTEN, a lipid and protein phosphatase and important tumor suppressor gene, acts as a key negative regulator of the PI3K pathway. Systemic injection of a locked-nucleic-acid (LNA) modified antagomiR against miR-21 diminished the pro-proliferative impact of down-regulated PTEN, leading to a significant increase in expansion of AAAs. Further down-regulation of aortic PTEN with a pre-miR-21-loaded lentivirus had significant protective effects on aneurysm expansion by inducing massive proliferation in the aortic wall in both murine models [59].

As mentioned above, smoking is considered to be the major modifiable risk factor for AAA disease. In our study, nicotine (a major constituent of tobacco smoke) accelerated AAA growth in both murine aneurysm models, and caused an augmented increase in miR-21 levels, which appeared to be a protective response to limit further aneurysm expansion and rupture. *In vitro* studies utilizing human aortic SMCs and ECs, as well as adventitial fibroblasts showed aortic SMCs to be the most responsive to miR-21 modulation. Our group also showed that miR-21 induction in nicotine, as well as AngII and interleukin-6 (IL-6) pre-treated SMCs, is dependent on NF-κB signaling. In support of these findings, we found increased expression of miR-21 and down-regulated PTEN in samples obtained from human AAA patients undergoing surgical repair of their enlarged infrarenal aorta compared to control abdominal aorta from organ donors. Notably, miR-21 was even further up-regulated (with PTEN being further decreased) in smokers with AAA disease compared with non-smokers [59].

4.2. miR-26a

Employing *in vitro* experiments with human aortic SMCs, Leeper and colleagues [60] found that miR-26a promotes the synthetic phenotype through regulation of SMAD1 and SMAD4, contributing to the regulatory circuit of TGF-β signaling-associated pathways. Overexpression of SMAD-1 and SMAD-4 was inducible with anti-miR-26a treatment. In two mouse models of aneurysm formation (PPE- and AngII-infusion), miR-26 levels were decreased, which might contribute to AAA formation through enhanced SMC apoptosis. Thus, miR-26 regulation in aneurysmal tissue with AAA development may in fact be causal, and not compensatory.

4.3. miR-29b

The miR-29 family of miRs contains three members (miR-29a, miR-29b, and miR-29c) that are encoded by two separate loci, giving rise to bi-cistronic precursor miRs (miR-29a/b1 and miR-29b2/c). This family targets numerous gene transcripts that encode ECM proteins involved in fibrotic responses, including several collagen isoforms (e.g., COL1A1, COL1A2, COL3A1), fibrillin-1, and elastin (ELN) [78], and is known to modulate gene expression during development and aging of the aorta [61] and during the progression of aortic aneurysms [61,62].

Other fibrosis-related responses and diseases, such as liver [79] and kidney fibrosis [80], systemic sclerosis [81], as well as cardiac fibrosis in response to myocardial ischemia [78], have all been linked to repressed levels of miR-29. TGF-β-associated pathways are important regulators of miR-29 expression, leading to triggering of the fibrotic response by decreasing miR-29 levels in cardiac fibroblasts, hepatic stellate cells, and dermal fibroblasts, and leading to a substantial increase in the aforementioned ECM target genes [78,81,82].

Based on these observations, miR-29 seems to be a crucial regulator of aortic aneurysm disease through modulating genes and pathways which are responsible for ECM composition and dynamics. We found that miR-29b was the only member of the miR-29 family to be significantly down-regulated at three different time points during murine AAA development and progression [82]. Further decreasing of miR-29b expression with a LNA-anti-miR-29b led to an acceleration of collagen encoding gene expression (COL1A1, COL2A1, COL3A1, COL5A1), as well as elastin (ELN). Furthermore, matrix-metalloproteinases-2 and -9 (MMP2 and MMP9) were down-regulated in LNA-anti-miR-29b-transduced mice. These results were reproducible in two independent mouse AAA models, (PPE-and AngII-infusion), and led to a significant decrease in aneurysm expansion compared to a scrambled-control-miR injected group.

Human AAA tissue samples displayed a similar pattern of reduced miR-29b expression with increased collagen gene expression in comparison to non-aneurysmal organ donor controls. These results suggest that the aortic wall, which weakens due to steadily increasing diameter, acts to induce expression of collagens by repressing miR-29b levels, providing additional support to the aortic wall in an attempt to limit the risk for rupture.

Aging is a well-established risk factor for aneurysm development. Boon *et al.* were the first to publish a study connecting miR regulation to aortic dilatation and aging. They discovered that expression of the miR-29 family was increased in the aging mouse aorta [61]. Rather than utilizing the more commonly employed ApoE$^{-/-}$ or LDL receptor$^{-/-}$ mice, Boon and colleagues studied AngII infusion in 18-month-old C57BL/6 (wild type) mice. In these mice, AngII infusion increased miR-29b expression in samples derived from the entire aorta, which would seem to suggest that with aging the protective role of miR-29b during AAA development may be diminished. In accordance with our aforementioned results, Boon *et al.* found that systemic treatment with an LNA-modified anti-miR-29b significantly increased the expression of collagen isoforms (COL1A1, COL3A1), as well as ELN, and decreased suprarenal aortic dilatation in aged AngII-treated mice.

5. miR-143/145

Probably the most extensively studied miR in VSMC pathology is the miR-143/145 cluster, which is transcribed as a bi-cistronic transcript from a common promoter, which in turn is regulated by serum response factor (SRF), myocardin, and myocardin-related transcription factor-A [83]. MiR-143/145 is dramatically reduced in several vascular disease models, e.g., carotid balloon/wire injury, carotid ligation in rats, and in ApoE$^{-/-}$ mice [83–85].

miR-143/145 alters SMC phenotypic switching in response to vascular injury, influencing both the synthetic/proliferative and the contractile/differentiated states [63,83–86]. Studies from several different groups have shown that these effects are partly mediated by targeting of multiple transcription factors, including KLF4, KLF5, and ELK-1 [84–86]. Further, down-regulation of miR-143/145 is sufficient to up-regulate PDGF receptor (PDGF-R), protein kinase C (PKC) epsilon, and fascin, an actin bundling protein of podosomes. These last are thought to be necessary for vascular wall matrix remodeling, potentially affecting the progression of aortic dilatation [87]. Interestingly, one of the first reports regarding the role of miRs in aneurysm disease showed that miR-143/145 expression is reduced in aortas from patients with thoracic aortic aneurysm, permitting dedifferentiation of aortic VSMC with a resultant decrease in contractile function [63].

Finally, miR-143/145 may be secreted in microvesicles derived from ECs (which otherwise do not usually express these miRs) [84]. It has been proposed that shear stress-induced KLF-2 may stimulate expression of miR-143/145 in ECs [88], leading to miR secretion in microvesicles and transfer into VSMCs [84]. EC-derived microvesicles containing miR-143/145 can reduce atherosclerotic lesions when injected into ApoE$^{-/-}$ mice [88].

6. Other miRs

A growing body of literature highlights the role of miRs in the regulation of angiogenesis and inflammation [12–14,89]. Smooth muscle degradation, along with decreased VSMC proliferation, decreased ECM synthesis and impaired ECM remodeling, have all previously been linked to AAA development. Clearly, these contributing mechanisms of aortic dilation may be regulated through miRs. However, the miRs described below have not yet been directly tied to aortic aneurysm initiation, propagation, or rupture.

6.1. miR-126

One of the most intriguing miRs as regards vascular inflammation is miR-126, an EC-enriched miR, which negatively regulates VCAM-1 expression [90,91]. Apoptotic bodies are released from ECs during atherosclerotic progression, and have been shown to contain miR-126. miR-126 decreases the expression of G-protein signaling 16 (RGS16) in ECs, thereby up-regulating the chemokine (C-X-C motif) ligand 12 (CXCL12) receptor. CXCL12 activation then decreases EC apoptosis and recruits progenitor cells at the lesion site, reducing the atherosclerotic burden *in vivo*, and contributing to plaque stabilization [92].

6.2. miR-146a

Alterations associated with aging in blood vessels include a decrease in compliance and an increase in vascular inflammatory response, which could promote AAA propagation. Several reports show dysregulation of miRs in the vasculature during aging. In particular, miR-146a expression is decreased in senescent ECs. It targets NADPH oxidase 4 (NOX4), decreasing reactive oxygen species (ROS) production. These data suggest that the reduction in miR-146 expression potentially enhances aging effects through NOX4-derived ROS [93]. In another study, miR-146a and KLF4 were found to form a feedback loop, regulating each other's expression and VSMC proliferation. The authors propose that miR-146a regulates KLF4, which competes with KLF5 binding to the miR-146a promoter to inhibit transcription [94].

6.3. miR-155

miR-155 is another miR of potential interest in AAA disease progression due to its effects on the renin-angiotensin-system (RAS). miR-155 is induced by TNF (which independently has been shown to contribute to AAA development) [95], and then negatively regulates the expression of the transcription factor "v-ets erythroblastosis virus E26 oncogene homolog 1" (Ets-1) [96]. AngII-induced overexpression of miR-155 results in a decrease in Ets-1, affecting expression of downstream targets such as VCAM-1, fms-related tyrosine kinase 1 (FLT1) and MCP1, and impairing lymphocyte adhesion to ECs [96]. miR-155 also has been shown to target the angiotensin II type 1 receptor (AT1R), resulting in decreased AngII-induced migration of ECs [96].

While the above-described effects of miR-155 might suggest an anti-inflammatory role, Nazari-Jahantigh *et al.* validated miR-155 in macrophages as a crucial component of atherosclerosis development. In these cells, miR-155 promoted the expression of MCP-1/CCL2, and directly suppressed Bcl-6, a transcription factor that inhibits NF-κB [97]. It has also been described that hematopoietic deficiency of miR-155 increases atherosclerotic plaque size and instability [98], possibly by inhibition of lipid uptake and inflammatory responses in monocytes. Clearly, findings thus far regarding the role of miR-155 have been somewhat ambiguous.

In addition to these miRs, Pahl *et al.* examined miR-regulation in human abdominal aortic tissue of patients undergoing elective open repair with samples collected at autopsy or obtained from a pre-existing tissue biobank [99], utilizing microRNA-array. Out of a total of 847 miRs, 3 miRs presented as significantly up- (miR-181a*, miR-146a, miR-21) and 5 miRs as down-regulated (miR-133b, miR133a, miR331-3p, miR30c-2*, miR-204) in patients with AAAs compared to controls. However, using an additional tissue set, qRT-PCR was only able to confirm the down-regulated miRs from the array.

7. Therapeutic Approaches Using miR Modulators

The identification of both the underlying causes of vascular disease, as well as appropriate interventions, remain great challenges to both basic vascular biology and everyday clinical practice. The traditional methods of drug design, involving enzymes, cell surface receptors, and other proteins, appear sometimes less effective in the treatment of cardiovascular diseases, due to the highly sensitive nature of the targeted systems.

In this dismaying scenario, the discovery of an entirely new method of gene regulation by miRs, and their recent validation as markers and modulators of vascular functionality during pathological conditions, provide new hope for innovative therapies. Research in recent years has recognized the crucial regulatory roles that miRs play in vascular diseases such as myocardial infarction, stroke, and aortic aneurysm [100].

Intriguingly, miRs also appear to represent valid therapeutic targets, because modulation of their expression *in vivo* with either antisense RNA molecules or miR-mimics/pre-miRs has been shown to effectively modulate cardiovascular disease in various animal models [101]. Inhibition or overexpression of a single miR can induce or attenuate pathological responses in the cardiovascular system, as a result of the regulated coordination of numerous target genes involved in complex physiological and disease phenotypes. The most important difference between modulating miRs, and the traditional therapeutic approach is that standard drugs typically interact with specific cellular targets, whereas miRs have the capability of modulating entire functional networks [102].

miR modulation is performed by supplying antagomiRs (or anti-miRs; synthetic reverse compliments of oligonucleotides) that bind to a target miR and silence it, or by using pre-miRs/miR-mimics that act similarly to the original miR [101]. Recent animal and even human efficacy data indicate that antagomiRs have the potential to become a whole new class of drugs. These inhibitors of miR expression have several significant advantages, which make them very attractive from a drug development standpoint, including small size, as well as frequent conservation of their target miRs across species. Using lessons learned from antisense technologies (e.g., siRNA), potent oligonucleotide chemistries to inhibit miRs are currently being investigated [103]. These efforts have given rise to candidates that bind to their putative miR targets with remarkable affinity and specificity, and which have desirable drug-like qualities, including increased stability and favorable pharmacokinetics.

The most common type of modification being utilized to protect antagomiRs from immediate degradation *in vivo* is the addition of a locked nucleic acid (LNA). LNA contains a class of bicyclic RNA analogs in which the furanose ring in the sugar-phosphate backbone is chemically locked in a RNA-mimicking N-type (C3'-endo) conformation by the introduction of a 2'-*O*,4'-*C*-methylene bridge. This modification leads to nuclease resistance, as well as an increase in binding affinity to the targeted miR, which is accomplished by Watson-Crick complementary base pairing [104]. Regarding the use of antagomiRs in humans, there have been no immunogenic or toxicological safety issues reported to date. However, the major drawback of these substances at this point seems to be the necessity of repeated delivery of doses for long-term therapeutic effects. This becomes a critical issue when the route of delivery is an invasive procedure, such as systemic injection [105]. The antagomiR that has advanced the farthest in clinical trials to date is Miravirsen (anti-122) for patients with chronic hepatitis-C (HCV) infections. Recently published data from a Phase 2a trial demonstrated that the drug was not only safe, but also well tolerated, providing prolonged antiviral activity well after the last dose of monotherapy [106].

Unlike antagomiRs, the prospect of delivery of injectable, naked miR-mimics and/or pre-miRs has remained problematic. For now, lenti- as well as adeno-associated viruses (AAV) represent efficacious delivery platforms for miRs, but these carry the risks common to most gene therapies. Lentiviral vectors, for example, are derived from HIV type 1 (HIV-1), and thus the production of wild-type HIV through homologous recombination of the virus remains a major safety concern. However, recent lentiviral vector developments permitting deletion of the U3 promoter region of the long terminal repeats from the virus, leading to self-inactivation, may resolve this issue, making them a promising vector for future applications [107].

miR-mimic and pre-miR development also present difficulties related to the need to deliver synthetic RNA duplexes in which one strand (the "guide" strand) is identical to the miR of interest, while the complementary strand ("passenger" strand) is modified to increase stability as well as cellular uptake. Apart from the problems involved in permitting cellular uptake of double-stranded miR-mimics, the passenger strand has the potential to counter-productively act as an antagomiR [105].

Given the above limitations, the development of miR mimics, which do not require a viral vector represents an important therapeutic goal. Some preclinical studies have achieved this in murine models by packaging synthetic miR duplexes within lipid nanoparticles [108,109].

In summary, the ability to modulate miR activity through systemic delivery of miR inhibitors or mimics without toxicity provides unprecedented opportunities for intervening in disease processes. While challenges such as potential off-target effects and the urgent need for local and/or cell-type specific delivery mechanisms remain, the pace of discovery in this field portends new, feasible clinical therapeutic approaches in patients.

8. miRs as Biomarkers in AAA Disease

At the outset, it is necessary to point out that, to date, no easily accessible and reproducibly measurable biomarker has been identified with prognostic value for AAA growth, or even for the potential to rupture [1,110].

Recently, miRs have received much attention regarding their suitability as biomarkers for vascular disease. Following pioneering work from the cancer field, several cardiovascular studies have found substantial variations in miR expression in numerous clinical specimen subtypes (e.g., blood, urine, saliva, etc.) [111–114]. Measuring levels of circulating miRs has several advantages and offers novel opportunities. For example, as with nucleic acids, miRs can be both amplified and detected with high sensitivity and specificity. Also, miR-microarrays and quantitative PCR (qPCR) methodology allows the quantification of many miRs in a single experiment. There is evidence that the combined analysis of many miRs and their co-expression patterns (miR networks) enhances their predictive power as biomarkers. Furthermore, miRs are relatively stable over time in human blood and appear to be protected from degradation through various mechanisms [115].

Despite this, the quantitative analysis of miRs in material such as blood and urine comes with certain disadvantages. Firstly, the concentrations of most circulating miRs are typically very low (with the exception of whole blood samples), making reliable quantitation and normalization a challenge with existing technology. Also, there exists no consensus for miR normalization controls. Beyond this, current qPCR and microarray technologies are still quite time-consuming (several hours) compared with some protein-based biomarker tests such as troponin or C-reactive protein, which can offer results within minutes [116]. For now, the added value of miR-based biomarkers remains to be established by more rigorous testing and optimization.

Despite these hurdles, several laboratories have already obtained profiles of circulating miRs in cardiovascular disease and explored their biomarker potential. Immediately apparent are certain inconsistencies between studies, where the same or highly similar settings have been studied. This is partially attributed to the current immaturity of the field, which still includes technical issues such as variability of RNA extraction protocols, different means of nucleic acid detection, and the aforementioned normalization procedures. However, many studies are also simply clinically underpowered, and/or do not use appropriate controls matched for potentially confounding factors such as age, sex, medication, comorbidities, and tissue source. Also, there has been minimal comparison of miRs to traditional reference biomarkers.

The first study to look at expression levels of circulating miRs in AAA disease was performed by Kin *et al.* The authors investigated a subset of miRs, which they identified to be significantly altered in abdominal aortic tissue samples from patients with AAA undergoing surgical repair when compared with non-aneurysmal thoracic aortic specimen from patients undergoing aortic valve replacement [117]. Interestingly, miRs that were up-regulated in AAA tissue samples appeared significantly down-regulated in plasma from patients with AAA compared to a small group of healthy volunteers. These included miRs-15a/b, -29b, -124a, -126, -146a, -155, and -223. Clearly, further studies in larger cohorts are necessary to explore the diagnostic, and, even more important, the predictive capabilities of miRs as biomarkers in AAA disease.

9. Summary and Perspectives

The demonstration that miRs play crucial roles in cardiovascular disease and can be easily regulated *in vitro* and *in vivo* by antagomiRs and pre-miRs/miR-mimics has tremendously accelerated miR research and nourished hopes that the agents used and verified in animal models could some day be employed in humans with AAA disease. miRs represent a relatively young, but rapidly advancing, field of basic biological and translational research with potentially new and innovative therapeutic applications. For vascular diseases in particular, the availability of local (coated stents and/or balloons) or cell type-specific delivery mechanisms would significantly increase the value of miR therapeutics in everyday clinical practice.

Acknowledgments

We would like to thank all past and current lab members of our laboratories at Stanford and Stockholm for their determination to generate the data for parts of the research presented in this present review. Our own research projects are supported by grants from the National Institutes of Health (1P50HL083800-01 to PST; 5K08 HL080567 to JMS), the Stanford Cardiovascular Institute (to JMS), the American Heart Association (0840172N to PST, 09POST2260118 to LM), the Karolinska Institute Cardiovascular Program Career Development Grant, and the Swedish Heart-Lung-Foundation (20120615 both to LM).

References

1. Golledge, J.; Muller, J.; Daugherty, A.; Norman, P. Abdominal aortic aneurysm: Pathogenesis and implications for management. *Arterioscler. Thromb. Vasc. Biol.* **2006**, *26*, 2605–2613.

2. Go, A.S.; Mozaffarian, D.; Roger, V.L.; Benjamin, E.J.; Berry, J.D.; Borden, W.B.; Bravata, D.M.; Dai, S.; Ford, E.S.; Fox, C.S.; Franco, S.; *et al.* Executive summary: Heart disease and stroke statistics—2013 update: A report from the American Heart Association. *Circulation* **2013**, *127*, 143–152.

3. Svensjo, S.; Martin Björck, M.; Gürtelschmid, M.; Gidlund, K.D.; Hellberg, A.; Wanhainen, A. Low prevalence of abdominal aortic aneurysm among 65-year-old Swedish men indicates a change in the epidemiology of the disease. *Circulation* **2011**, *124*, 1118–1123.

4. Golledge, J.; Norman, P.E. Current status of medical management for abdominal aortic aneurysm. *Atherosclerosis* **2011**, *217*, 57–63.

5. Thom, T.; Haase, N.; Rosamond, W.; Howard, V.J.; Rumsfeld, J.; Manolio, T.; Zheng, Z.J.; Flegal, K.; O'Donnell, C.; Kittner, S.; *et al.* Heart disease and stroke statistics—2006 update: A report from the American Heart Association Statistics Committee and Stroke Statistics Subcommittee. *Circulation* **2006**, *113*, e85–e151.

6. Golledge, J.; Tsao, P.S.; Dalman, R.L.; Norman, P.E. Circulating markers of abdominal aortic aneurysm presence and progression. *Circulation* **2008**, *118*, 2382–2392.

7. Jones, D.W.; Easton, J.D.; Halperin, J.L.; Hirsch, A.T.; Matsumoto, A.H.; O'Gara, P.T.; Safian, R.D.; Schwartz, G.L.; Spittell, J.A.; American Heart Association. Atherosclerotic Vascular Disease Conference: Writing Group V: Medical decision making and therapy. *Circulation* **2004**, *109*, 2634–2642.

8. Weintraub, N.L. Understanding abdominal aortic aneurysm. *N. Engl. J. Med.* **2009**, *361*, 1114–1116.

9. Franks, P.J.; Edwards, R.J.; Greenhalgh, R.M.; Powell, J.T. Smoking as a risk factor for abdominal aortic aneurysm. *Ann. N. Y. Acad. Sci.* **1996**, *800*, 246–248.

10. Norman, P.E.; Powell, J.T. Abdominal aortic aneurysm: The prognosis in women is worse than in men. *Circulation* **2007**, *115*, 2865–2869.

11. Powell, J.T.; Greenhalgh, R.M. Clinical practice. Small abdominal aortic aneurysms. *N. Engl. J. Med.* **2003**, *348*, 1895–1901.

12. Lu, H.; Rateri, D.L.; Bruemmer, D.; Cassis, L.A.; Daugherty, A. Novel mechanisms of abdominal aortic aneurysms. *Curr. Atheroscler. Rep.* **2012**, *14*, 402–412.

13. Lu, H.; Rateri, D.L.; Bruemmer, D.; Cassis, L.A.; Daugherty, A. Involvement of the renin-angiotensin system in abdominal and thoracic aortic aneurysms. *Clin. Sci.* **2012**, *123*, 531–543.

14. Daugherty, A.; Cassis, L.A. Mouse models of abdominal aortic aneurysms. *Arterioscler. Thromb. Vasc. Biol.* **2004**, *24*, 429–434.

15. Milewicz, D.M. MicroRNAs, fibrotic remodeling, and aortic aneurysms. *J. Clin. Invest.* **2012**, *122*, 490–493.

16. Lindsay, M.E.; Dietz, H.C. Lessons on the pathogenesis of aneurysm from heritable conditions. *Nature* **2011**, *473*, 308–316.

17. Guo, D.; Pannu, H.; Tran-Fadulu, V.; Papke, C.L.; Yu, R.K.; Avidan, N.; Bourgeois, S.; Estrera, A.L.; Safi, H.J.; Sparks, E. Mutations in smooth muscle alpha-actin (ACTA2) lead to thoracic aortic aneurysms and dissections. *Nat. Genet.* **2007**, *39*, 1488–1493.

18. Loeys, B.L.; Schwarze, U.; Holm, T.; Callewaert, B.L.; Thomas, G.H.; Pannu, H.; de Backer, J.F.; Oswald, G.L.; Symoens, S.; Manouvrier, S. Aneurysm syndromes caused by mutations in the TGF-beta receptor. *N. Engl. J. Med.* **2006**, *355*, 788–798.

19. Regalado, E.S.; Guo, D.; Villamizar, C.; Avidan, N.; Gilchrist, D.; McGillivray, B.; Clarke, L.; Bernier, F.; Santos-Cortez, R.L.; Leal, S.M. Exome sequencing identifies SMAD3 mutations as a cause of familial thoracic aortic aneurysm and dissection with intracranial and other arterial aneurysms. *Circ. Res.* **2011**, *109*, 680–686.

20. Van de Laar, I.M.B.H.; Oldenburg, R.A.; Pals, G.; Roos-Hesselink, J.W.; de Graaf, B.M.; Verhagen, J.M.A.; Hoedemaekers, Y.M.; Willemsen, R.; Severijnen, L.; Venselaar, H. Mutations in SMAD3 cause a syndromic form of aortic aneurysms and dissections with early-onset osteoarthritis. *Nat. Genet.* **2011**, *43*, 121–126.

21. Dai, J.; Losy, F.; Guinault, A.M.; Pages, C.; Anegon, I.; Desgranges, P.; Becquemin, J.P.; Allaire, E. Overexpression of transforming growth factor-beta1 stabilizes already-formed aortic aneurysms: A first approach to induction of functional healing by endovascular gene therapy. *Circulation* **2005**, *112*, 1008–1015.

22. Biros, E.; Walker, P.J.; Nataatmadja, M.; West, M.; Golledge, J. Downregulation of transforming growth factor, beta receptor 2 and Notch signaling pathway in human abdominal aortic aneurysm. *Atherosclerosis* **2012**, *221*, 383–386.

23. Golledge, J.; Clancy, P.; Jones, G.T.; Cooper, M.; Palmer, L.J.; van Rij, A.M.; Norman, P.E. Possible association between genetic polymorphisms in transforming growth factor beta receptors, serum transforming growth factor beta1 concentration and abdominal aortic aneurysm. *Br. J. Surg.* **2009**, *96*, 628–632.

24. Wang, L.; Guo, D.; Cao, J.; Gong, L.; Kamm, K.E.; Regalado, E.; Li, L.; Shete, S.; He, W.; Zhu, M.; *et al.* Mutations in myosin light chain kinase cause familial aortic dissections. *Am. J. Hum. Genet.* **2010**, *87*, 701–707.

25. Zhu, L.; Vranckx, R.; van Kien, P.K.; Lalande, A.; Boisset, N.; Mathieu, F.; Wegman, M.; Glancy, L.; Gasc, J.; Brunotte, F.; *et al.* Mutations in myosin heavy chain 11 cause a syndrome associating thoracic aortic aneurysm/aortic dissection and patent ductus arteriosus. *Nat. Genet.* **2006**, *38*, 343–349.

26. Takagi, H.; Manabe, H.; Kawai, N.; Goto, S.; Umemoto, T. Circulating matrix metalloproteinase-9 concentrations and abdominal aortic aneurysm presence: A meta-analysis. *Interact. Cardiovasc. Thorac. Surg.* **2009**, *9*, 437–440.

27. Freestone, T.; Turner, R.J.; Higman, D.J.; Lever, M.J.; Powell, J.T. Influence of hypercholesterolemia and adventitial inflammation on the development of aortic aneurysm in rabbits. *Arterioscler. Thromb. Vasc. Biol.* **1997**, *17*, 10–17.

28. Anidjar, S.; Salzmann, J.L.; Gentric, D.; Lagneau, P.; Camilleri, J.P.; Michel, J.B. Elastase-induced experimental aneurysms in rats. *Circulation* **1990**, *82*, 973–981.

29. Saraff, K.; Babamusta, F.; Cassis, L.A.; Daugherty, A. Aortic dissection precedes formation of aneurysms and atherosclerosis in angiotensin II-infused, apolipoprotein E-deficient mice. *Arterioscler. Thromb. Vasc. Biol.* **2003**, *23*, 1621–1626.

30. Rateri, D.L.; Howatt, D.A.; Moorleghen, J.J.; Charnigo, R.; Cassis, L.A.; Daugherty, A. Prolonged infusion of angiotensin II in apoE(−/−) mice promotes macrophage recruitment with continued expansion of abdominal aortic aneurysm. *Am. J. Pathol.* **2011**, *179*, 1542–1548.

31. Daugherty, A.; Rateri, D.L.; Charo, I.F.; Phillip Owens, A., III; Howatt, D.A.; Cassis, L.A. Angiotensin II infusion promotes ascending aortic aneurysms: Attenuation by CCR2 deficiency in apoE−/− mice. *Clin. Sci.* **2010**, *118*, 681–689.

32. MacTaggart, J.N.; Xiong, W.; Knispel, R.; Baxter, B.T. Deletion of CCR2 but not CCR5 or CXCR3 inhibits aortic aneurysm formation. *Surgery* **2007**, *142*, 284–288.

33. Phillip Owens, A., III; Rateri, D.L.; Howatt, D.A.; Moore, K.J.; Tobias, P.S.; Curtiss, L.K.; Lu, H.; Cassis, L.A.; Daugherty, A. MyD88 deficiency attenuates angiotensin II-induced abdominal aortic aneurysm formation independent of signaling through Toll-like receptors 2 and 4. *Arterioscler. Thromb. Vasc. Biol.* **2011**, *31*, 2813–2819.

34. Daugherty, A.; Manning, M.W.; Cassis, L.A. Angiotensin II promotes atherosclerotic lesions and aneurysms in apolipoprotein E-deficient mice. *J. Clin. Invest.* **2000**, *105*, 1605–1612.

35. Yin, M.; Zhang, J.; Wang, Y.; Wang, S.; Böckler, D.; Duan, Z.; Xin, S. Deficient CD4+CD25+ T regulatory cell function in patients with abdominal aortic aneurysms. *Arterioscler. Thromb. Vasc. Biol.* **2010**, *30*, 1825–1831.

36. Eliason, J.L.; Hannawa, K.K.; Ailawadi, G.; Sinha, I.; Ford, J.W.; Deogracias, M.P.; Roelofs, K.J.; Woodrum, D.T.; Ennis, T.L.; Henke, P.K.; *et al.* Neutrophil depletion inhibits experimental abdominal aortic aneurysm formation. *Circulation* **2005**, *112*, 232–240.

37. Hannawa, K.K.; Eliason, J.L.; Woodrum, D.T.; Pearce, C.G.; Roelofs, K.J.; Grigoryants, V.; Eagleton, M.J.; Henke, P.K.; Wakefield, T.W.; Myers, D.D. *et al.* L-selectin-mediated neutrophil recruitment in experimental rodent aneurysm formation. *Circulation* **2005**, *112*, 241–247.

38. Middleton, R.K.; Lloyd, G.M.; Bown, M.J.; Cooper, N.J.; London, N.J.; Sayers, R.D. The pro-inflammatory and chemotactic cytokine microenvironment of the abdominal aortic aneurysm wall: A protein array study. *J. Vasc. Surg.* **2007**, *45*, 574–580.

39. Kaneko, H.; Anzai, T.; Horiuchi, K.; Kohno, T.; Nagai, T.; Anzai, A.; Takahashi, T.; Sasaki, A.; Shimoda, M.; Maekawa, Y.; *et al.* Tumor necrosis factor-alpha converting enzyme is a key mediator of abdominal aortic aneurysm development. *Atherosclerosis* **2011**, *218*, 470–478.

40. Satoh, H.; Nakamura, M.; Satoh, M.; Nakajima, T.; Izumoto, H.; Maesawa, C.; Kawazoe, K.; Masuda, T.; Hiramori, K. Expression and localization of tumour necrosis factor-alpha and its converting enzyme in human abdominal aortic aneurysm. *Clin. Sci.* **2004**, *106*, 301–306.

41. Juvonen, J.; Surcel, H.; Satta, J.; Teppo, A.; Bloigu, A.; Syrjälä, H.; Airaksinen, J.; Leinonen, M.; Saikku, P.; Juvonen, T. Elevated circulating levels of inflammatory cytokines in patients with abdominal aortic aneurysm. *Arterioscler. Thromb. Vasc. Biol.* **1997**, *17*, 2843–2847.

42. Koole, D.; Hurks, R.; Schoneveld, A.; Vink, A.; Golledge, J.; Moran, C.S.; de Kleijn, D.P.; van Herwaarden, J.A.; de Vries, J.; Laman, J.D.; *et al.* Osteoprotegerin is associated with aneurysm diameter and proteolysis in abdominal aortic aneurysm disease. *Arterioscler. Thromb. Vasc. Biol.* **2012**, *32*, 1497–1504.

43. Xiong, W.; MacTaggart, J.; Knispel, R.; Worth, J.; Persidsky, Y.; Baxter, B.T. Blocking TNF-α attenuates aneurysm formation in a murine model. *J. Immunol.* **2009**, *183*, 2741–2746.

44. Kloosterman, W.P.; Plasterk, R.H. The diverse functions of microRNAs in animal development and disease. *Dev. Cell* **2006**, *11*, 441–450.

45. Cai, X.; Hagedorn, C.H.; Cullen, B.R. Human microRNAs are processed from capped, polyadenylated transcripts that can also function as mRNAs. *RNA* **2004**, *10*, 1957–1966.

46. Yeom, K.; Lee, Y.; Han, J.; Suh, M.R.; Kim, V.N. Characterization of DGCR8/Pasha, the essential cofactor for Drosha in primary miRNA processing. *Nucleic. Acids Res.* **2006**, *34*, 4622–4629.

47. Krol, J.; Loedige, I.; Filipowicz, W. The widespread regulation of microRNA biogenesis, function and decay. *Nat. Rev. Genet.* **2010**, *11*, 597–610.

48. Kim, V.N. MicroRNA precursors in motion: Exportin-5 mediates their nuclear export. *Trends Cell Biol.* **2004**, *14*, 156–159.

49. Hutvagner, G. Small RNA asymmetry in RNAi: Function in RISC assembly and gene Regulation. *FEBS Lett.* **2005**, *579*, 5850–5857.

50. Okamura, K.; Phillips, M.D.; Tyler, D.M.; Duan, H.; Chou, Y.T.; Lai, E.C. The regulatory activity of microRNA* species has substantial influence on microRNA and 3' UTR evolution. *Nat. Struct. Mol. Biol.* **2008**, *15*, 354–363.

51. Guo, L.; Lu, Z. The fate of miRNA* strand through evolutionary analysis: Implication for degradation as merely carrier strand or potential regulatory molecule? *PLoS One* **2010**, *5*, e11387.

52. Chang, K.W.; Kao, S.Y.; Wu, Y.H.; Tsai, M.M.; Tu, H.F.; Liu, C.J.; Lui, M.T.; Lin, S.C. Passenger strand miRNA miR-31* regulates the phenotypes of oral cancer cells by targeting RhoA. *Oral. Oncol.* **2013**, *49*, 27–33.

53. Yang, J.; Phillips, M.D.; Betel, D.; Mu, P.; Ventura, A.; Siepel, A.C.; Chen, K.C.; Lai, E.C. Widespread regulatory activity of vertebrate microRNA* species. *RNA* **2011**, *17*, 312–326.

54. Zhou, H.; Huang, X.; Cui, H.; Luo, X.; Tang, Y.; Chen, S.; Wu, L.; Shen, N. miR-155 and its star-form partner miR-155* cooperatively regulate type I interferon production by human plasmacytoid dendritic cells. *Blood* **2010**, *116*, 5885–5894.

55. Doench, J.G.; Sharp, P.A. Specificity of microRNA target selection in translational repression. *Genes Dev.* **2004**, *18*, 504–511.

56. Lewis, B.P.; Burge, C.B.; Bartel, D.P. Conserved seed pairing, often flanked by adenosines, indicates that thousands of human genes are microRNA targets. *Cell* **2005**, *120*, 15–20.

57. Huntzinger, E.; Izaurralde, E. Gene silencing by microRNAs: Contributions of translational repression and mRNA decay. *Nat. Rev. Genet.* **2011**, *12*, 99–110.

58. Pillai, R.S.; Bhattacharyya, S.N.; Artus, C.G.; Zoller, T.; Cougot, N.; Basyuk, E.; Bertrand, E.; Filipowicz, W. Inhibition of translational initiation by Let-7 MicroRNA in human cells. *Science* **2005**, *309*, 1573–1576.

59. Maegdefessel, L.; Azuma, J.; Toh, R.; Deng, A.; Merk, D.R.; Raiesdana, A.; Leeper, N.J.; Raaz, U.; Schoelmerich, A.M.; McConnell, M.V.; *et al.* MicroRNA-21 blocks abdominal aortic aneurysm development and nicotine-augmented expansion. *Sci. Transl. Med.* **2012**, *4*, 122ra22.

60. Leeper, N.J.; Raiesdana, A.; Kojima, Y.; Chun, H.J.; Azuma, J.; Maegdefessel, L.; Kundu, R.K.; Quertermous, T.; Tsao, P.S.; Spin, J.M. MicroRNA-26a is a novel regulator of vascular smooth muscle cell function. *J. Cell Physiol.* **2011**, *226*, 1035–1043.

61. Boon, R.A.; Seeger, T.; Heydt, S.; Fischer, A.; Hergenreider, E.; Horrevoets, A.J.G.; Vinciguerra, M.; Rosenthal, N.; Sciacca, S.; Pilato, M.; *et al.* MicroRNA-29 in aortic dilation: Implications for aneurysm formation. *Circ. Res.* **2011**, *109*, 1115–1119.

62. Maegdefessel, L.; Azuma, J.; Toh, R.; Merk, D.R.; Deng, A.; Chin, J.T.; Raaz, U.; Schoelmerich, A.M.; Raiesdana, A.; Leeper, N.J.; *et al.* Inhibition of microRNA-29b reduces murine abdominal aortic aneurysm development. *J. Clin. Invest.* **2012**, *122*, 497–506.

63. Elia, L.; Quintavalle, M.; Zhang, J.; Contu, R.; Cossu, L.; Latronico, M.V.G.; Peterson, K.L.; Indolfi, C.; Catalucci, D.; Chen, J.; *et al.* The knockout of miR-143 and -145 alters smooth muscle cell maintenance and vascular homeostasis in mice: Correlates with human disease. *Cell Death Differ.* **2009**, *16*, 1590–1598.

64. Lee, Y.S.; Dutta, A. MicroRNAs in cancer. *Annu. Rev. Pathol.* **2009**, *4*, 199–227.

65. Jazbutyte, V.; Thum, T. MicroRNA-21: From cancer to cardiovascular disease. *Curr. Drug Targets* **2010**, *11*, 926–35.

66. Cheng, Y.; Zhang, C. MicroRNA-21 in cardiovascular disease. *J. Cardiovasc. Transl. Res.* **2010**, *3*, 251–255.

67. Davis, B.N.; Hilyard, A.C.; Lagna, G.; Hata, A. SMAD proteins control DROSHA-mediated microRNA maturation. *Nature* **2008**, *454*, 56–61.

68. Ji, R.; Cheng, Y.; Yue, J.; Yang, J.; Liu, X.; Chen, H.; Dean, D.B.; Zhang, C. MicroRNA expression signature and antisense-mediated depletion reveal an essential role of MicroRNA in vascular neointimal lesion formation. *Circ. Res.* **2007**, *100*, 1579–1588.

69. Kang, H.; Hata, A. MicroRNA regulation of smooth muscle gene expression and phenotype. *Curr. Opin. Hematol.* **2012**, *19*, 224–231.

70. Sarkar, J.; Gou, D.; Turaka, P.; Viktorova, E.; Ramchandran, R.; Usha Raj, J. MicroRNA-21 plays a role in hypoxia-mediated pulmonary artery smooth muscle cell proliferation and migration. *Am. J. Physiol. Lung Cell Mol. Physiol.* **2010**, *299*, L861–L871.

71. Wang, M.; Li, W.; Chang, G.; Ye, C.; Ou, J.; Li, X.; Liu, Y.; Cheang, T.; Huang, X.; Wang, S. MicroRNA-21 regulates vascular smooth muscle cell function via targeting tropomyosin 1 in arteriosclerosis obliterans of lower extremities. *Arterioscler. Thromb. Vasc. Biol.* **2011**, *31*, 2044–2053.

72. Song, J.T.; Hu, B.; Qu, H.Y.; Bi, C.L.; Huang, X.Z.; Zhang, M. Mechanical stretch modulates MicroRNA 21 expression, participating in proliferation and apoptosis in cultured human aortic smooth muscle cells. *PLoS One* **2012**, *7*, e47657.

73. Chapman, G.B.; Durante, W.; Hellums, J.D.; Schafer, A.I. Physiological cyclic stretch causes cell cycle arrest in cultured vascular smooth muscle cells. *Am. J. Physiol. Heart Circ. Physiol.* **2000**, *278*, H748–H754.

74. Cheng, W.; Wang, B.; Chen, S.; Chang, H.; Shyu, K. Mechanical stretch induces the apoptosis regulator PUMA in vascular smooth muscle cells. *Cardiovasc. Res.* **2012**, *93*, 181–189.

75. Li, C.; Wernig, F.; Leitges, M.; Hu, Y.; Xu, Q. Mechanical stress-activated PKCdelta regulates smooth muscle cell migration. *FASEB J.* **2003**, *17*, 2106–2108.

76. Weber, M.; Baker, M.B.; Moore, J.P.; Searles, C.D. MiR-21 is induced in endothelial cells by shear stress and modulates apoptosis and eNOS activity. *Biochem. Biophys. Res. Commun.* **2010**, *393*, 643–648.

77. Zhu, S.; Deng, S.; Ma, Q.; Zhang, T.; Jia, C.; Zhuo, D.; Yang, F.; Wei, J.; Wang, L.; Dykxhoorn, D.M.; *et al.* microRNA-10A* and microRNA-21 modulate endothelial progenitor cell senescence via suppressing Hmga2. *Circ. Res.* **2013**, *112*, 152–164.

78. van Rooij, E.; Sutherland, L.B.; Liu, N.; Williams, A.H.; McAnally, J.; Gerard, R.D.; Richardson, J.A.; Olson, E.N. A signature pattern of stress-responsive microRNAs that can evoke cardiac hypertrophy and heart failure. *Proc. Natl. Acad. Sci. USA* **2006**, *103*, 18255–18260.

79. Kwiecinski, M.; Noetel, A.; Elfimova, N.; Trebicka, J.; Schievenbusch, S.; Strack, I.; Molnar, L.; von Brandenstein, M.; Töx, U.; Nischt, R. *et al.* Hepatocyte growth factor (HGF) inhibits collagen I and IV synthesis in hepatic stellate cells by miRNA-29 induction. *PLoS One* **2011**, *6*, e24568.

80. Wang, B.; Komers, R.; Carew, R.; Winbanks, C.E.; Xu, B.; Herman-Edelstein, M.; Koh, P.; Thomas, M.; Jandeleit-Dahm, K.; Gregorevic, P.; *et al.* Suppression of microRNA-29 expression by TGF-β1 promotes collagen expression and renal fibrosis. *J. Am. Soc. Nephrol.* **2012**, *23*, 252–265.

81. Maurer, B.; Stanczyk, J.; Jüngel, A.; Akhmetshina, A.; Trenkmann, M.; Brock, M.; Kowal-Bielecka, O.; Gay, R.E.; Michel, B.A.; Distler, J.H.; *et al.* MicroRNA-29, a key regulator of collagen expression in systemic sclerosis. *Arthritis Rheum* **2010**, *62*, 1733–1743.

82. Ogawa, T.; Iizuka, M.; Sekiya, Y.; Yoshizato, K.; Ikeda, K.; Kawada, N. Suppression of type I collagen production by microRNA-29b in cultured human stellate cells. *Biochem. Biophys. Res. Commun.* **2010**, *391*, 316–321.

83. Boettger, T.; Beetz, N.; Kostin, S.; Schneider, J.; Krüger, M.; Hein, L.; Braun, T. Acquisition of the contractile phenotype by murine arterial smooth muscle cells depends on the Mir143/145 gene cluster. *J. Clin. Invest.* **2009**, *119*, 2634–2647.

84. Cheng, Y.; Liu, X.; Yang, J.; Lin, Y.; Xu, D.Z.; Lu, Q.; Deitch, E.A.; Huo, Y.; Delphin, E.S.; Zhang, C. MicroRNA-145, a novel smooth muscle cell phenotypic marker and modulator, controls vascular neointimal lesion formation. *Circ. Res.* **2009**, *105*, 158–166.

85. Cordes, K.R.; Sheehy, N.T.; White, M.P.; Berry, E.C.; Morton, S.U.; Muth, A.N.; Lee, T.H.; Miano, J.M.; Ivey, K.N.; Srivastava, D. miR-145 and miR-143 regulate smooth muscle cell fate and plasticity. *Nature* **2009**, *460*, 705–710.

86. Xin, M.; Small, E.M.; Sutherland, L.B.; Qi, X.; McAnally, J.; Plato, C.F.; Richardson, J.A.; Bassel-Duby, R.; Olson, E.N. MicroRNAs miR-143 and miR-145 modulate cytoskeletal dynamics and responsiveness of smooth muscle cells to injury. *Genes Dev.* **2009**, *23*, 2166–2178.

87. Quintavalle, M.; Elia, L.; Condorelli, G.; Courtneidge, S.A. MicroRNA control of podosome formation in vascular smooth muscle cells *in vivo* and *in vitro*. *J. Cell Biol.* **2010**, *189*, 13–22.

88. Hergenreider, E.; Heydt, S.; Tréguer, K.; Boettger, T.; Horrevoets, A.J.; Zeiher, A.M.; Scheffer, M.P.; Frangakis, A.S.; Yin, X.; Mayr, M.; *et al.* Atheroprotective communication between endothelial cells and smooth muscle cells through miRNAs. *Nat. Cell Biol.* **2012**, *14*, 249–256.

89. Golledge, A.L.; Walker, P.; Norman, P.E.; Golledge, J. A systematic review of studies examining inflammation associated cytokines in human abdominal aortic aneurysm samples. *Dis. Markers* **2009**, *26*, 181–188.

90. Harris, T.A.; Yamakuchi, M.; Ferlito, M.; Mendell, J.T.; Lowenstein, C.J. MicroRNA-126 regulates endothelial expression of vascular cell adhesion molecule 1. *Proc. Natl. Acad. Sci. USA* **2008**, *105*, 1516–1521.

91. Asgeirsdóttir, S.A.; van Solingen, C.; Kurniati, N.F.; Zwiers, P.J.; Heeringa, P.; van Meurs, M.; Satchell, S.C.; Saleem, M.A.; Mathieson, P.W.; Banas, B.; *et al.* MicroRNA-126 contributes to renal microvascular heterogeneity of VCAM-1 protein expression in acute inflammation. *Am. J. Physiol. Renal Physiol.* **2012**, *302*, F1630–F1639.

92. Zernecke, A.; Bidzhekov, K.; Noels, H.; Shagdarsuren, E.; Gan, L.; Denecke, B.; Hristov, M.; Köppel, T.; Jahantigh, M.N.; Lutgens, E.; *et al.* Delivery of microRNA-126 by apoptotic bodies induces CXCL12-dependent vascular protection. *Sci. Signal.* **2009**, *2*, ra81.

93. Vasa-Nicotera, M.; Chen, H.; Tucci, P.; Yang, A.L.; Saintigny, G.; Menghini, R.; Mahè, C.; Agostini, M.; Knight, R.A.; Melino, G.; *et al.* miR-146a is modulated in human endothelial cell with aging. *Atherosclerosis* **2011**, *217*, 326–330.

94. Sun, S.G.; Zheng, B.; Han, M.; Fang, X.M.; Li, H.X.; Miao, S.B.; Su, M.; Han, Y.; Shi, H.J.; Wen, J.K. miR-146a and Kruppel-like factor 4 form a feedback loop to participate in vascular smooth muscle cell proliferation. *EMBO Rep.* **2011**, *12*, 56–62.

95. Suárez, Y.; Wang, C.; Manes, T.D.; Pober, J.S. Cutting edge: TNF-induced microRNAs regulate TNF-induced expression of E-selectin and intercellular adhesion molecule-1 on human endothelial cells: Feedback control of inflammation. *J. Immunol.* **2010**, *184*, 21–25.

96. Zhu, N.; Zhang, D.; Chen, S.; Liu, X.; Lin, L.; Huang, X.; Guo, Z.; Liu, J.; Wang, Y.; Yuan, W.; Qin, Y. Endothelial enriched microRNAs regulate angiotensin II-induced endothelial inflammation and migration. *Atherosclerosis* **2011**, *215*, 286–293.

97. Nazari-Jahantigh, M.; Wei, Y.; Noels, H.; Akhtar, S.; Zhou, Z.; Koenen, R.R.; Heyll, K.; Gremse, F.; Kiessling, F.; Grommes, J.; *et al.* MicroRNA-155 promotes atherosclerosis by repressing Bcl6 in macrophages. *J. Clin. Invest.* **2012**, *122*, 4190–4202.

98. Donners, M.M.; Wolfs, I.M.; Stöger, L.J.; van der Vorst, E.P.; Pöttgens, C.C.; Heymans, S.; Schroen, B.; Gijbels, M.J.; de Winther, M.P. Hematopoietic miR155 deficiency enhances atherosclerosis and decreases plaque stability in hyperlipidemic mice. *PLoS One* **2012**, *7*, e35877.

99. Pahl, M.C.; Derr, K.; Gäbel, G.; Hinterseher, I.; Elmore, J.R.; Schworer, C.M.; Peeler, T.C.; Franklin, D.P.; Gray, J.L.; Carey, D.J. *et al.* MicroRNA expression signature in human abdominal aortic aneurysms. *BMC Med. Genomics* **2012**, *5*, 25.

100. Small, E.M.; Olson, E.N. Pervasive roles of microRNAs in cardiovascular biology. *Nature* **2011**, *469*, 336–342.

101. Van Rooij, E.; Olson, E.N. MicroRNA therapeutics for cardiovascular disease: Opportunities and obstacles. *Nat. Rev. Drug Discov.* **2012**, *11*, 860–872.

102. Small, E.M.; Frost, R.J.; Olson, E.N. MicroRNAs add a new dimension to cardiovascular disease. *Circulation* **2010**, *121*, 1022–1032.

103. Van Rooij, E.; Purcell, A.L.; Levin, A.A. Developing microRNA therapeutics. *Circ. Res.* **2012**, *110*, 496–507.

104. Stenvang, J.; Petri, A.; Lindow, M.; Obad, S.; Kauppinen, S. Inhibition of microRNA function by antimiR oligonucleotides. *Silence* **2012**, *3*, 1.

105. Mendell, J.T.; Olson, E.N. MicroRNAs in stress signaling and human disease. *Cell* **2012**, *148*, 1172–1187.

106. Janssen, H.L.A.; Reesink, H.W.; Lawitz, E.J.; Zeuzem, S.; Rodriguez-Torres, M.; Patel, K.; van der Meer, A.J.; Patick, A.K.; Chen, A.; Zhou, Y.; *et al.* Treatment of HCV infection by targeting microRNA. *N. Engl. J. Med.* **2013**, *368*, 1685–1694.

107. Mishra, P.K.; Tyagi, N.; Kumar, M.; Tyagi, S.C. MicroRNAs as a therapeutic target for cardiovascular diseases. *J. Cell Mol. Med.* **2009**, *13*, 778–789.

108. Pramanik, D.; Campbell, N.R.; Karikari, C.; Chivukula, R.; Kent, O.A.; Mendell, J.T.; Maitra, A. Restitution of tumor suppressor microRNAs using a systemic nanovector inhibits pancreatic cancer growth in mice. *Mol. Cancer Ther.* **2011**, *10*, 1470–1480.

109. Trang, P.; Wiggins, J.F.; Daige, C.L.; Cho, C.; Omotola, M.; Brown, D.; Weidhaas, J.B.; Bader, A.G.; Slack, F.J. Systemic delivery of tumor suppressor microRNA mimics using a neutral lipid emulsion inhibits lung tumors in mice. *Mol. Ther.* **2011**, *19*, 1116–1122.

110. Moxon, J.V.; Parr, A.; Emeto, T.I.; Walker, P.; Norman, P.E.; Golledge, J. Diagnosis and monitoring of abdominal aortic aneurysm: Current status and future prospects. *Curr. Probl. Cardiol.* **2010**, *35*, 512–548.

111. D'Alessandra, Y.; Devanna, P.; Limana, F.; Straino, S.; di Carlo, A.; Brambilla, P.G.; Rubino, M.; Carena, M.C.; Spazzafumo, L.; de Simone, M.; *et al.* Circulating microRNAs are new and sensitive biomarkers of myocardial infarction. *Eur. Heart J.* **2010**, *31*, 2765–2773.

112. Fichtlscherer, S.; de Rosa, S.; Fox, H.; Schwietz, T.; Fischer, A.; Liebetrau, C.; Weber, M.; Hamm, C.W.; Röxe, T.; Müller-Ardogan, M.; *et al.* Circulating microRNAs in patients with coronary artery disease. *Circ. Res.* **2010**, *107*, 677–684.

113. Tijsen, A.J.; Creemers, E.E.; Moerland, P.D.; de Windt, L.J.; van der Wal, A.C.; Kok, W.E.; Pinto, Y.M. MiR423-5p as a circulating biomarker for heart failure. *Circ. Res.* **2010**, *106*, 1035–1039.

114. Zampetaki, A.; Willeit, P.; Tilling, L.; Drozdov, I.; Prokopi, M.; Renard, J.M.; Mayr, A.; Weger, S.; Schett, G.; Shah, A.; *et al.* Prospective study on circulating MicroRNAs and risk of myocardial infarction. *J. Am. Coll Cardiol.* **2012**, *60*, 290–299.

115. Engelhardt, S. Small RNA biomarkers come of age. *J. Am. Coll Cardiol.* **2012**, *60*, 300–303.

116. Zampetaki, A.; Mayr, M. Analytical challenges and technical limitations in assessing circulating miRNAs. *Thromb. Haemost.* **2012**, *108*, 592–598.

117. Kin, K.; Miyagawa, S.; Fukushima, S.; Shirakawa, Y.; Torikai, K.; Shimamura, K.; Daimon, T.; Kawahara, Y.; Kuratani, T.; Sawa, Y. Tissue- and plasma-specific microRNA signatures for atherosclerotic abdominal aortic aneurysm. *J. Am. Heart Assoc.* **2012**, *1*, e000745.

8

Principles of miRNA-Target Regulation in Metazoan Models

Epaminondas Doxakis

Basic Neurosciences Division, Biomedical Research Foundation of the Academy of Athens, Soranou Efesiou 4, Athens 11527, Greece; E-Mail: edoxakis@bioacademy.gr;

Abstract: MicroRNAs (miRs) are key post-transcriptional regulators that silence gene expression by direct base pairing to target sites of RNAs. They have a wide variety of tissue expression patterns and are differentially expressed during development and disease. Their activity and abundance is subject to various levels of control ranging from transcription and biogenesis to miR response elements on RNAs, target cellular levels and miR turnover. This review summarizes and discusses current knowledge on the regulation of miR activity and concludes with novel non-canonical functions that have recently emerged.

Keywords: miR; miR biogenesis; miR targets; miR turnover; isomiR; ceRNA

1. Introduction

Mature microRNAs (miRs) are a class of highly conserved small non-coding RNA molecules, about 22 *nucleotides* in length, that act to inhibit protein expression by partially hybridizing to complementary sequences, mainly in the 3' UTR, of target RNA transcripts. Each miR is estimated to regulate multiple functionally-related target mRNAs, and the combinatorial action of miRs is expected to regulate the expression of hundreds of mRNAs. Currently, over 1100 and 1800 miRs have been annotated and categorized in mice and humans, respectively (miRBase 20, [1]). However, these numbers are likely to be inflated by mistakenly identified miRs [2]. In addition, the high rate of miR family turnover in mammals—with many newly emerged miR families being lost soon after their formation—indicate that many more of the truly-identified miRs are likely to have little functional significance [3]. It is now predicted that more than half of human genes are regulated by miRs [4]. miRs have a wide variety of tissue expression patterns and are differentially expressed during development [5–8]. They are deregulated in most human diseases and the profiles they generate carry more diagnostic information than those of mRNAs or proteins [9]. Moreover, the therapeutic potential of miRs, already demonstrated in numerous studies, has further heightened the importance of research that seeks to understand both their mechanism of action and their biological significance [10–13].

This paper aims at reviewing the latest information on miR biogenesis and the factors that determine the efficacy of miR-mediated repression and miR endogenous levels. It concludes with novel atypical functions that stand-out from the canonical repression activity of miRs.

2. miR Biogenesis

miRs are transcribed as part of longer primary transcripts (pri-miRs) by, mainly, RNA polymerase II (Pol II) and only few by RNA polymerase III (Pol III) [14–19]. The majority of miR genes are transcribed from introns of protein-coding genes. The remaining are transcribed as part of long non-coding RNAs that are often arranged as clusters that lead to one pri-miR being subsequently, processed into several mature miRs (Figure 1) [17,20]. Similar to other Pol II transcripts, pri-miRs possess a 5' 7-methyl-guanosine cap and a 3' poly (A) tail, the use of which is currently poorly understood. Within pri-miR long transcripts, mature miR sequences form hairpin structures that contain imperfect double-stranded stems of ~30 bp connected by a terminal loop at the top and single-stranded RNA segments at the base [21,22]. In the canonical miR biogenesis pathway, these hairpin structures are recognized in the nucleus and cleaved by a multi-protein microprocessor complex that is composed of two core components, Drosha (a RNase III ribonuclease) and DGCR8 (also known as Pasha in invertebrates which is a double-stranded RNA binding protein). Mechanistically, DGCR8, initially, recognizes the base of the miR hairpin structure and then guides Drosha to cleave the pri-miR at a distance of ~11 bp from the base generating a ~70 nucleotide (nt) hairpin RNA (named precursor miR or pre-miR) with a 2 nt 3' overhang [21–26]. This 3' overhang and the double-stranded hairpin structure of the pre-miR are subsequently recognized by exportin-5, which together with its cofactor RAN-GTP, shuttle pre-miR from the nucleus into the cytosol. The hydrolysis of GTP bound to RAN in the cytosol triggers the dissociation of the complex, allowing the pre-miR to bind Dicer, a double-stranded ribonuclease III. Dicer cleaves the pre-miR terminal loop in concert with its cofactors TRBP (also known as Loqs in *Drosophila*) and PACT. In this process, Dicer binds to the pre-miR 2 nt 3' overhang and cuts two helical turns (~22 nt) away to produce a double-stranded RNA with 3' overhangs of 2 nt at both ends. TRBP and PACT regulate Dicer's substrate recognition and RNA processing power but are not essential for Dicer's slicing activity [27–33]. After cleavage, the strand with the 5' terminus that has less stable base-pairing (the "guide strand") is transferred onto an Argonaute (Ago) protein, which is part of a poorly defined multi-protein miR-induced silencing complex (miRISC) that includes Dicer, TRBP and TNRC6 (also known as GW182) whereas the other strand (the "passenger" or "star strand") is degraded [34–37]. Additional features on the miR duplex may also play a role in the strand selection process and there are several miRs where both strands are incorporated, to varying degrees, onto Ago proteins [38–41]. Once in place, the miR nucleotide sequence serves as a guide for RNA interference (RNAi) based on the partial complementarity with the various RNA substrates, a process which is largely attained by random diffusion of miRISC into the cytosol [42]. TNRC6 proteins are essential for RNAi as they interact with poly(A)-binding protein (PABP) and the PAN2-PAN3 and CCR4-NOT deadenylase complexes to induce translation repression, deadenylation and decay of the mRNA targets [43–45].

Figure 1. miR biogenesis. Monocistronic or polycistronic miRs are transcribed by RNA polymerase II into long pri-miR transcripts. These pri-miRs are, subsequently, processed by RNase III Drosha complex to ~70 nt pre-miRs that are exported out of the nucleus and into the cytosol by Exportin-5. In the cytoplasm, the RNase III Dicer complex cleaves pre-miRs to double-stranded ~22 nt miRs. One strand is then selected and loaded onto an Argonaute protein, which is part of the miRISC complex. The single-stranded miR then serves as a guide for RNA interference based on the partial complementarity with the various RNA substrates.

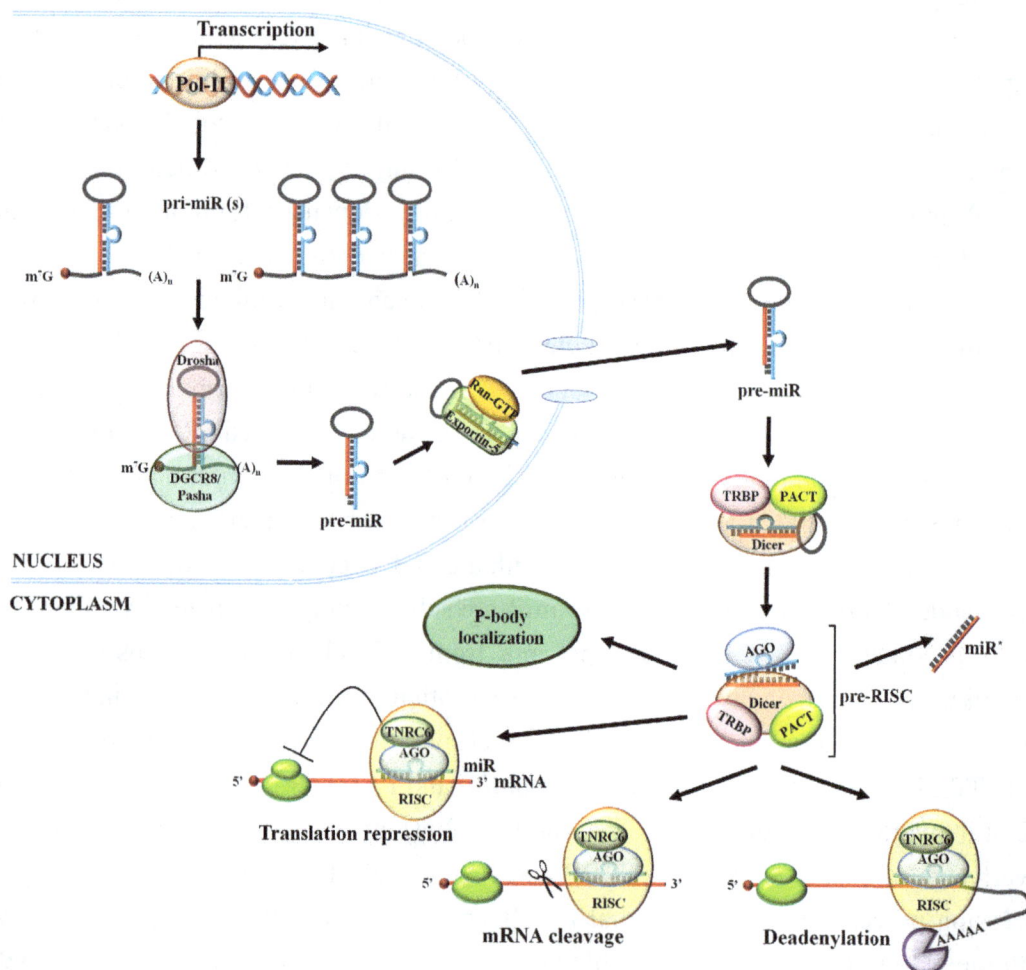

3. Efficacy of miR Repression

Despite a wealth of genome-wide and biochemical data on the role of miRs in the regulation of their targets, we do not yet have a clear understanding of the factors that determine which mRNAs will be targeted by miRs or by which mechanism individual mRNAs will be silenced, that is, translation repression or mRNA destabilization. Likely, this reflects on the vast repertoire of context-specific determinants that modulate miR-target interactions (Figure 2).

Figure 2. miR repression determinants. Multiple factors determine repression effectiveness of miRs. These include: (**a**) Sequence complementarity at positions 2–7 or 2–8 of the 5' end of the mature miR; (**b**) Target site features: binding site location near the edges of 3' UTR or multiple binding sites for miRs; (**c**) Alternative cleavage and polyadenylation to maintain miR binding sites; (**d**) Relatively high miR levels; (**e**) Relatively low target levels; and (**f**) Presence of stabilizing and absence of destabilizing RNA binding protein sites.

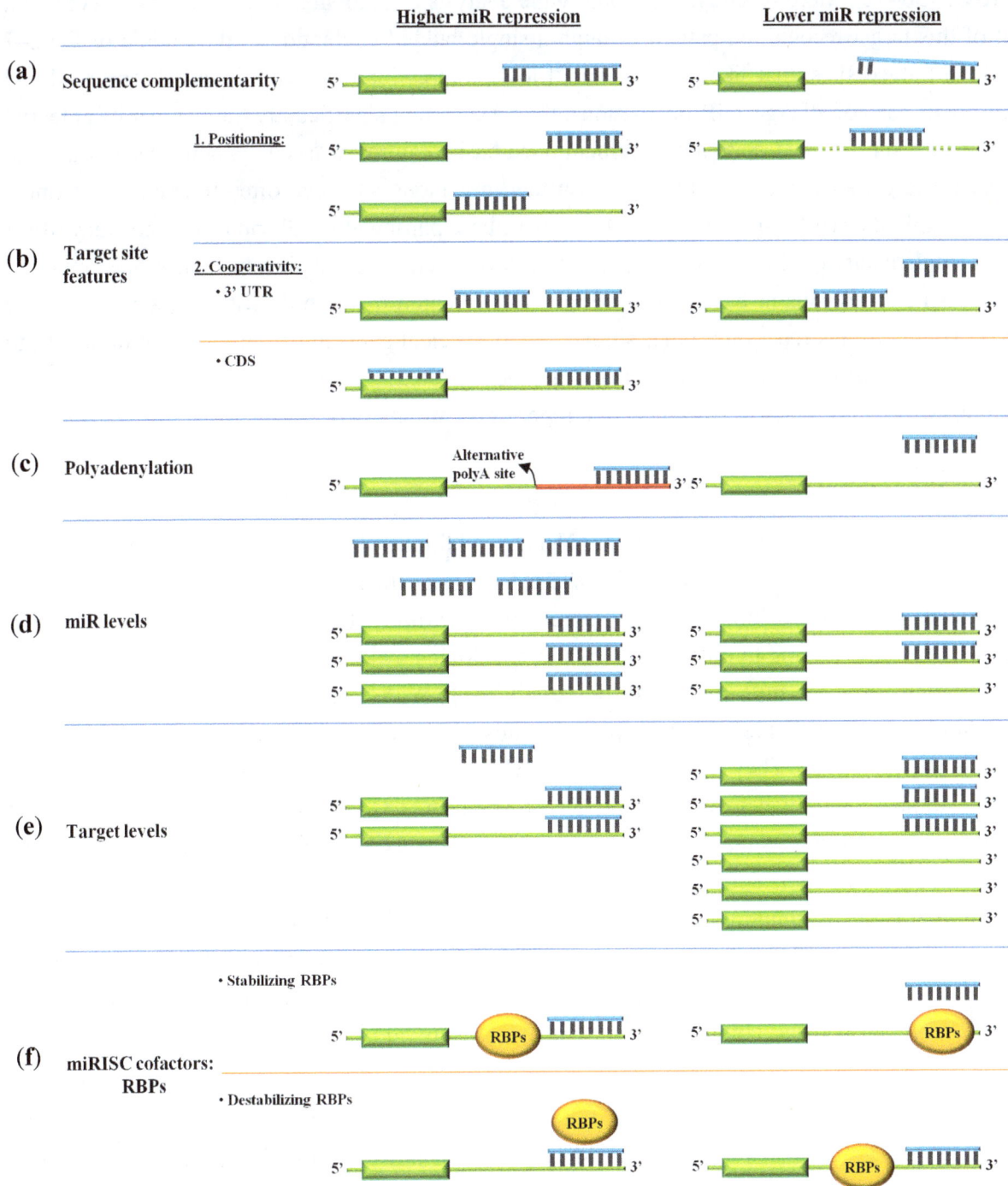

3.1. The miR Nucleotide Sequence

Large-scale transcriptomic and proteomic studies have revealed that the primary determinant for miR binding is perfect consecutive Watson-Crick base-pairing between the target RNA and the miR at positions 2–7 or 2–8 of the 5' end of the mature miR, often denoted as the "seed" region [46–49]. This signature has been reaffirmed with crystallographic studies of ribonucleoprotein Ago-miR complexes showing that the seed region is organized in a helical conformation that exposes it to base-pair with the target RNA [50–52]. More recently, a genome-wide analysis of Ago sites in murine brain revealed a variant of this target recognition pattern through a single bulged nucleotide in the middle of the 2–7 seed. These bulged sites, that likely yield overall lower repression, are evolutionarily conserved and comprise over 15% of all Ago-miR interactions, thus, expanding significantly the number of potential miR regulatory sites [53]. Despite the aforementioned basic features, a "seed" is neither necessary nor sufficient for target silencing. It has been shown that miR target sites can often tolerate G:U wobble base pairs within the seed region [54,55] and extensive base pairing at the 3' end of the miR may offset missing complementarity at the seed region [46,56]. Moreover, centered sites have also been reported showing 11–12 contiguous nt base-pairing to the central region of the miR without pairing to either end [57]. To add to this repertoire, other studies report efficient silencing from sites that do not fit to any of the above patterns, seemingly appearing random [58,59], and even sites with extensive 5' complementarity can be inactive when tested in reporter constructs [60].

3.2. Target Site Features

Considerable progress has been made to identify additional features that could predict target regulation with more precision. Grimson *et al.* have reported that local sequence context, such as AU-rich nucleotide composition near the target site, proximity to sites for co-expressed miRs, proximity to residues pairing to miR nt 13–16, and positioning away from the center of long 3' UTRs can all promote efficient miR repression of targets [61]. With respect to these findings, several different studies have reaffirmed that multiple miR sites in the same 3' UTR can potentiate the degree of translational repression. They reported that optimal downregulation is obtained when two sites are closely positioned, usually between 13 and 35 nt apart [62,63]. However, target sites spaced at substantially longer distances may still cooperate to lower the expression of proteins [64,65]. In this context, miR cooperativity is defined as the positive interaction of two or more individual miRs or one individual miR acting on multiple target sites on the same 3' UTR for target regulation. Recently, it was estimated that the miR site density of brain synaptic mRNAs is twice higher than that of the rest of cellular mRNAs, indicating that miR cooperativity may be a prevalent mechanism for physiological processes that require precise control, such as synaptic transmission [65].

An additional feature that has also emerged is that miR target sites tend to be less evolutionary preserved in the first ~15 nt downstream of the stop codon, presumably, to avoid being in the path of the translational machinery that could displace the miRISC complex [61]. However, both computational and biochemical approaches have later identified that nearly half of miR sites are located in open reading frame (ORF) sequences [66–70]. Experimental analysis indicated that the sites in coding regions and to a lesser extent 5' UTRs can confer miR repression, albeit at lower levels than

3' UTRs [71,72]. Recently, it was reported that coding region-located sites induce more rapid reduction in mRNA translation than 3' UTR-located sites in a process that does not involve mRNA degradation, however, the effect may only be transient. The authors elaborate that this type of response may be suited for the regulation of cell cycle proteins [73]. Further, there are several families of paralogous genes that contain multiple repeat sequences in their coding regions, arisen through evolutionary duplications, that are miR targets [74]. Like for miR cooperativity in 3' UTRs, it was shown that miR sites in the coding region potentiate the repression activity of miRs acting on 3' UTR [75].

To add a twist to miR regulation, it has been reported that individual miRs may also display distinct preference for binding to different regions of an mRNA. For instance, neuronal miR-124 seed sequences are preferentially located in the 3' UTR, while miR-107 seed sequences are enriched in the coding region of the mRNAs. Further, mRNA targets of neuronal miR-128 and miR-320 are less enriched for 6-mer seed sequences than miR-124 and mir-107 [76]. The reason for these differences is unknown but they, evidently, enrich the heterogeneity of miR-mediated repression.

3.3. Target Accessibility and Polyadenylation

Another important determinant of efficient silencing is target RNA folding with several reports indicating that miR sites are preferentially positioned in highly accessible/unstructured regions at the start and end of 3' UTRs [61–79]. Experimentally, target sites in the middle of 3' UTR have been found to be less efficient for RNAi regulation [80] while those positioned near the end of 3' UTRs are associated with highest repression [62].

Another contributing factor is the length of the 3' UTR. Approximately, half of human genes undergo alternative cleavage and polyadenylation (pA) to generate transcripts with variable 3' UTR lengths [53]. Given that 3' UTRs are the main targets of miRs, alternative pA is expected to modify target RNA translation. Consequently, a close connection between gene transcription and pA site choice has been demonstrated, in which highly expressed genes contain shorter 3' UTRs while transcripts that are expressed in lower levels are associated with longer 3' UTR isoforms [81]. Along this, higher gene expression is tightly linked to cell division where short 3' UTR isoforms with fewer miR sites are abundant in proliferating cells [82]. In contrast, differentiated cells possess longer 3' UTRs [81]. A noteworthy consequence of alternative splicing was observed in transformed cells where the loss of miR target sites by pA contributed to oncogene activation without any apparent mutagenesis [82].

3.4. miR and Target RNA Levels

An additional critical determinant for miR repression effectiveness is the cellular concentrations of (a) the target RNA, (b) the miR and (c) the miRISC complex. miRs that have multiple targets and are not highly expressed are expected to downregulate individual target genes to a lesser extent than those with a lower number of targets. Similarly, highly abundant target transcripts, that may act as decoys, dilute the effect of miRs under differential conditions [83–85] and this effect is more pronounced when the decoys are capable of perfect base-pairing with the miR [86]. Along these lines, it has been observed that lower levels of a miR may fail to regulate its target mRNA, however, it retains the ability to promote inhibition in conjunction with another miR, indicating that cooperative silencing requires lower concentration of miRs [65]. Going beyond, it is predicted that imbalances in the relative concentrations of miRs and their gene targets may exaggerate or compensate for sequence mismatches between miR and target RNA pairs. miRISC stability has emerged as an additional level at which miR activity can be controlled. Specifically, LIN41, an E3 ubiquitin ligase, has been shown to target Ago2 for ubiquitination and proteasome degradation. Because Ago proteins are limiting factors for the activity of the miRISC complex, alterations in the levels of LIN41 result in global attenuation of miR-mediated repression [87].

3.5. RNA Binding Proteins

RNA binding proteins (RBPs) regulate key aspects of gene expression including pre-mRNA splicing, nuclear-cytosolic shuttling, cytosolic transport and storage, local translation and turnover. Although most RBPs have housekeeping functions, a subset of RBPs controls the expression of specific labile mRNAs by binding to U- and AU-rich elements (AREs) on either 5' UTR or 3' UTR. These include HuR, TIAR, TIA-1, AUF1, TTP and KSRP, collectively known as translation and turnover regulatory (TTR)-RBPs. They primarily modulate mRNA levels in response to external stimuli and have been shown to influence all aspects of cellular activities that include proliferation and differentiation [88]. Recently, a link between RBPs and miRs has emerged. Initially, it has been observed that destabilization mediated by a transfected miR is generally attenuated by the presence of destabilizing AU-rich motifs and augmented by stabilizing U-rich motifs, the binding sites of TTR-RBPs [89,90]. Subsequently, transcriptome-wide analysis for the best characterized ubiquitous RBP, HuR revealed that most miR sites were found in the immediate vicinity of HuR sites [91,92] (reviewed in [93]). The authors elaborated that where miR and HuR sites overlapped the transcripts were preferentially regulated by HuR, but when they were not overlapping the transcripts were regulated by miR. Interestingly, HuR transcript is itself a direct target of miRs and of itself, and at the same time, directly regulates stability and/or maturation of other miRs pointing to the vast repertoire of the different regulatory loops [91–96].

4. Availability of miRs

It has become increasingly evident that miR activity is determined not only by target site features but also miR levels, target abundance and the presence of multiple RNA decoys (Figure 3). It is the summation of all these inputs that ultimately shapes miR function.

4.1. Transcription

Most miRs are transcribed by RNA polymerase II (Pol II) and only few by RNA polymerase III (Pol III) [14–19,97]. Pol III-mediated transcription is usually restricted to housekeeping non-coding genes, such as tRNAs and snRNAs, that require ubiquitous expression under all conditions [98], whereas Pol II-mediated transcription permits tight control of expression during all types of regulatory conditions [99]. Nonetheless, there is evidence that the same promoter elements can be used by both polymerases in humans [100–102] and transcription factors can also regulate RNA Pol III activity to some degree [103]. Furthermore, whole genome analysis has revealed that miR promoters are, in general, very similar to protein-coding promoters containing proportionally similar levels of CpG islands, TATA boxes, TFIIB recognition elements (BRE) and initiators (Inr) [16].

Figure 3. miR availability determinants. Multiple factors determine availability of miRs. These include: (**a**) High transcription rates; (**b**) Enhanced Drosha processing; (**c**) Lower levels of isomiRs that result from RNA editing, sloppy Drosha/Dicer cleavage, exoribonuclease trimming and nucleotidyl transferase additions; (**d**) Lower levels of miR sequestering ceRNAs; and (**e**) Lower levels of exoribonucleases.

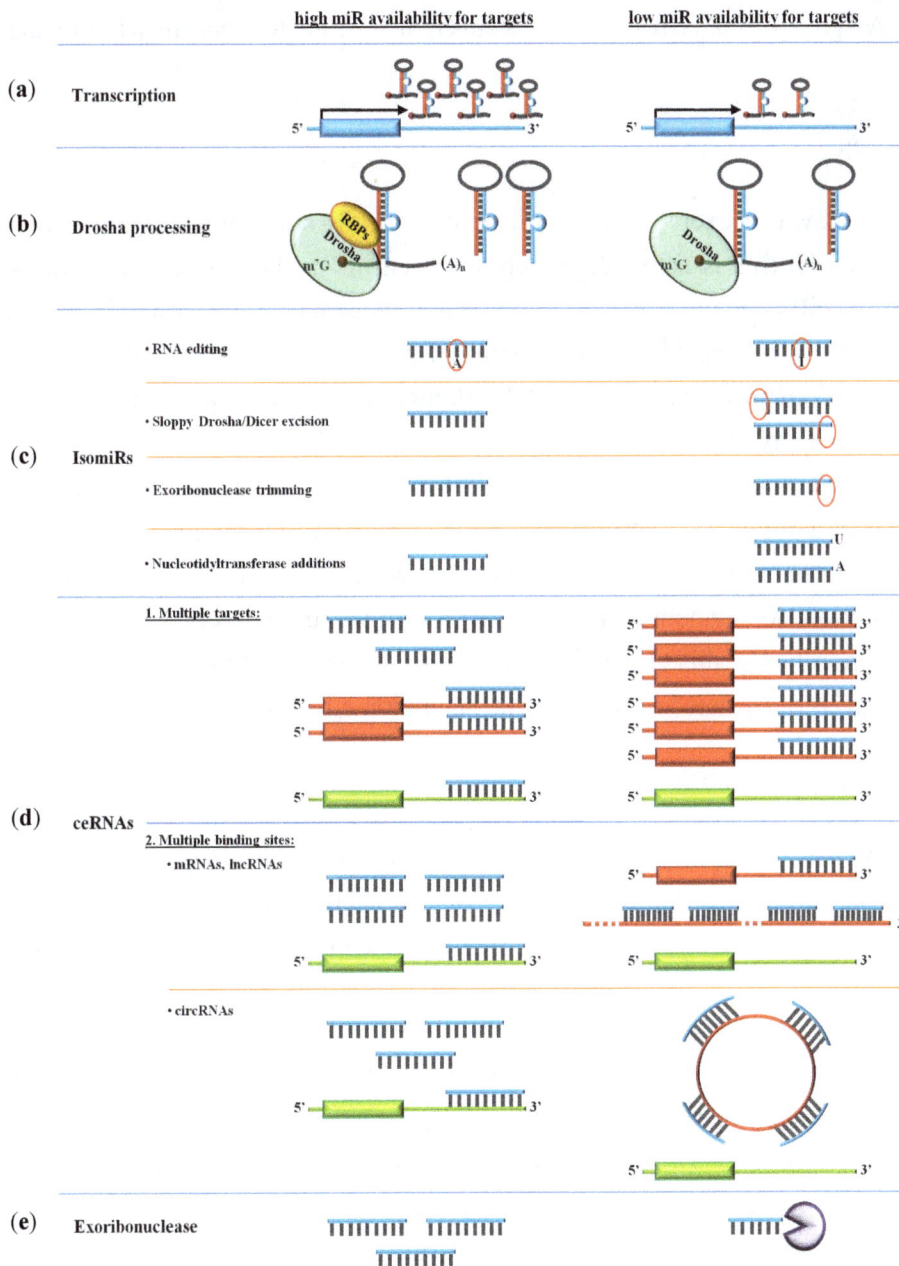

The majority of miR genes are transcribed from introns (and to lesser extent exons) of protein-coding genes [17,20]. As a consequence, miRs, more often than not, are co-expressed with host genes [104]. Nevertheless, increasingly, there have been reports that showed that intragenic miRs could, independently, initiate transcription from own promoters [105–107]. It is now estimated that about a third of hosted miRs use their own promoters for more efficient and tailored transcription [14,16,108]. With respect to the miRs that are located in intergenic regions, these are often arranged in clusters that lead to one pri-miR being subsequently processed into several mature miRs [104]. Using microarray profiling, Baskerville and Bartel have proposed that miRs separated by <50 kb are typically derived from a common transcript [104]. Accordingly, the latest miRBase release (Release 20) groups human and murine miRs in 153 (containing 465 miRs) and 92 (containing 366 miRs) clusters, respectively, using a default of <10 kb inter-miR distance. Clusters provide an effective mechanism to express cooperative miRs, simultaneously. Many clusters contain representatives from different miR families that together regulate specific protein networks by co-targeting downstream mRNAs [109]. This provides another layer of coordinated system-wide regulation of gene output in cells [110].

4.2. Drosha Processing

Drosha has been shown to exert selectivity over its pri-miR substrates compared to other RNAs. The mechanism by which this is achieved differs between miRs. Thus far, microarray profiling has shown that subsets of miRs contain a Smad binding RNA sequence (R-SBE) within the stem region of the pri-miR that resembles the Smad binding element in DNA. Smad proteins bind to these motifs on the miRs with one (MH1) domain while another (MH2) domain binds p68, a protein that is integral part of the microprocessor complex in the nucleus and is known to induce Drosha processing [111,112]. Similarly, DNA damage induces p53 association with p68, promoting the processing of specific miRs that subsequently exert a tumor suppressor function via repression of c-myc [113]. A different mode of regulation is demonstrated by the RNA binding proteins KHSRP and hnRNPA1 that bind to specific single- and double- stranded segments on the pre-miR hairpin, respectively, inducing microprocessor complex cleavage. Importantly, this targeted processing of the pri-miR has been shown to uncouple the uniform expression levels of clustered miRs from the maturation efficiency of individual miRs [114–116].

4.3. miR Polymorphism and isomiRs

Computational predictions have strongly suggested that miRs may have shaped the evolution of their targets based on the fact that the conservation of predicted miR target sites in mRNAs is higher than that of other conserved 3' UTR motifs [117]. Consequently, polymorphisms in miR sequences were presumed rare. Towards this, bioinformatic analysis has revealed that the density of single nucleotide polymorphisms (SNPs) in miRs is 4.5-times lower than in protein coding sequences [118] and from these polymorphisms, only 1/10 or less are located in the seed region [60,118–120]. As expected, miR SNPs in the seed region would ultimately result in the regulation of a completely different set of mRNA targets. An increasing number of epidemiological reports have now linked several of these miR SNPs to pathology and, in particular, cancer susceptibility. miR-146a-3p and miR-499-3p, for instance, have so far been associated with the largest variety of cancer pathologies affecting all organ systems (for review see [120,121]).

Recent advances in high-throughput small RNA sequencing technologies have revealed novel post-transcriptional processing mechanisms that increase mature miR sequence heterogeneity from single genomic locus in cells. It is estimated that as many as 90% of miRs are presented with some sort of modification mainly in the form of trimming and/or nucleotide addition in the 3' terminus [122].

Thus far, four mechanisms that generate functionally distinct miR isoforms, annotated as isomiRs, have been identified [123]. These are RNA editing, inexact Drosha and Dicer processing, exonuclease ribonucleotide trimming and template-independent ribonucleotide addition.

RNA editing is a chemical alteration in the primary nucleotide sequence of double-stranded RNAs. The most common RNA editing modification involves the hydrolytic deamination of adenosine-to-inosine (A-to-I) catalyzed by the adenosine deaminase acting on RNA (ADAR) enzymes [124]. Because inosine preferentially base pairs with cytidine, this conversion is equivalent to an adenosine to guanosine change. Although earlier reports identified widespread A-to-I editing in pri-miRs, more recent studies have revealed that RNA editing is rather rare for mature miRs [125,126]. A comprehensive profiling of human RNA editome revealed only 44 edited miR sites of which 11 were in the seed region [127]. This indicates that miRs exhibit low frequency of editing and that the primary biological function of miR editing in animals is the regulation of the miR maturation pathway, rather than the specificity of miR targeting [125]. Nevertheless, editing of mature miRs at seed region, such as for the most thoroughly studied mir-376, resulted in changes in the targeting profile that subsequently altered biological function in a tissue-specific manner [126] promoting carcinogenesis [128]. For another miR, mir-142, pri-miR editing resulted in impaired Drosha processing and enhanced degradation by the specific I-U nuclease Tudor-SN [129]. Recently, the adenosine deaminase ADAR1 was shown to differentiate from its deaminase activity and participate in RNAi when in heteroduplex with Dicer. Hence, when in complex with Dicer, it increased the rate of pre-miR cleave and facilitated miRISC loading of mature miRs, while in homodimer form, it mediated RNA editing [130].

Multiple isomiRs with various 5' and/or 3' ends are thought to be the result of sloppy Drosha and Dicer excision [131,132]. More recently, mammalian TRBP and its *Drosophila* ortholog Loqs have been shown to fine-tune Dicer cleavage sites for a subset of miRs generating longer miR isoforms by one nucleotide at either 5' or 3' ends [133,134]. In addition, it was shown that the hairpin loop and stem structure of the pre-miR affected Dicer-TRBP processing with different sensitivity compared with Dicer alone. The authors proposed that TRBP might induce a Dicer conformational change influencing Dicer substrate specificity and kinetics [134].

Post-Dicer processing by exoribonucleases modulates 3' shortening in miRs. Nibbler, a 3' to 5' exoribonuclease has been shown to trim Ago-bound miRs in *Drosophila*; depletion of Nibbler resulted in the loss of about a quarter of 3' isomiRs; unexpectedly, Han *et al.* also found that miRs are frequently produced by Dicer as intermediates that are longer than ~22 nt, and are subsequently trimmed to appropriate size by exoribonucleases [129,135]. It remains to be seen whether similar mechanisms exist in mammals.

Besides nucleotide excisions, post-Dicer 3' additions are widespread and conserved. These are mediated by several nucleotidyl transferases that catalyze the addition of ribonucleotides, most often adenine and uridine, to the ends of mature miR molecules [136,137]. Interestingly, these isomiRs are differentially expressed across development and different tissues. For instance, adenines are highly abundant in early *Drosophila* development, while a subset of miRs with uridines is expressed in adult tissues [138]. With respect to function, these 3' ribonucleotide additions have been shown to alter (enhance or lower) miR stability in some cases [138,139] and/or effectiveness in others [140,141]. However, the authors concluded that these effects are likely to be restricted to only a small subset of isomiRs in animals [140]. Like for differentially expressed splice mRNA variants, several isomiRs with 3' additions have been associated with human diseases. Thus far, significant alterations have been reported in cancer, Huntington's disease and pre-eclampsia [122,142,143].

4.4. ceRNAs and miR Degrading Enzymes

Recently, a new model of post-transcriptional regulation has emerged in which RNA targets are not merely passive substrates of miR repression, but cross-talk with each other in distinct networks by competing for shared miRs. Such competing endogenous RNAs (ceRNAs) ultimately determine mature miR availability and function within cells (reviewed in [144,145]). This reverse reasoning compels a redefinition of the idea that miRs stand at the top of mRNA networks to regulate protein output, by considering that any RNA that shares same target sequences actively regulate each other and miRs through direct competition for miR binding. Initial reports that provided proof of principle to this concept have shown that exogenous overexpression of 3' UTR sequences alone titrated cellular miR abundance and inactivated miR functions by freeing target mRNAs from repression [83,146,147]. Subsequently, it was shown that tenths of protein-coding mRNAs that share multiple miR target sites with dose-sensitive phosphatase and tensin homolog (PTEN) act as decoys to modulate PTEN levels [85,148]. An implication of these studies is that any RNA with miR target sites can potentially function as ceRNA. Thus, long non-coding RNAs (lncRNAs), due to their length, may be good candidates for sequestering miRs within cells. Hence, muscle-specific lncRNA, linc-MD1, was shown to sponge out two miRs to regulate the expression of transcription factors that activate muscle-specific gene expression [149]. Similarly, the PTENP1 pseudogene that is highly homologous to PTEN regulated cellular levels of PTEN (and the reverse) by sponging out common miRs [150]. Very recently, the repertoire of ceRNAs has been expanded by the identification of a new subclass of circular RNAs (circRNAs) [151,152]. Like other ceRNAs, these circRNAs serve as miR reservoirs. Distinctly, however, circRNAs may have multiple binding sites for specific miRs and therefore, are dedicated to sequestering particular miRs. Furthermore, being circulized, they possess enhanced stability by avoiding RNA exoribonuclease enzymes that act on 3' and 5' RNA ends and hence, maintain their effects for longer. An extreme case is characterized by human circRNA, ciRS-7, that harbors 74 mismatched mir-7 seed matches of which 63 are conserved in at least one other species [152]. This circRNA acts as a mir-7 sponge; it is resistant to miR-dependent destabilization and strongly suppresses miR-7 activity [151].

Exoribonucleases and the exosome have also been implicated in miR turnover. Using microarrays, Bail et al. have found that most miRs are remarkably stable (half-life over 8hrs), but some, including miR-382, were short-lived and were degraded to a modest extent (1.5-fold) by XRN1, a 5' to 3' exoribonuclease, and exosome, but not by XRN2 [153]. Moreover, overexpression of polynucleotide phosphorylase hPNPase(old-35), a 3' to 5' exoribonuclease, resulted in the downregulation of specific mature miRs in human melanoma cells without affecting their pri- or pre- miR levels [154].

5. Non-Canonical miR Activities

A relatively small number of studies have demonstrated that miRs can stimulate gene expression along their assigned repressive roles. These reports indicated that miR-mediated effects via gene promoters, extracellular receptors and 3' or 5' UTRs can be selective and controlled, ordained by either the miR sequence, associated proteins and/or cellular context.

5.1. Promoter Activation

Earlier studies have shown that exogenous application of small duplex RNAs, that are complementary to promoters, activate gene expression just like proteins and hormones, a phenomenon referred to as RNA activation (RNAa) [155,156]. Soon later, Dahiya's group discovered mir-373 target sites in the promoters of e-cadherin and cold shock domain containing protein C2 (CSDC2). miR-373 overexpression readily induced transcription of these two genes and this concurrent induction required mir-373 target sites in both promoters [157]. Subsequently, they showed that mir-205 sites are present in the promoter of interleukin (IL) tumor suppressor genes IL-24 and IL-32 and, similar to mir-373, mir-205 induced gene expression [157,158].

5.2. Target Activation

Several reports have shown that miRs can induce translation by binding to 5' or 3' UTR. In the brain, a target sequence of mir-346 was found in the 5' UTR of a splice variant of receptor-interacting protein 140 (RIP140). Gain- and loss- of-function studies established that mir-346 elevated RIP140 protein levels by facilitating association of its mRNA with the polysome fraction. Furthermore, the activity of the mir-346 did not require Ago2 indicating that other RNPs in complex with the miR or different RIP140 mRNA conformation induced by the miR mediated the effect [159]. In another study, mir-145 was shown to regulate smooth muscle cell fate and plasticity via upregulation of myocardin (*Myocd*). *Myocd* bears mir-145 sites in 3' UTR and mir-145 expression specifically upregulated luciferase expression by 150-fold; at the same time other mir-145 targets were repressed. It remains to be seen whether miR-145 interferes with binding of a destabilizing RBP to 3' UTR [160]. Along this, miR-466l, a miR discovered in mouse embryonic stem cells, upregulated IL-10 expression in TLR-triggered macrophages by antagonizing the RBP tristetraprolin (TTP)-mediated IL-10 mRNA degradation [161].

5.3. Receptors' Ligands

Members of the Toll-like receptor (TLR) family, mouse TLR7 and human TLR8, expressed on dendritic cells and B lymphocytes, physiologically recognize and bind ~20 nt viral single-stranded RNAs leading to their activation [162,163]. Because miRs are secreted in exosomes and are of similar size, it was predicted that they may also serve as TLR7/8 ligands. Indeed, Fabbri *et al.* identified that the tumor-secreted mir-21 and mir-29a were ligands for TLR7/8 and were capable of triggering a TLR-mediated prometastatic inflammatory response [164].

6. Conclusions

Over the past years, significant advances have been made into understanding how miRs interact with their RNA targets, and several key features, such as base-pair complementarity, local context factors and de/stabilization signals have been identified and finely analyzed as a result. The ultimate goal of all these studies has been to predict miR function through the identification of their targets. More recent analyses, however, demonstrated that local mRNA determinants could only explain a fraction of the miR repression activity and system level factors such as isomiRs, RBPs, and ceRNAs have been brought into attention. The very recent discovery that miRs can both regulate and be regulated by their RNA targets has presented a completely new twist into understanding the role of miRs in development and disease. It remains to be seen how miRs and RNA targets communicate using the miR nt sequence as a "language" to deliver large-scale concerted instructions in cells.

Acknowledgments

The Author wishes to thank Maria Paschou for assistance with the preparation of the figures. Work in the author's laboratory is funded by grants from the Greek General Secretariat for Research and Development (GSRT), Ministry of Education.

References

1.   Kozomara, A.; Griffiths-Jones, S. miRBase: Integrating microRNA annotation and deep-sequencing data. *Nucleic Acids Res.* **2011**, *39*, D152–D157.

2.   Chiang, H.R.; Schoenfeld, L.W.; Ruby, J.G.; Auyeung, V.C.; Spies, N.; Baek, D.; Johnston, W.K.; Russ, C.; Luo, S.; Babiarz, J.E.; *et al.* Mammalian microRNAs: Experimental evaluation of novel and previously annotated genes. *Genes Dev.* **2010**, *24*, 992–1009.

3.   Meunier, J.; Lemoine, F.; Soumillon, M.; Liechti, A.; Weier, M.; Guschanski, K.; Hu, H.; Khaitovich, P.; Kaessmann, H. Birth and expression evolution of mammalian microRNA genes. *Genome Res.* **2013**, *23*, 34–45.

4.   Friedman, R.C.; Farh, K.K.; Burge, C.B.; Bartel, D.P. Most mammalian mRNAs are conserved targets of microRNAs. *Genome Res.* **2009**, *19*, 92–105.

5.   Beuvink, I.; Kolb, F.A.; Budach, W.; Garnier, A.; Lange, J.; Natt, F.; Dengler, U.; Hall, J.; Filipowicz, W.; Weiler, J. A novel microarray approach reveals new tissue-specific signatures of known and predicted mammalian microRNAs. *Nucleic Acids Res.* **2007**, *35*, e52.

6.   Landgraf, P.; Rusu, M.; Sheridan, R.; Sewer, A.; Iovino, N.; Aravin, A.; Pfeffer, S.; Rice, A.; Kamphorst, A.O.; Landthaler, M.; *et al.* A mammalian microRNA expression atlas based on small RNA library sequencing. *Cell* **2007**, *129*, 1401–1414.

7.   Liang, Y.; Ridzon, D.; Wong, L.; Chen, C. Characterization of microRNA expression profiles in normal human tissues. *BMC Genomics* **2007**, *8*, 166.

8.   Moreau, M.P.; Bruse, S.E.; Jornsten, R.; Liu, Y.; Brzustowicz, L.M. Chronological changes in microRNA expression in the developing human brain. *PLoS One* **2013**, *8*, e60480.

9.   Lu, J.; Getz, G.; Miska, E.A.; Alvarez-Saavedra, E.; Lamb, J.; Peck, D.; Sweet-Cordero, A.; Ebert, B.L.; Mak, R.H.; Ferrando, A.A.; *et al.* microRNA expression profiles classify human cancers. *Nature* **2005**, *435*, 834–838.

10. Aldrich, B.T.; Frakes, E.P.; Kasuya, J.; Hammond, D.L.; Kitamoto, T. Changes in expression of sensory organ-specific microRNAs in rat dorsal root ganglia in association with mechanical hypersensitivity induced by spinal nerve ligation. *Neuroscience* **2009**, *164*, 711–723.

11. Elmen, J.; Lindow, M.; Schutz, S.; Lawrence, M.; Petri, A.; Obad, S.; Lindholm, M.; Hedtjarn, M.; Hansen, H.F.; Berger, U.; *et al.* LNA-mediated microRNA silencing in non-human primates. *Nature* **2008**, *452*, 896–899.

12. Elmen, J.; Lindow, M.; Silahtaroglu, A.; Bak, M.; Christensen, M.; Lind-Thomsen, A.; Hedtjarn, M.; Hansen, J.B.; Hansen, H.F.; Straarup, E.M.; *et al.* Antagonism of microRNA-122 in mice by systemically administered LNA-antimiR leads to up-regulation of a large set of predicted target mRNAs in the liver. *Nucleic Acids Res.* **2008**, *36*, 1153–1162.

13. Krutzfeldt, J.; Rajewsky, N.; Braich, R.; Rajeev, K.G.; Tuschl, T.; Manoharan, M.; Stoffel, M. Silencing of microRNAs *in vivo* with "antagomirs". *Nature* **2005**, *438*, 685–689.

14. Corcoran, D.L.; Pandit, K.V.; Gordon, B.; Bhattacharjee, A.; Kaminski, N.; Benos, P.V. Features of mammalian microRNA promoters emerge from polymerase II chromatin immunoprecipitation data. *PLoS One* **2009**, *4*, e5279.

15. Marson, A.; Levine, S.S.; Cole, M.F.; Frampton, G.M.; Brambrink, T.; Johnstone, S.; Guenther, M.G.; Johnston, W.K.; Wernig, M.; Newman, J.; *et al.* Connecting microRNA genes to the core transcriptional regulatory circuitry of embryonic stem cells. *Cell* **2008**, *134*, 521–533.

16. Ozsolak, F.; Poling, L.L.; Wang, Z.; Liu, H.; Liu, X.S.; Roeder, R.G.; Zhang, X.; Song, J.S.; Fisher, D.E. Chromatin structure analyses identify miRNA promoters. *Genes Dev.* **2008**, *22*, 3172–3183.

17. Saini, H.K.; Griffiths-Jones, S.; Enright, A.J. Genomic analysis of human microRNA transcripts. *Proc. Natl. Acad. Sci. USA* **2007**, *104*, 17719–17724.

18. Wang, X.; Xuan, Z.; Zhao, X.; Li, Y.; Zhang, M.Q. High-resolution human core-promoter prediction with CoreBoost_HM. *Genome Res.* **2009**, *19*, 266–275.

19. Zhou, X.; Ruan, J.; Wang, G.; Zhang, W. Characterization and identification of microRNA core promoters in four model species. *PLoS Comput. Biol.* **2007**, *3*, e37.

20. Rodriguez, A.; Griffiths-Jones, S.; Ashurst, J.L.; Bradley, A. Identification of mammalian microRNA host genes and transcription units. *Genome Res.* **2004**, *14*, 1902–1910.

21. Han, J.; Lee, Y.; Yeom, K.H.; Nam, J.W.; Heo, I.; Rhee, J.K.; Sohn, S.Y.; Cho, Y.; Zhang, B.T.; Kim, V.N. Molecular basis for the recognition of primary microRNAs by the Drosha-DGCR8 complex. *Cell* **2006**, *125*, 887–901.

22. Zeng, Y.; Cullen, B.R. Efficient processing of primary microRNA hairpins by Drosha requires flanking nonstructured RNA sequences. *J. Biol. Chem.* **2005**, *280*, 27595–27603.

23. Basyuk, E.; Suavet, F.; Doglio, A.; Bordonne, R.; Bertrand, E. Human let-7 stem-loop precursors harbor features of RNase III cleavage products. *Nucleic Acids Res.* **2003**, *31*, 6593–6597.

24. Denli, A.M.; Tops, B.B.; Plasterk, R.H.; Ketting, R.F.; Hannon, G.J. Processing of primary microRNAs by the Microprocessor complex. *Nature* **2004**, *432*, 231–235.

25. Gregory, R.I.; Yan, K.P.; Amuthan, G.; Chendrimada, T.; Doratotaj, B.; Cooch, N.; Shiekhattar, R. The Microprocessor complex mediates the genesis of microRNAs. *Nature* **2004**, *432*, 235–240.

26. Lee, Y.; Ahn, C.; Han, J.; Choi, H.; Kim, J.; Yim, J.; Lee, J.; Provost, P.; Radmark, O.; Kim, S.; *et al.* The nuclear RNase III Drosha initiates microRNA processing. *Nature* **2003**, *425*, 415–419.

27. Bernstein, E.; Caudy, A.A.; Hammond, S.M.; Hannon, G.J. Role for a bidentate ribonuclease in the initiation step of RNA interference. *Nature* **2001**, *409*, 363–366.

28. Chakravarthy, S.; Sternberg, S.H.; Kellenberger, C.A.; Doudna, J.A. Substrate-specific kinetics of Dicer-catalyzed RNA processing. *J. Mol. Biol.* **2010**, *404*, 392–402.

29. Haase, A.D.; Jaskiewicz, L.; Zhang, H.; Laine, S.; Sack, R.; Gatignol, A.; Filipowicz, W. TRBP, a regulator of cellular PKR and HIV-1 virus expression, interacts with Dicer and functions in RNA silencing. *EMBO Rep.* **2005**, *6*, 961–967.

30. Hutvagner, G.; McLachlan, J.; Pasquinelli, A.E.; Balint, E.; Tuschl, T.; Zamore, P.D. A cellular function for the RNA-interference enzyme Dicer in the maturation of the let-7 small temporal RNA. *Science* **2001**, *293*, 834–838.

31. Lee, H.Y.; Zhou, K.; Smith, A.M.; Noland, C.L.; Doudna, J.A. Differential roles of human Dicer-binding proteins TRBP and PACT in small RNA processing. *Nucleic Acids Res.* **2013**, doi:10.1093/nar/gkt361.

32. Lee, Y.; Hur, I.; Park, S.Y.; Kim, Y.K.; Suh, M.R.; Kim, V.N. The role of PACT in the RNA silencing pathway. *EMBO J.* **2006**, *25*, 522–532.

33. MacRae, I.J.; Zhou, K.; Doudna, J.A. Structural determinants of RNA recognition and cleavage by Dicer. *Nat. Struct. Mol. Biol.* **2007**, *14*, 934–940.

34. Chendrimada, T.P.; Gregory, R.I.; Kumaraswamy, E.; Norman, J.; Cooch, N.; Nishikura, K.; Shiekhattar, R. TRBP recruits the Dicer complex to Ago2 for microRNA processing and gene silencing. *Nature* **2005**, *436*, 740–744.

35. Khvorova, A.; Reynolds, A.; Jayasena, S.D. Functional siRNAs and miRNAs exhibit strand bias. *Cell* **2003**, *115*, 209–216.

36. Krol, J.; Sobczak, K.; Wilczynska, U.; Drath, M.; Jasinska, A.; Kaczynska, D.; Krzyzosiak, W.J. Structural features of microRNA (miRNA) precursors and their relevance to miRNA biogenesis and small interfering RNA/short hairpin RNA design. *J. Biol. Chem.* **2004**, *279*, 42230–42239.

37. Schwarz, D.S.; Hutvagner, G.; Du, T.; Xu, Z.; Aronin, N.; Zamore, P.D. Asymmetry in the assembly of the RNAi enzyme complex. *Cell* **2003**, *115*, 199–208.

38. Czech, B.; Zhou, R.; Erlich, Y.; Brennecke, J.; Binari, R.; Villalta, C.; Gordon, A.; Perrimon, N.; Hannon, G.J. Hierarchical rules for Argonaute loading in *Drosophila*. *Mol. Cell* **2009**, *36*, 445–456.

39. Hu, H.Y.; Yan, Z.; Xu, Y.; Hu, H.; Menzel, C.; Zhou, Y.H.; Chen, W.; Khaitovich, P. Sequence features associated with microRNA strand selection in humans and flies. *BMC Genomics* **2009**, *10*, 413.

40. Noland, C.L.; Doudna, J.A. Multiple sensors ensure guide strand selection in human RNAi pathways. *RNA* **2013**, *19*, 639–648.

41. Okamura, K.; Liu, N.; Lai, E.C. Distinct mechanisms for microRNA strand selection by *Drosophila* Argonautes. *Mol. Cell* **2009**, *36*, 431–444.

42. Ameres, S.L.; Martinez, J.; Schroeder, R. Molecular basis for target RNA recognition and cleavage by human RISC. *Cell* **2007**, *130*, 101–112.

43. Fabian, M.R.; Sonenberg, N. The mechanics of miRNA-mediated gene silencing: A look under the hood of miRISC. *Nat. Struct. Mol. Biol.* **2012**, *19*, 586–593.

44. Huntzinger, E.; Kuzuoglu-Ozturk, D.; Braun, J.E.; Eulalio, A.; Wohlbold, L.; Izaurralde, E. The interactions of GW182 proteins with PABP and deadenylases are required for both translational repression and degradation of miRNA targets. *Nucleic Acids Res.* **2013**, *41*, 978–994.

45. Zekri, L.; Kuzuoglu-Ozturk, D.; Izaurralde, E. GW182 proteins cause PABP dissociation from silenced miRNA targets in the absence of deadenylation. *EMBO J.* **2013**, *32*, 1052–1065.

46. Brennecke, J.; Stark, A.; Russell, R.B.; Cohen, S.M. Principles of microRNA-target recognition. *PLoS Biol.* **2005**, *3*, e85.

47. Krek, A.; Grun, D.; Poy, M.N.; Wolf, R.; Rosenberg, L.; Epstein, E.J.; MacMenamin, P.; da Piedade, I.; Gunsalus, K.C.; Stoffel, M.; *et al.* Combinatorial microRNA target predictions. *Nat. Genet.* **2005**, *37*, 495–500.

48. Lewis, B.P.; Burge, C.B.; Bartel, D.P. Conserved seed pairing, often flanked by adenosines, indicates that thousands of human genes are microRNA targets. *Cell* **2005**, *120*, 15–20.

49. Lewis, B.P.; Shih, I.H.; Jones-Rhoades, M.W.; Bartel, D.P.; Burge, C.B. Prediction of mammalian microRNA targets. *Cell* **2003**, *115*, 787–798.

50. Elkayam, E.; Kuhn, C.D.; Tocilj, A.; Haase, A.D.; Greene, E.M.; Hannon, G.J.; Joshua-Tor, L. The structure of human argonaute-2 in complex with miR-20a. *Cell* **2012**, *150*, 100–110.

51. Schirle, N.T.; MacRae, I.J. The crystal structure of human Argonaute 2. *Science* **2012**, *336*, 1037–1040.

52. Wang, Y.; Sheng, G.; Juranek, S.; Tuschl, T.; Patel, D.J. Structure of the guide-strand-containing argonaute silencing complex. *Nature* **2008**, *456*, 209–213.

53. Chi, S.W.; Hannon, G.J.; Darnell, R.B. An alternative mode of microRNA target recognition. *Nat. Struct. Mol. Biol.* **2012**, *19*, 321–327.

54. Miranda, K.C.; Huynh, T.; Tay, Y.; Ang, Y.S.; Tam, W.L.; Thomson, A.M.; Lim, B.; Rigoutsos, I. A pattern-based method for the identification of MicroRNA binding sites and their corresponding heteroduplexes. *Cell* **2006**, *126*, 1203–1217.

55. Vella, M.C.; Choi, E.Y.; Lin, S.Y.; Reinert, K.; Slack, F.J. The C. elegans microRNA let-7 binds to imperfect let-7 complementary sites from the lin-41 3' UTR. *Genes Dev.* **2004**, *18*, 132–137.

56. Reinhart, B.J.; Slack, F.J.; Basson, M.; Pasquinelli, A.E.; Bettinger, J.C.; Rougvie, A.E.; Horvitz, H.R.; Ruvkun, G. The 21-nucleotide let-7 RNA regulates developmental timing in Caenorhabditis elegans. *Nature* **2000**, *403*, 901–906.

57. Shin, C.; Nam, J.W.; Farh, K.K.; Chiang, H.R.; Shkumatava, A.; Bartel, D.P. Expanding the microRNA targeting code: functional sites with centered pairing. *Mol. Cell* **2010**, *38*, 789–802.

58. Lal, A.; Navarro, F.; Maher, C.A.; Maliszewski, L.E.; Yan, N.; O'Day, E.; Chowdhury, D.; Dykxhoorn, D.M.; Tsai, P.; Hofmann, O.; *et al.* miR-24 inhibits cell proliferation by targeting E2F2, MYC, and other cell-cycle genes via binding to "seedless" 3' UTR microRNA recognition elements. *Mol. Cell* **2009**, *35*, 610–625.

59. Tay, Y.; Zhang, J.; Thomson, A.M.; Lim, B.; Rigoutsos, I. microRNAs to Nanog, Oct4 and Sox2 coding regions modulate embryonic stem cell differentiation. *Nature* **2008**, *455*, 1124–1128.

60. Didiano, D.; Hobert, O. Perfect seed pairing is not a generally reliable predictor for miRNA-target interactions. *Nat. Struct. Mol. Biol.* **2006**, *13*, 849–851.

61. Grimson, A.; Farh, K.K.; Johnston, W.K.; Garrett-Engele, P.; Lim, L.P.; Bartel, D.P. microRNA targeting specificity in mammals: Determinants beyond seed pairing. *Mol. Cell* **2007**, *27*, 91–105.

62. Hon, L.S.; Zhang, Z. The roles of binding site arrangement and combinatorial targeting in microRNA repression of gene expression. *Genome Biol.* **2007**, *8*, R166.

63. Saetrom, P.; Heale, B.S.; Snove, O., Jr.; Aagaard, L.; Alluin, J.; Rossi, J.J. Distance constraints between microRNA target sites dictate efficacy and cooperativity. *Nucleic Acids Res.* **2007**, *35*, 2333–2342.

64. Doxakis, E. Post-transcriptional regulation of alpha-synuclein expression by mir-7 and mir-153. *J. Biol. Chem.* **2010**, *285*, 12726–12734.

65. Paschou, M.; Doxakis, E. Neurofibromin 1 is a miRNA target in neurons. *PLoS One* **2012**, *7*, e46773.

66. Chi, S.W.; Zang, J.B.; Mele, A.; Darnell, R.B. Argonaute HITS-CLIP decodes microRNA-mRNA interaction maps. *Nature* **2009**, *460*, 479–486.

67. Hafner, M.; Landthaler, M.; Burger, L.; Khorshid, M.; Hausser, J.; Berninger, P.; Rothballer, A.; Ascano, M., Jr.; Jungkamp, A.C.; Munschauer, M.; *et al.* Transcriptome-wide identification of RNA-binding protein and microRNA target sites by PAR-CLIP. *Cell* **2010**, *141*, 129–141.

68. Leung, A.K.; Vyas, S.; Rood, J.E.; Bhutkar, A.; Sharp, P.A.; Chang, P. Poly(ADP-ribose) regulates stress responses and microRNA activity in the cytoplasm. *Mol. Cell* **2011**, *42*, 489–499.

69. Rigoutsos, I. New tricks for animal microRNAS: Targeting of amino acid coding regions at conserved and nonconserved sites. *Cancer Res.* **2009**, *69*, 3245–3248.

70. Zisoulis, D.G.; Lovci, M.T.; Wilbert, M.L.; Hutt, K.R.; Liang, T.Y.; Pasquinelli, A.E.; Yeo, G.W. Comprehensive discovery of endogenous Argonaute binding sites in Caenorhabditis elegans. *Nat. Struct. Mol. Biol.* **2010**, *17*, 173–179.

71. Baek, D.; Villen, J.; Shin, C.; Camargo, F.D.; Gygi, S.P.; Bartel, D.P. The impact of microRNAs on protein output. *Nature* **2008**, *455*, 64–71.

72. Selbach, M.; Schwanhausser, B.; Thierfelder, N.; Fang, Z.; Khanin, R.; Rajewsky, N. Widespread changes in protein synthesis induced by microRNAs. *Nature* **2008**, *455*, 58–63.

73. Hausser, J.; Syed, A.P.; Bilen, B.; Zavolan, M. Analysis of CDS-located miRNA target sites suggests that they can effectively inhibit translation. *Genome Res.* **2013**, *23*, 604–615.

74. Schnall-Levin, M.; Rissland, O.S.; Johnston, W.K.; Perrimon, N.; Bartel, D.P.; Berger, B. Unusually effective microRNA targeting within repeat-rich coding regions of mammalian mRNAs. *Genome Res.* **2011**, *21*, 1395–1403.

75. Fang, Z.; Rajewsky, N. The impact of miRNA target sites in coding sequences and in 3' UTRs. *PLoS One* **2011**, *6*, e18067.

76. Wang, W.X.; Wilfred, B.R.; Xie, K.; Jennings, M.H.; Hu, Y.H.; Stromberg, A.J.; Nelson, P.T. Individual microRNAs (miRNAs) display distinct mRNA targeting "rules". *RNA Biol.* **2010**, *7*, 373–380.

77. Kertesz, M.; Iovino, N.; Unnerstall, U.; Gaul, U.; Segal, E. The role of site accessibility in microRNA target recognition. *Nat. Genet.* **2007**, *39*, 1278–1284.

78. Long, D.; Lee, R.; Williams, P.; Chan, C.Y.; Ambros, V.; Ding, Y. Potent effect of target structure on microRNA function. *Nat. Struct. Mol. Biol.* **2007**, *14*, 287–294.

79. Paschou, M.; Paraskevopoulou, M.D.; Vlachos, I.S.; Koukouraki, P.; Hatzigeorgiou, A.G.; Doxakis, E. miRNA regulons associated with synaptic function. *PLoS One* **2012**, *7*, e46189.

80. Bergauer, T.; Krueger, U.; Lader, E.; Pilk, S.; Wolter, I.; Bielke, W. Analysis of putative miRNA binding sites and mRNA 3' ends as targets for siRNA-mediated gene knockdown. *Oligonucleotides* **2009**, *19*, 41–52.

81. Ji, Z.; Lee, J.Y.; Pan, Z.; Jiang, B.; Tian, B. Progressive lengthening of 3' untranslated regions of mRNAs by alternative polyadenylation during mouse embryonic development. *Proc. Natl. Acad. Sci. USA* **2009**, *106*, 7028–7033.

82. Sandberg, R.; Neilson, J.R.; Sarma, A.; Sharp, P.A.; Burge, C.B. Proliferating cells express mRNAs with shortened 3' untranslated regions and fewer microRNA target sites. *Science* **2008**, *320*, 1643–1647.

83. Arvey, A.; Larsson, E.; Sander, C.; Leslie, C.S.; Marks, D.S. Target mRNA abundance dilutes microRNA and siRNA activity. *Mol. Syst. Biol.* **2010**, *6*, 363.

84. Karreth, F.A.; Tay, Y.; Perna, D.; Ala, U.; Tan, S.M.; Rust, A.G.; de Nicola, G.; Webster, K.A.; Weiss, D.; Perez-Mancera, P.A.; *et al. In vivo* identification of tumor-suppressive PTEN ceRNAs in an oncogenic BRAF-induced mouse model of melanoma. *Cell* **2011**, *147*, 382–395.

85. Sumazin, P.; Yang, X.; Chiu, H.S.; Chung, W.J.; Iyer, A.; Llobet-Navas, D.; Rajbhandari, P.; Bansal, M.; Guarnieri, P.; Silva, J.; *et al.* An extensive microRNA-mediated network of RNA-RNA interactions regulates established oncogenic pathways in glioblastoma. *Cell* **2011**, *147*, 370–381.

86. Ameres, S.L.; Horwich, M.D.; Hung, J.H.; Xu, J.; Ghildiyal, M.; Weng, Z.; Zamore, P.D. Target RNA-directed trimming and tailing of small silencing RNAs. *Science* **2010**, *328*, 1534–1539.

87. Rybak, A.; Fuchs, H.; Hadian, K.; Smirnova, L.; Wulczyn, E.A.; Michel, G.; Nitsch, R.; Krappmann, D.; Wulczyn, F.G. The let-7 target gene mouse lin-41 is a stem cell specific E3 ubiquitin ligase for the miRNA pathway protein Ago2. *Nat. Cell Biol.* **2009**, *11*, 1411–1420.

88. Pullmann, R., Jr.; Kim, H.H.; Abdelmohsen, K.; Lal, A.; Martindale, J.L.; Yang, X.; Gorospe, M. Analysis of turnover and translation regulatory RNA-binding protein expression through binding to cognate mRNAs. *Mol. Cell. Biol.* **2007**, *27*, 6265–6278.

89. Jacobsen, A.; Wen, J.; Marks, D.S.; Krogh, A. Signatures of RNA binding proteins globally coupled to effective microRNA target sites. *Genome Res.* **2010**, *20*, 1010–1019.

90. Larsson, E.; Sander, C.; Marks, D. mRNA turnover rate limits siRNA and microRNA efficacy. *Mol. Syst. Biol.* **2010**, *6*, 433.

91. Lebedeva, S.; Jens, M.; Theil, K.; Schwanhausser, B.; Selbach, M.; Landthaler, M.; Rajewsky, N. Transcriptome-wide analysis of regulatory interactions of the RNA-binding protein HuR. *Mol. Cell* **2011**, *43*, 340–352.

92. Mukherjee, N.; Corcoran, D.L.; Nusbaum, J.D.; Reid, D.W.; Georgiev, S.; Hafner, M.; Ascano, M., Jr.; Tuschl, T.; Ohler, U.; Keene, J.D. Integrative regulatory mapping indicates that the RNA-binding protein HuR couples pre-mRNA processing and mRNA stability. *Mol. Cell* **2011**, *43*, 327–339.

93. Srikantan, S.; Tominaga, K.; Gorospe, M. Functional interplay between RNA-binding protein HuR and microRNAs. *Curr. Protein Pept. Sci.* **2012**, *13*, 372–379.

94. Abdelmohsen, K.; Srikantan, S.; Kuwano, Y.; Gorospe, M. miR-519 reduces cell proliferation by lowering RNA-binding protein HuR levels. *Proc. Natl. Acad. Sci. USA* **2008**, *105*, 20297–20302.

95. Guo, X.; Wu, Y.; Hartley, R.S. microRNA-125a represses cell growth by targeting HuR in breast cancer. *RNA Biol.* **2009**, *6*, 575–583.

96. Young, L.E.; Moore, A.E.; Sokol, L.; Meisner-Kober, N.; Dixon, D.A. The mRNA stability factor HuR inhibits microRNA-16 targeting of COX-2. *Mol. Cancer Res.* **2012**, *10*, 167–180.

97. Borchert, G.M.; Lanier, W.; Davidson, B.L. RNA polymerase III transcribes human microRNAs. *Nat. Struct. Mol. Biol.* **2006**, *13*, 1097–1101.

98. Dieci, G.; Fiorino, G.; Castelnuovo, M.; Teichmann, M.; Pagano, A. The expanding RNA polymerase III transcriptome. *Trends Genet.* **2007**, *23*, 614–622.

99. Lee, Y.; Kim, M.; Han, J.; Yeom, K.H.; Lee, S.; Baek, S.H.; Kim, V.N. microRNA genes are transcribed by RNA polymerase II. *EMBO J.* **2004**, *23*, 4051–4060.

100. Carlson, D.P.; Ross, J. Human beta-globin promoter and coding sequences transcribed by RNA polymerase III. *Cell* **1983**, *34*, 857–864.

101. Chung, J.; Sussman, D.J.; Zeller, R.; Leder, P. The c-myc gene encodes superimposed RNA polymerase II and III promoters. *Cell* **1987**, *51*, 1001–1008.

102. Listerman, I.; Bledau, A.S.; Grishina, I.; Neugebauer, K.M. Extragenic accumulation of RNA polymerase II enhances transcription by RNA polymerase III. *PLoS Genet.* **2007**, *3*, e212.

103. Felton-Edkins, Z.A.; Kenneth, N.S.; Brown, T.R.; Daly, N.L.; Gomez-Roman, N.; Grandori, C.; Eisenman, R.N.; White, R.J. Direct regulation of RNA polymerase III transcription by RB, p53 and c-Myc. *Cell Cycle* **2003**, *2*, 181–184.

104. Baskerville, S.; Bartel, D.P. Microarray profiling of microRNAs reveals frequent coexpression with neighboring miRNAs and host genes. *RNA* **2005**, *11*, 241–247.

105. Fujita, S.; Ito, T.; Mizutani, T.; Minoguchi, S.; Yamamichi, N.; Sakurai, K.; Iba, H. miR-21 gene expression triggered by AP-1 is sustained through a double-negative feedback mechanism. *J. Mol. Biol.* **2008**, *378*, 492–504.

106. He, L.; Thomson, J.M.; Hemann, M.T.; Hernando-Monge, E.; Mu, D.; Goodson, S.; Powers, S.; Cordon-Cardo, C.; Lowe, S.W.; Hannon, G.J.; *et al.* A microRNA polycistron as a potential human oncogene. *Nature* **2005**, *435*, 828–833.

107. O'Donnell, K.A.; Wentzel, E.A.; Zeller, K.I.; Dang, C.V.; Mendell, J.T. c-Myc-regulated microRNAs modulate E2F1 expression. *Nature* **2005**, *435*, 839–843.

108. Monteys, A.M.; Spengler, R.M.; Wan, J.; Tecedor, L.; Lennox, K.A.; Xing, Y.; Davidson, B.L. Structure and activity of putative intronic miRNA promoters. *RNA* **2010**, *16*, 495–505.

109. Yuan, X.; Liu, C.; Yang, P.; He, S.; Liao, Q.; Kang, S.; Zhao, Y. Clustered microRNAs' coordination in regulating protein-protein interaction network. *BMC Syst. Biol.* **2009**, *3*, 65.

110. Gusev, Y. Computational methods for analysis of cellular functions and pathways collectively targeted by differentially expressed microRNA. *Methods* **2008**, *44*, 61–72.

111. Davis, B.N.; Hilyard, A.C.; Lagna, G.; Hata, A. SMAD proteins control DROSHA-mediated microRNA maturation. *Nature* **2008**, *454*, 56–61.

112. Davis, B.N.; Hilyard, A.C.; Nguyen, P.H.; Lagna, G.; Hata, A. Smad proteins bind a conserved RNA sequence to promote microRNA maturation by Drosha. *Mol. Cell* **2010**, *39*, 373–384.

113. Suzuki, H.I.; Yamagata, K.; Sugimoto, K.; Iwamoto, T.; Kato, S.; Miyazono, K. Modulation of microRNA processing by p53. *Nature* **2009**, *460*, 529–533.

114. Guil, S.; Caceres, J.F. The multifunctional RNA-binding protein hnRNP A1 is required for processing of miR-18a. *Nat. Struct. Mol. Biol.* **2007**, *14*, 591–596.

115. Michlewski, G.; Caceres, J.F. Antagonistic role of hnRNP A1 and KSRP in the regulation of let-7a biogenesis. *Nat. Struct. Mol. Biol.* **2010**, *17*, 1011–1018.

116. Trabucchi, M.; Briata, P.; Garcia-Mayoral, M.; Haase, A.D.; Filipowicz, W.; Ramos, A.; Gherzi, R.; Rosenfeld, M.G. The RNA-binding protein KSRP promotes the biogenesis of a subset of microRNAs. *Nature* **2009**, *459*, 1010–1014.

117. Chen, K.; Rajewsky, N. Natural selection on human microRNA binding sites inferred from SNP data. *Nat. Genet.* **2006**, *38*, 1452–1456.

118. Muinos-Gimeno, M.; Montfort, M.; Bayes, M.; Estivill, X.; Espinosa-Parrilla, Y. Design and evaluation of a panel of single-nucleotide polymorphisms in microRNA genomic regions for association studies in human disease. *Eur. J. Hum. Genet.* **2010**, *18*, 218–226.

119. Saunders, M.A.; Liang, H.; Li, W.H. Human polymorphism at microRNAs and microRNA target sites. *Proc. Natl. Acad. Sci. USA* **2007**, *104*, 3300–3305.

120. Zorc, M.; Skok, D.J.; Godnic, I.; Calin, G.A.; Horvat, S.; Jiang, Z.; Dovc, P.; Kunej, T. Catalog of microRNA seed polymorphisms in vertebrates. *PLoS One* **2012**, *7*, e30737.

121. Ryan, B.M.; Robles, A.I.; Harris, C.C. Genetic variation in microRNA networks: The implications for cancer research. *Nat. Rev. Cancer* **2010**, *10*, 389–402.

122. Marti, E.; Pantano, L.; Banez-Coronel, M.; Llorens, F.; Minones-Moyano, E.; Porta, S.; Sumoy, L.; Ferrer, I.; Estivill, X. A myriad of miRNA variants in control and Huntington's disease brain regions detected by massively parallel sequencing. *Nucleic Acids Res.* **2010**, *38*, 7219–7235.

123. Morin, R.D.; O'Connor, M.D.; Griffith, M.; Kuchenbauer, F.; Delaney, A.; Prabhu, A.L.; Zhao, Y.; McDonald, H.; Zeng, T.; Hirst, M.; *et al*. Application of massively parallel sequencing to microRNA profiling and discovery in human embryonic stem cells. *Genome Res.* **2008**, *18*, 610–621.

124. Nishikura, K. Functions and regulation of RNA editing by ADAR deaminases. *Annu. Rev. Biochem.* **2010**, *79*, 321–349.

125. De Hoon, M.J.; Taft, R.J.; Hashimoto, T.; Kanamori-Katayama, M.; Kawaji, H.; Kawano, M.; Kishima, M.; Lassmann, T.; Faulkner, G.J.; Mattick, J.S.; *et al*. Cross-mapping and the identification of editing sites in mature microRNAs in high-throughput sequencing libraries. *Genome Res.* **2010**, *20*, 257–264.

126. Kawahara, Y.; Megraw, M.; Kreider, E.; Iizasa, H.; Valente, L.; Hatzigeorgiou, A.G.; Nishikura, K. Frequency and fate of microRNA editing in human brain. *Nucleic Acids Res.* **2008**, *36*, 5270–5280.

127. Peng, Z.; Cheng, Y.; Tan, B.C.; Kang, L.; Tian, Z.; Zhu, Y.; Zhang, W.; Liang, Y.; Hu, X.; Tan, X.; *et al*. Comprehensive analysis of RNA-Seq data reveals extensive RNA editing in a human transcriptome. *Nat. Biotechnol.* **2012**, *30*, 253–260.

128. Choudhury, Y.; Tay, F.C.; Lam, D.H.; Sandanaraj, E.; Tang, C.; Ang, B.T.; Wang, S. Attenuated adenosine-to-inosine editing of microRNA-376a* promotes invasiveness of glioblastoma cells. *J. Clin. Invest.* **2012**, *122*, 4059–4076.

129. Liu, N.; Abe, M.; Sabin, L.R.; Hendriks, G.J.; Naqvi, A.S.; Yu, Z.; Cherry, S.; Bonini, N.M. The exoribonuclease Nibbler controls 3' end processing of microRNAs in *Drosophila*. *Curr. Biol.* **2011**, *21*, 1888–1893.

130. Ota, H.; Sakurai, M.; Gupta, R.; Valente, L.; Wulff, B.E.; Ariyoshi, K.; Iizasa, H.; Davuluri, R.V.; Nishikura, K. ADAR1 Forms a complex with Dicer to spromote microRNA processing and RNA-induced gene silencing. *Cell* **2013**, *153*, 575–589.

131. Ruby, J.G.; Jan, C.; Player, C.; Axtell, M.J.; Lee, W.; Nusbaum, C.; Ge, H.; Bartel, D.P. Large-scale sequencing reveals 21U-RNAs and additional microRNAs and endogenous siRNAs in *C. elegans*. *Cell* **2006**, *127*, 1193–1207.

132. Ruby, J.G.; Stark, A.; Johnston, W.K.; Kellis, M.; Bartel, D.P.; Lai, E.C. Evolution, biogenesis, expression, and target predictions of a substantially expanded set of *Drosophila* microRNAs. *Genome Res.* **2007**, *17*, 1850–1864.

133. Fukunaga, R.; Han, B.W.; Hung, J.H.; Xu, J.; Weng, Z.; Zamore, P.D. Dicer partner proteins tune the length of mature miRNAs in flies and mammals. *Cell* **2012**, *151*, 533–546.

134. Lee, H.Y.; Doudna, J.A. TRBP alters human precursor microRNA processing *in vitro*. *RNA* **2012**, *18*, 2012–2019.

135. Han, B.W.; Hung, J.H.; Weng, Z.; Zamore, P.D.; Ameres, S.L. The 3'-to-5' exoribonuclease Nibbler shapes the 3' ends of microRNAs bound to *Drosophila* Argonaute1. *Curr. Biol.* **2011**, *21*, 1878–1887.

136. Martin, G.; Keller, W. RNA-specific ribonucleotidyl transferases. *RNA* **2007**, *13*, 1834–1849.

137. Wyman, S.K.; Knouf, E.C.; Parkin, R.K.; Fritz, B.R.; Lin, D.W.; Dennis, L.M.; Krouse, M.A.; Webster, P.J.; Tewari, M. Post-transcriptional generation of miRNA variants by multiple nucleotidyl transferases contributes to miRNA transcriptome complexity. *Genome Res.* **2011**, *21*, 1450–1461.

138. Fernandez-Valverde, S.L.; Taft, R.J.; Mattick, J.S. Dynamic isomiR regulation in *Drosophila* development. *RNA* **2010**, *16*, 1881–1888.

139. Katoh, T.; Sakaguchi, Y.; Miyauchi, K.; Suzuki, T.; Kashiwabara, S.; Baba, T. Selective stabilization of mammalian microRNAs by 3' adenylation mediated by the cytoplasmic poly(A) polymerase GLD-2. *Genes Dev.* **2009**, *23*, 433–438.

140. Burroughs, A.M.; Ando, Y.; de Hoon, M.J.; Tomaru, Y.; Nishibu, T.; Ukekawa, R.; Funakoshi, T.; Kurokawa, T.; Suzuki, H.; Hayashizaki, Y.; *et al.* A comprehensive survey of 3' animal miRNA modification events and a possible role for 3' adenylation in modulating miRNA targeting effectiveness. *Genome Res.* **2010**, *20*, 1398–1410.

141. Jones, M.R.; Quinton, L.J.; Blahna, M.T.; Neilson, J.R.; Fu, S.; Ivanov, A.R.; Wolf, D.A.; Mizgerd, J.P. Zcchc11-dependent uridylation of microRNA directs cytokine expression. *Nat. Cell Biol.* **2009**, *11*, 1157–1163.

142. Guo, L.; Yang, Q.; Lu, J.; Li, H.; Ge, Q.; Gu, W.; Bai, Y.; Lu, Z. A comprehensive survey of miRNA repertoire and 3' addition events in the placentas of patients with pre-eclampsia from high-throughput sequencing. *PLoS One* **2011**, *6*, e21072.

143. Kuchenbauer, F.; Morin, R.D.; Argiropoulos, B.; Petriv, O.I.; Griffith, M.; Heuser, M.; Yung, E.; Piper, J.; Delaney, A.; Prabhu, A.L.; *et al.* In-depth characterization of the microRNA transcriptome in a leukemia progression model. *Genome Res.* **2008**, *18*, 1787–1797.

144. Cesana, M.; Daley, G.Q. Deciphering the rules of ceRNA networks. *Proc. Natl. Acad. Sci. USA* **2013**, *110*, 7112–7113.

145. Salmena, L.; Poliseno, L.; Tay, Y.; Kats, L.; Pandolfi, P.P. A ceRNA hypothesis: The Rosetta Stone of a hidden RNA language? *Cell* **2011**, *146*, 353–358.

146. Jeyapalan, Z.; Deng, Z.; Shatseva, T.; Fang, L.; He, C.; Yang, B.B. Expression of CD44 3'-untranslated region regulates endogenous microRNA functions in tumorigenesis and angiogenesis. *Nucleic Acids Res.* **2011**, *39*, 3026–3041.

147. Lee, D.Y.; Shatseva, T.; Jeyapalan, Z.; Du, W.W.; Deng, Z.; Yang, B.B. A 3'-untranslated region (3' UTR) induces organ adhesion by regulating miR-199a* functions. *PLoS One* **2009**, *4*, e4527.

148. Tay, Y.; Kats, L.; Salmena, L.; Weiss, D.; Tan, S.M.; Ala, U.; Karreth, F.; Poliseno, L.; Provero, P.; di Cunto, F.; *et al.* Coding-independent regulation of the tumor suppressor PTEN by competing endogenous mRNAs. *Cell* **2011**, *147*, 344–357.

149. Cesana, M.; Cacchiarelli, D.; Legnini, I.; Santini, T.; Sthandier, O.; Chinappi, M.; Tramontano, A.; Bozzoni, I. A long noncoding RNA controls muscle differentiation by functioning as a competing endogenous RNA. *Cell* **2011**, *147*, 358–369.

150. Poliseno, L.; Salmena, L.; Zhang, J.; Carver, B.; Haveman, W.J.; Pandolfi, P.P. A coding-independent function of gene and pseudogene mRNAs regulates tumour biology. *Nature* **2010**, *465*, 1033–1038.

151. Hansen, T.B.; Jensen, T.I.; Clausen, B.H.; Bramsen, J.B.; Finsen, B.; Damgaard, C.K.; Kjems, J. Natural RNA circles function as efficient microRNA sponges. *Nature* **2013**, *495*, 384–388.

152. Memczak, S.; Jens, M.; Elefsinioti, A.; Torti, F.; Krueger, J.; Rybak, A.; Maier, L.; Mackowiak, S.D.; Gregersen, L.H.; Munschauer, M.; *et al.* Circular RNAs are a large class of animal RNAs with regulatory potency. *Nature* **2013**, *495*, 333–338.

153. Bail, S.; Swerdel, M.; Liu, H.; Jiao, X.; Goff, L.A.; Hart, R.P.; Kiledjian, M. Differential regulation of microRNA stability. *RNA* **2010**, *16*, 1032–1039.

154. Das, S.K.; Sokhi, U.K.; Bhutia, S.K.; Azab, B.; Su, Z.Z.; Sarkar, D.; Fisher, P.B. Human polynucleotide phosphorylase selectively and preferentially degrades microRNA-221 in human melanoma cells. *Proc. Natl. Acad. Sci. USA* **2010**, *107*, 11948–11953.

155. Janowski, B.A.; Younger, S.T.; Hardy, D.B.; Ram, R.; Huffman, K.E.; Corey, D.R. Activating gene expression in mammalian cells with promoter-targeted duplex RNAs. *Nat. Chem. Biol.* **2007**, *3*, 166–173.

156. Li, L.C.; Okino, S.T.; Zhao, H.; Pookot, D.; Place, R.F.; Urakami, S.; Enokida, H.; Dahiya, R. Small dsRNAs induce transcriptional activation in human cells. *Proc. Natl. Acad. Sci. USA* **2006**, *103*, 17337–17342.

157. Place, R.F.; Li, L.C.; Pookot, D.; Noonan, E.J.; Dahiya, R. microRNA-373 induces expression of genes with complementary promoter sequences. *Proc. Natl. Acad. Sci. USA* **2008**, *105*, 1608–1613.

158. Majid, S.; Dar, A.A.; Saini, S.; Yamamura, S.; Hirata, H.; Tanaka, Y.; Deng, G.; Dahiya, R. microRNA-205-directed transcriptional activation of tumor suppressor genes in prostate cancer. *Cancer* **2010**, *116*, 5637–5649.

159. Tsai, N.P.; Lin, Y.L.; Wei, L.N. microRNA mir-346 targets the 5'-untranslated region of receptor-interacting protein 140 (RIP140) mRNA and up-regulates its protein expression. *Biochem. J.* **2009**, *424*, 411–418.

160. Cordes, K.R.; Sheehy, N.T.; White, M.P.; Berry, E.C.; Morton, S.U.; Muth, A.N.; Lee, T.H.; Miano, J.M.; Ivey, K.N.; Srivastava, D. miR-145 and miR-143 regulate smooth muscle cell fate and plasticity. *Nature* **2009**, *460*, 705–710.

161. Ma, F.; Liu, X.; Li, D.; Wang, P.; Li, N.; Lu, L.; Cao, X. microRNA-466l upregulates IL-10 expression in TLR-triggered macrophages by antagonizing RNA-binding protein tristetraprolin-mediated IL-10 mRNA degradation. *J. Immunol.* **2010**, *184*, 6053–6059.

162. Heil, F.; Hemmi, H.; Hochrein, H.; Ampenberger, F.; Kirschning, C.; Akira, S.; Lipford, G.; Wagner, H.; Bauer, S. Species-specific recognition of single-stranded RNA via Toll-like receptor 7 and 8. *Science* **2004**, *303*, 1526–1529.

163. Lund, J.M.; Alexopoulou, L.; Sato, A.; Karow, M.; Adams, N.C.; Gale, N.W.; Iwasaki, A.; Flavell, R.A. Recognition of single-stranded RNA viruses by Toll-like receptor 7. *Proc. Natl. Acad. Sci. USA* **2004**, *101*, 5598–5603.

164. Fabbri, M.; Paone, A.; Calore, F.; Galli, R.; Gaudio, E.; Santhanam, R.; Lovat, F.; Fadda, P.; Mao, C.; Nuovo, G.J.; *et al.* microRNAs bind to Toll-like receptors to induce prometastatic inflammatory response. *Proc. Natl. Acad. Sci. USA* **2012**, *109*, E2110–E2116.

9

ADAR Enzyme and miRNA Story: A Nucleotide that can Make the Difference

Sara Tomaselli [1,†], Barbara Bonamassa [1,†], Anna Alisi [2,*], Valerio Nobili [2], Franco Locatelli [1,3] and Angela Gallo [1,*]

[1] Laboratory of RNA Editing, Onco-haematology Department, Bambino Gesù Children's Hospital, IRCCS, Piazza S. Onofrio 4, Rome 00165, Italy; E-Mails: sara.tomaselli@opbg.net (S.T.); barbara.bonamassa@opbg.net (B.B.); franco.locatelli@opbg.net (F.L.)

[2] Hepato-Metabolic Disease Unit and Liver Research Unit, Bambino Gesù Children's Hospital, IRCCS, Piazza S. Onofrio 4, Rome 00165, Italy; E-Mail: valerio.nobili@opbg.net

[3] Department of Pediatric Science, Università di Pavia, Strada Nuova 65, Pavia 27100, Italy

[†] These authors contributed equally to this work.

[*] Authors to whom correspondence should be addressed; E-Mails: anna.alisi@opbg.net (A.A.); angela.gallo@opbg.net (A.G.);

Abstract: Adenosine deaminase acting on RNA (ADAR) enzymes convert adenosine (A) to inosine (I) in double-stranded (ds) RNAs. Since Inosine is read as Guanosine, the biological consequence of ADAR enzyme activity is an A/G conversion within RNA molecules. A-to-I editing events can occur on both coding and non-coding RNAs, including microRNAs (miRNAs), which are small regulatory RNAs of ~20–23 nucleotides that regulate several cell processes by annealing to target mRNAs and inhibiting their translation. Both miRNA precursors and mature miRNAs undergo A-to-I RNA editing, affecting the miRNA maturation process and activity. ADARs can also edit 3' UTR of mRNAs, further increasing the interplay between mRNA targets and miRNAs. In this review, we provide a general overview of the ADAR enzymes and their mechanisms of action as well as miRNA processing and function. We then review the more recent findings about the impact of ADAR-mediated activity on the miRNA pathway in terms of biogenesis, target recognition, and gene expression regulation.

Keywords: microRNA; Adenosine deaminase acting on RNA (ADAR); A-to-I RNA editing; double-stranded RNA (dsRNA); non-coding sequence

---

1. Introduction

Protein-coding genes account for approximately 1% of the mammalian genome and 70%–90% of the rest can be transcribed but not translated [1]. Therefore, a large part of the human transcriptome consists of non-coding RNA sequences, (*i.e.*, UTRs, introns of protein-coding genes and non-coding RNAs, such as transfer RNAs (tRNAs), ribosomal RNAs (rRNAs), microRNAs (miRNAs), small interfering RNAs (siRNAs), piwi-interacting RNAs (piRNAs), long non-coding RNAs (lncRNAs), small nuclear RNAs (snRNAs), and small nucleolar RNAs (snoRNAs)). Both protein-coding and non-coding RNAs undergo several post-transcriptional modifications, which (partially) account for the complexity of both the transcriptome and proteome that characterizes the high level of gene regulation in higher eukaryotes [2]. Among these post-transcriptional mechanisms, RNA editing is an ubiquitous and crucial modification event that alters RNA molecules by nucleotide modification bypassing the genomic information [3,4]. There are different types of RNA editing [3], but the best characterized and frequent editing event in higher eukaryotes involves the conversion of adenosine (A) to inosine (I) in double-stranded RNA (dsRNA) regions through the action of the Adenosine Deaminase Acting on RNA (ADAR) enzymes [4–6].

Computational analysis combined with next generation sequencing (NGS) has recently been used to identify A-to-I RNA editing sites [7–10]. Reverse transcriptase recognizes Inosine as Guanosine. Therefore, an A-to-I RNA editing site can be identified when a cDNA sequence and the corresponding genomic DNA (gDNA) sequence are aligned. Surprisingly, several editing sites were found in non-coding regions of the human transcriptome (~15,000 sites, mapped in ~2000 different genes) and most of them are clustered within inversely oriented repetitive Alu elements (~90%). On the basis of this analysis, it is predicted that >85% of pre-mRNAs are possibly edited, with the vast majority being targeted in introns (~90%) and UTRs [10].

As Inosine is interpreted as Guanosine by splicing and translational machineries, A-to-I editing can change the informational content of the RNA coding molecules by altering splicing and translation processes. Moreover, Inosine has different base-pairing properties compared to Adenosine and differs from Guanosine by the loss of the N_2 amino group (due to the ADAR deamination event), which accounts for the less strong interaction with Cytosine (two H-bound instead of three). Thus, A-to-I RNA editing has the potential to alter RNA structure by introducing bulges/mismatches or creating different base pairs (for examples, A-U base pairs can change into I:U mismatches in dsRNAs). The final picture is that ADARs can alter splicing, translation, and the dsRNA structure. It was originally thought that the main function of ADAR enzymes was their re-coding capacity. However, A-to-I editing most frequently targets non-coding sequences [9,11,12] and, recently, numerous interactions between ADARs and miRNA/siRNA pathways [13] have been discovered, which suggests a role of ADARs and A-to-I editing in RNA-mediated regulation of gene expression. In this review, we first provide a general overview of the ADAR enzymes and their mechanisms of action. We then focus on the miRNA pathway and the effects of ADAR-mediated modifications on the biogenesis and functions of miRNAs.

2. ADAR Family

ADAR-mediated A-to-I RNA editing converts A to I by hydrolytic deamination of adenine bases. Three ADARs (ADAR1, ADAR2, and ADAR3) are present in vertebrates (Figure 1). ADARs contain a highly conserved catalytic deaminase domain (DM) at their *C*-terminal. Crystallography structure of the DM showed that the surface of this domain contains a positively-charged cleft for the binding of negatively-charged dsRNA and that it catalyses the hydrolytic deamination of Adenosine via a catalytic zinc ion [14]. Moreover, an inositol hexakisphosphate (IP6) was found buried within the enzyme core that contributes to the protein fold [14]. A nucleotide "flip-out" mechanism is necessary to force the targeted Adenosine into the catalytic pocket in the correct orientation for the deamination reaction [15].

The second key domain of all ADAR enzymes is the dsRNA-binding domain (dsRBD) at the *N*-terminus. Each dsRBD (three for ADAR1 and two for ADAR2-3) has an α-β-β-β-α topology consisting of approximately 70 amino acids, with the two α helices packing against a three-stranded anti-parallel β-sheets. Multiple dsRBDs are thought to act synergistically, which, as a consequence, increases both the affinity and specificity for dsRNA targets [16].

At the *N*-terminus, ADAR1 carries a Z-DNA binding domains (Zα plus Zβ) that suggests its localization at highly transcribed DNA sites. Moreover, as these domains can also bind Z-RNA, ADAR1 is also able to localize to underwound dsRNAs in RNA virus [17,18].

Figure 1. Structure of ADAR family proteins: ADAR1, ADAR2, and ADAR3. The ADAR enzymes contain a *C*-terminal conserved catalytic deaminase domain (DM), two or three dsRBDs in the *N*-terminal portion. ADAR1 full-length protein also contains a *N*-terminal Zα domain with a nuclear export signal (NES) and a Zβ domain, while ADAR3 has a R-domain. A nuclear localization signal is also indicated.

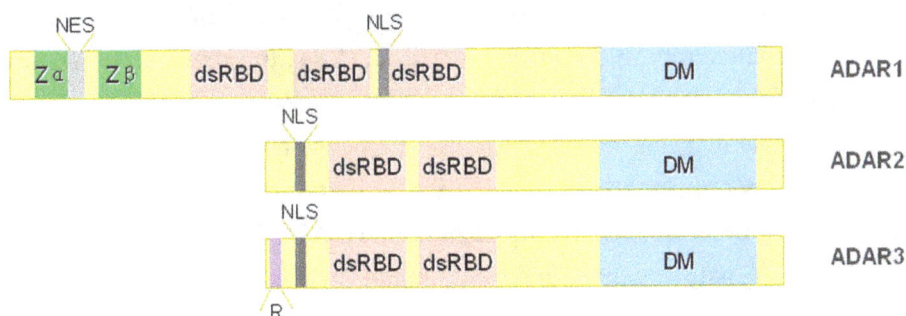

2.1. ADAR1

The *ADAR1* human gene is located on the long arm of chromosome 1 (1q21.3) spanning ~30Kbp [19]. The protein was discovered to exist in two isoforms of different size, *i.e.*, the interferon (IFN-α, -β, and -γ)-inducible long (ADAR1L, 150 KDa) and the constitutive short (ADAR1S, 110 KDa) isoform, which result from the use of alternative start codons and promoters [20]. While

ADAR1S promoter is constitutively active, IFN can induce ADAR1L, suggesting a role in the cellular response to stress factors such as viral infections [21]. In addition to the finding of regulatory elements within the IFN-inducible ADAR1 promoter, recent studies revealed distinct tissue-specific expression features for different ADAR1 transcripts [22]. Both transcripts contain three dsRBDs but the N-terminus of ADAR1L has several domains that are absent in ADAR1S, including an arginine-glycine-enriched domain (RG domain) and a nuclear export signal (NES) within the Zα domain. Thus, ADAR1L is found both in the cytoplasm and nucleus since it also has a nuclear/nucleolar localization signal (NLS/NoLS) [23]. Consequently, the intracellular distribution of the various ADAR1 isoforms is determined by the export/import regulatory proteins available in a cell. On the contrary, ADAR1S localizes mainly to the nucleus since it carries only the NLS/NoLS signal. However, it has been shown that ADAR1S can also localize to the cytoplasm thanks to the cooperative action of all three dsRBDs, with dsRNAs able to interact with exportin-5 [24].

The small ubiquitin-like modifier 1 (SUMO-1) binds ADAR1 at lysine 418, decreasing the editing activity of the enzyme [25]. ADAR enzymes can form homo- and hetero-dimers and dimerization is essential for their editing activity [26,27]. Several studies have shown that ADAR1 (and ADAR2) can work as homodimer, whereas other investigations have demonstrated that also heterodimers can be formed, which may be necessary for the ADARs to act as active deaminases [27–29]. Other ADAR-interacting proteins include the nuclear factor 90 (NF90) proteins [30], the protein-kinase RNA-activated protein (PKR) [31], the adenovirus-associated (VAI) RNA [32], and the Vaccinia virus E3L protein [33].

2.2. ADAR2

The *ADAR2* human gene is located on the long arm of chromosome 21 (21q22.3), spanning ~153 Kbp [34]. The promoter that directs ADAR2 expression has not been functionally characterized, although a putative promoter region upstream of a newly identified exon was described for both the human and the mouse *Adar2* gene [35]. This promoter includes a TATA box sequence and the consensus binding sites for the Nuclear factor kappa-light-chain-enhancer of activated B cells (Nf-κB) and for the Specificity Protein 1 (SP1) [35]. While it has to be established whether ADAR2 possesses multiple promoters to produce multiple transcripts like ADAR1, the regulatory mechanism(s) driving the transcriptional control of ADAR2 in a tissue- and cell type-specific fashion have been partially unveiled. Indeed, it was shown that cAMP response element-binding (CREB) can indirectly induce ADAR2 expression [36]. More recently, Yang *et al.* [37] demonstrated that JNK1 serves as a crucial component in mediating glucose-responsive up-regulation of ADAR2 expression in pancreatic β-cells, suggesting that the JNK1 pathway may be functionally linked to the nutrient-sensing actions of ADAR2-mediated RNA editing in professional secretory cells.

ADAR2 N-terminus has an arginine-enriched domain (R-domain) (similar to that identified in the ADAR3 protein, Figure 1) that contains a NLS [35,38], while an extra NLS is located before the first dsRBD [35,39]. Consequently, ADAR2 localizes into the cell nuclei thanks to the action of importin α1, α4, and α5 [39].

ADAR2 can form homodimers and heterodimers with ADAR1 [27–29]. ADAR2 dimerization seems to be essential for editing activity, although it is not clear whether the interaction is or not dsRNA-mediated [27,40,41].

2.3. ADAR3

The *ADAR3* human gene is located on the short arm of chromosome 10 (10p15) in proximity of the telomere [42]. Although ADAR3 has conserved all the key catalytic residues of the ADAR family members, no deaminase activity has been found for this enzyme so far [43]. All the editing sites have been, thus, attributed to ADAR1 and 2 activity. ADAR3 protein carries two dsRBDs and, additionally, an R-domain that binds single-stranded RNAs (ssRNAs) [43,44], suggesting that both ss- and dsRNAs can be bound by the enzyme. ADAR3 is localized in the nucleus of the cell and interacts with the importin α1 through the R-motif [35].

Differently from the ubiquitously expressed ADAR1 and 2, ADAR3 is expressed at detectable levels only in certain post-mitotic cells in the central nervous system (CNS) [43]. Furthermore, ADAR3 remains in the monomeric form, which may explain the lack of editing activity, at least in part [28]. Thus, ADAR3 function is unknown so far, although its ability to bind both ss- and dsRNAs would suggest a regulatory activity over ADAR1 and 2. Indeed, ADAR3 can compete for dsRNA substrates preventing the binding of the other ADAR enzymes [43,45].

3. ADAR Substrates

Any dsRNAs of ≥20 bp can be an ADAR substrate [6]. ADAR substrates were originally identified by chance, comparing cDNAs to their genomic counterparts and finding editing events as a mixture of A/G instead of A only. Different editing sites have been identified over the years, particularly in transcripts coding for proteins expressed in the CNS [4,5,46], *i.e.*, those coding for subunits of the glutamate receptor super-family GluR, the serotonin 5-hydroxytryptamine 2C (5-HT2C)-receptor, the potassium voltage-gated channel (Kv1.1), and the a3 subunit of the γ-aminobutyric acid (GABAA) receptor. These editing events have a major impact on protein properties.

More recently, bio-computational studies and innovative sequencing techniques have demonstrated that A-to-I RNA editing mainly affects non-coding RNAs [6]. Importantly, the majority of editing events occur in introns and 5'–3' UTRs enriched with Alu repeat-mediated dsRNAs. Recently, a database collecting the identified (validated or not) editing RNAs has become available (http://darned.ucc.ie) [47].

ADAR-mediated editing levels range from 2% to 100% [5,13,46], depending on cell and tissue type [48] as well as developmental stages [49]. How ADAR chooses the target adenosine is still not completely clear. ADARs show slight sequence preferences [50]. However, dsRNA length and structure seem to play an important role. For example, dsRNAs of 15–40 bp are edited selectively at very few sites, whereas those longer than 50 bp are extensively or non-selectively deaminated (with 50%–60% of adenosines being edited) [51]. Similarly, selective deamination is also observed in dsRNAs with bulges, loops, and mismatches [52]. It has been suggested that ADAR substrate specificity may also depend on editor modulators (such as snoRNAs) [53] and on the different dsRBD number and spacing of ADAR proteins that allow discrimination between dsRNA structures and stabilities. While the importance of site-specific editing (within coding sequence genes or microRNAs) has been explored and was found to affect the final protein or miRNA maturation/targeting, the role of the non-specific/promiscuous editing (within non-coding RNA portions such as introns and

5'–3' UTRs) is still poorly understood. However, recent studies would point out their involvement in modulation of gene expression, which may occur by changing the splicing enhancers/silencers recognition sites [54–57], by perturbing/inducing the binding of RBPs for RNA nuclear localization/ retention [58] or inducing inosine-specific degradation (Tudor-SN nuclease) [59].

4. miRNA World Machinery Overview

miRNAs are short (~20–23 nucleotides) ssRNAs that regulate, at post-transcriptional level, several genes playing crucial roles in various cellular processes such as cell cycle, apoptosis, differentiation and, when deregulated, neoplastic transformation [60]. Mammalian miRNA genes (in cluster or as single unit) are located either in introns/exons of protein-coding genes, in non-coding genes, or in intra-genic regions of the genome [61–63] (Figure 2). Intronic/exonic miRNAs are often transcribed by the RNA polymerase (Pol) II and co-expressed with their host gene, while intergenic miRNAs are independently transcribed by either RNA Pol II or III [64,65]. Usually, miRNA promoters located in the inter-genic or non-coding regions of the genome are regulated by transcriptional or epigenetic factors like protein-coding genes [66].

Figure 2. Schematic diagram of the miRNA genes. (**A**) Monocistronic intergenic miRNA gene; (**B**) Monocistronic exonic/intronic miRNA gene.

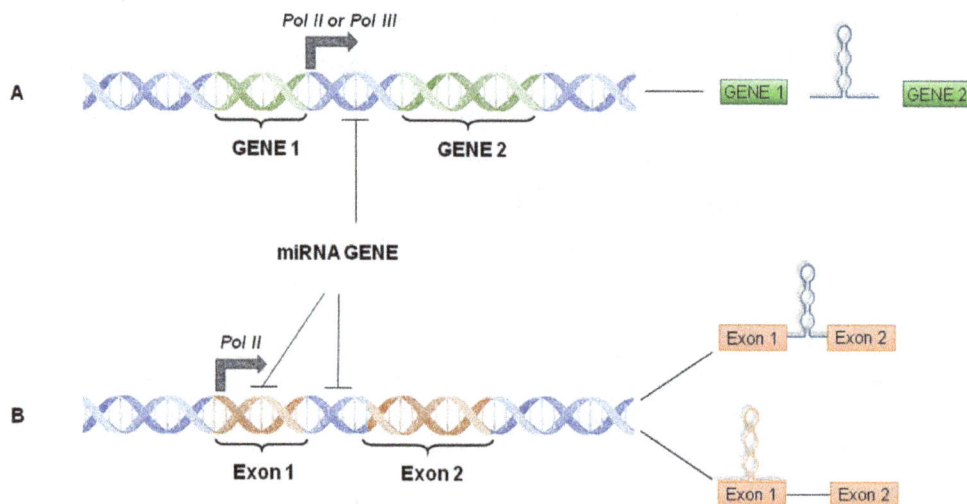

Each miRNA may regulate several mRNAs post-transcriptionally, while a single mRNA can be targeted by several miRNAs via base-pairing to the mRNA 3' UTRs [67,68].

The conventional theory assumes that the "seed sequence" (~6–8 nucleotides in length) at miRNA 5' end is crucial for target specificity and mediates its binding to 3' UTRs of target mRNAs, causing their translational repression or degradation [69]. However, recent studies suggest that miRNAs can exert their action over specific targets using alternative mechanisms, including the binding to specific proteins or to non-coding RNAs [70,71]. The biogenesis and processing of miRNAs occur in the nucleus/cytoplasm due to the action of multiple proteins. Some of these have a well-known role(s) in miRNA processing, including Drosha, exportin 5, Argonaute (Ago), and Dicer, while others have partially been explored such as ADAR1 [72].

4.1. miRNA Biogenesis and Processing into the Nucleus

The early step of miRNA biogenesis in the nucleus is the transcription of a miRNA precursor (Figure 3). Mature miRNAs are generated from long, hairpin-shaped primary transcripts (pri-miRNA) that are usually several thousand nucleotides long [66]. After transcription, pri-miRNAs undergo multiple steps of processing into the nucleus. Conventional nuclear processing of pri-miRNAs happen due to their cleavage by a large microprocessor complex (650 kDa in humans) consisting of the RNase III enzyme Drosha and the DiGeorge syndrome Critical Region gene 8 (DGCR8) protein [73,74]. Specifically, Drosha, a nuclear protein of 130–160 kDa, cuts the 5' and 3' ends of the pri-miRNA molecule with its RNase domain, giving a short hairpin of 60–70 nucleotides long (pre-miRNA) [66]. Although DGCR8-Drosha microprocessor is involved in the cropping of many miRNAs, Drosha may also form larger complexes with other proteins (e.g., RNA helicases, dsRNA binding proteins, heterogeneous nuclear ribonucleoproteins, *etc.*) to regulate the processing of specific pri-miRNAs [75]. A recent study provides evidence that certain mature miRNAs combined with Ago proteins may re-enter the nucleus and inhibit the pri-miRNA processing [76].

Figure 3. miRNA biogenesis and processing. Canonical biogenesis of pri-miRNA transcription is mediated by Pol II. Next, the microprocessor complex composed of Drosha and DGCR8 mediates the nuclear cleavage of pri-miRNA into pre-miRNA. The nuclear export of pre-miRNA is subsequently mediated by exportin-5/Ran-GTP61. Cytoplasmic pre-miRNA is processed by Dicer into a duplex microRNA. The next step is the unwinding of the duplex into a mature ~22 nucleotide miRNA and a miRNA* by the RISC complex. The mature miRNA is generally conveyed by the RISC on the targeted mRNA, whilst miRNA* can be degraded or alternatively perform a different targeting.

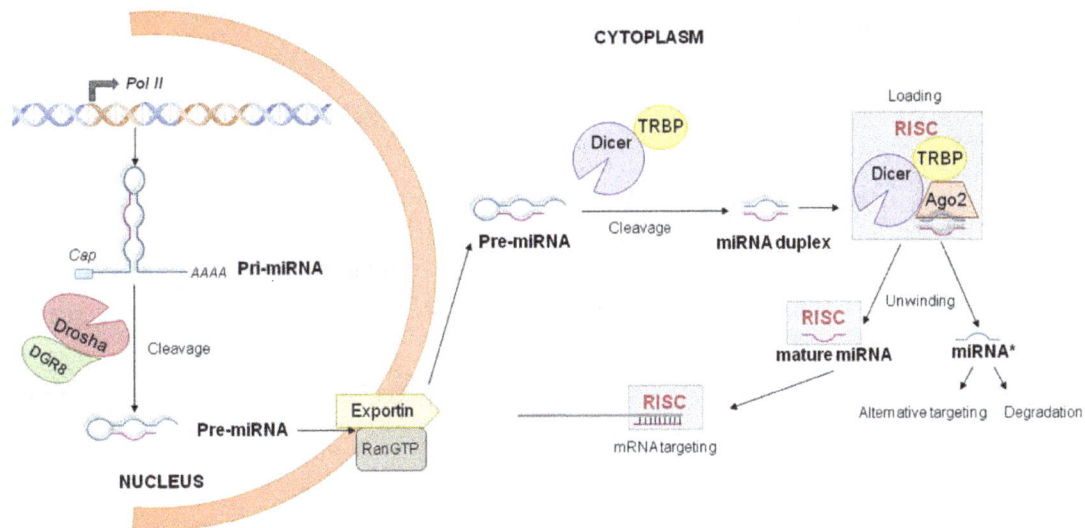

Following the nuclear processing, pre-miRNAs are exported to the cytoplasm by an energy-dependent mechanism involving the exportin-5/Ran-GTP61 complex. Exportin-5 binds pre-miRNA molecules and Ran-GTP61, which catalyses GTP hydrolysis and the consequent release of pre-miRNA short precursors into the cytoplasm. Interestingly, Exportin-5 also hampers pre-miRNA nuclear accumulation, protecting them from a potential nuclear digestion and retention [77,78]. In addition to the nuclear-to-cytoplasm pre-miRNA flux, the presence of functional mature miRNAs into the nucleus suggests a retrograde transport regulated by other carriers such as Importin 8 [79].

4.2. miRNA Processing into the Cytoplasm

Once exported from the nucleus, the cytoplasmic pre-miRNA duplex is further processed by Dicer and other accessory proteins, including the transactivation response RNA binding protein (TRBP), the protein activator of the dsRNA-dependent protein kinase (PACT), and the Ago proteins (Figure 3). Together they form the RNA-induced silencing complex (RISC) [80–83].

For miRNAs displaying a high degree of complementarity along the hairpin stem, a preliminary Ago 2-dependent cleavage is required before Dicer action. This Ago 2 slicer activity generates a nicked hairpin, producing a precursor miRNA or ac-pre-miRNA that is further processed by Dicer [84]. Dicer typically cleaves pre-miRNA duplexes near the terminal loop, releasing a small RNA duplex of ~22 nucleotides [66].

After Dicer-mediated cleavage, the small RNA duplex is loaded onto an Ago protein (Ago 1–4 in mammals) of the RISC to generate the microRNA containing ribonucleoprotein complex, i.e., miRNP or miRISC. Usually one single-strand (named guide) of the duplex (which is complementary to the target mRNA) is charged on Ago 2 as a mature miRNA, while the other strand of the duplex (named passenger or miRNA*) is usually degraded. miRNA guide (or in some cases miRNA* [85,86]) is selected to associate with Ago proteins by their thermodynamic stability [87]. There are at least two other hypotheses to explain duplex unwinding into guide and passenger strand. Dicer could cleave the miRNA*, releasing the miRNA guide that is subsequently captured by Ago 2. Alternatively, the miRNA* of a loaded duplex could be cleaved by the slicer activity of Ago 2, which simultaneously retains the miRNA guide. The activated RISC can bind the target mRNA, and direct its degradation, or repress its translation [88]. However, it has been reported that in some cases, miRNAs can also up-regulate the expression of their targets [85,89].

5. ADAR-Dependent Effects on miRNA Pathway

As ADARs can bind to and edit any dsRNA, the discovery that these enzymes are able to modify dsRNA substrates that enter the miRNA-mediated gene silencing and RNA interference (RNAi) pathways, i.e., miRNA and siRNA precursors [13], does not come as a surprise. It has been shown that mammalian pri-miRNAs undergo A-to-I RNA editing in adult brain [86,90–92]. Furthermore, NGS analysis has shown that ADARs can alter miRNA processing and sequence in C. elegans, mouse embryos, human and mouse brain [93–96]. Moreover, a more recent study showed that ADAR1 forms a complex with Dicer, promoting miRNA processing, RISC loading of miRNAs and silencing of target RNAs independently of its deaminase activity [72], as previously suggested [97].

In summary, several miRNA precursors (pri- and pre-miRs) undergo specific A-to-I RNA editing that may inhibit their maturation process and, thus, the production of mature miRNAs, affecting the loading of the edited miRNA to the RISC complex, or redirecting the edited miRNA to a new set of target mRNAs (Figure 4). Considering that A-to-I editing can also occur within the 3' UTR regions of mRNAs, the picture of miRNA-ADAR interaction becomes even more complex, underlining the high level of regulation of the miRNA world.

Figure 4. Editing-dependent effects of ADARs on miRNA pathway. miRNA precursors (pri- and pre-miRs) undergo specific A-to-I RNA editing that (i) may block their maturation process at either Drosha or Dicer step; (ii) may affect the loading of the edited miRNA to the RISC complex; (iii) may redirect the edited miRNA to a new set of target mRNAs.

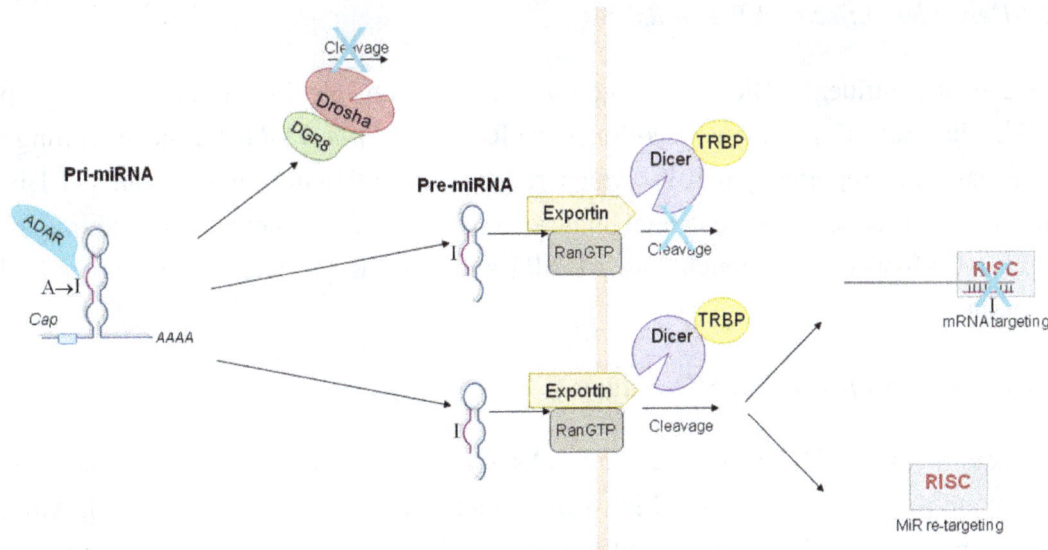

5.1. ADAR-Dependent Effects on Pri-miRs

The first report of RNA editing events in a miRNA precursor dates back to almost ten years ago, when Maas and co-authors detected a low level (~5% in human brain) of A-to-I changes within the pri-miR-22 [86]. Using human cell lines (HEK293T), ectopically expressing ADAR1 or ADAR2, they found that pri-miR-22 is mainly edited by ADAR1, although the physiological role of this editing was not elucidated.

A couple of years later, Yang *et al.* confirmed that ADARs can interact with pri-miRNAs using RNA editing assays and data from *Adar1* and *Adar2* null mice [98]. Four out of the eight analyzed miRNA precursors displayed A-to-I editing *in vitro* (*i.e.*, pri-miR-142, -223, -1-1, -143), with pri-miR-142 harboring the highest editing levels. Both ADAR1S and ADAR2 are able to edit pri-miR-142 at 11 specific sites, nine of which lie within the mature miRNA sequence. Transfecting edited pri-miR-142 in HEK293 cells, the authors determined that editing at the +4 and +5 sites destroys the integrity of the stem-loop structure, inhibiting the maturation of the pri- to pre-miRs. The consequence is a reduced production of mature miR-142. Indeed, the levels of endogenous miR-142 were lower in wild-type mouse spleens than those in *Adar1* and *Adar2* null mouse spleens. However, some editing sites (such as the one at site +40) seem not to affect pri-miR-142 processing. Editing-mediated inhibition of miRNA maturation at the pri-miR step does not cause accumulation of the edited pri-miR-142, as it may be degraded by Tudor-SN, a component of the RISC complex [99], known to mediate the degradation of inosine-containing dsRNAs (IU-dsRNAs) *in vitro* [59].

The discovery that edited pri-miRs can undergo rapid degradation by Tudor-SN suggests that the amount of edited pri-miRNAs into a cell could be higher than previously hypothesized. A recent study showed that ADAR1L (the ADAR1 nucleus/cytoplasmatic shuttling isoform) and Tudor-SN co-localize in the cytoplasm within stress granules (SGs) in HeLa cells under various stress conditions [100]. The authors speculated that ADAR1 may edit target dsRNAs in the cytoplasm and the resultant IU-dsRNA may recruit Tudor-SN to form SGs during cell stress responses. However, further experiments are needed to better define the role of ADAR1 in this context and the importance of the ADAR1-mediated SG formation.

5.2. ADAR-Dependent Effects on Pre-miRs

Editing can also influence Dicer cleavage, which is responsible for the processing of pre- into miRNAs. This has been first demonstrated for pri-miR-151 [101]. ADAR1-dependent editing at the -1 and +3 site has been reported [90,101], which reduces the efficiency of the Dicer-TRBP activity and results in the production of unedited mature miR-151 [101]. Interestingly, editing of mouse pri-miR-151 is CNS-specific, although both ADAR1 and pri-miR-151 were found expressed in many non-brain tissues.

5.3. ADAR-Dependent Effects on RISC-Loading

Epstein-Barr virus (EBV) encodes 23 miRNAs that are implicated in the attenuation of host antiviral immune response and the transition from latent to lytic replication [102,103]. Among EBV miRNAs, four primary miRNAs were found to undergo site-specific A-to-I editing events [104]. The authors focused on pri-miR-BART6, which showed high editing levels at the +20 in EBV latently infected cell lines. This editing reduces the correct loading of miR-BART6-5p into the RISC complex. Remarkably, this is the first report of pri-miRNA A-to-I editing that suppresses RISC loading [104]. Editing of pri-miR-BART6 reduces the activity of mature miR-BART6, playing a crucial role in the regulation of EBV life cycle and cell immune response.

Recently, new A-to-I editing events have been reported within another EBV miRNA, *i.e.*, pri-miR-BART3. Editing was found at four sites in EBV-infected epithelial carcinoma cells and in nasopharyngeal carcinoma samples, affecting both the biogenesis and targeting of mature miR-BART3 [105].

5.4. ADAR-Dependent Effects on Retargeting

A specific ADAR-mediated A-to-I change has been reported in Kaposi's sarcoma-associated herpesvirus (KSHV) transcripts [106,107]. This alteration modifies the seed sequence of the mature miR-K10, potentially affecting its target mRNAs [106,107]. ADAR1S heavily edits the K12 transcript in a specific site, as shown by *in vitro* editing assays [106]. Importantly, the authors observed that this editing event has a functional significance, playing a key role in the replication strategy of HHV-8 and in its tumorigenic potential. This was the first evidence that ADAR-mediated editing can also affect the target specificity of a mature miRNA.

Subsequently, Nishikura and colleagues demonstrated that edited mature miRNAs play a biological function *in vivo* [91]. The human pri-miR-376a1, previously showed to be edited [90], is situated in a cluster of 6 pri-miRNAs. The authors disclosed that five out of these six miRNAs are edited in human tissues (*i.e.*, pri-miR-367a1, -367a2, -367b, -368, -B2). Several adenosines within the miR-376 cluster members undergo A-to-I editing, with two positions showing the highest editing levels (nearly 100% in certain tissues), *i.e.*, the +4 site, which is preferentially edited by ADAR2, and the +44 site, which is selectively edited by ADAR1. These editing events do not affect the primary transcript maturation steps. However, both +4 and +44 sites lay within the seed sequences of miR-376a* and miR-376a respectively, suggesting that the edited miRNAs could have a different target mRNA profile. In particular, the authors demonstrated that a single ADAR2-mediated base change (at the +4 site) is able to modulate the expression of phosphoribosyl pyrophosphate synthetase 1 (PRPS1), a mouse protein involved in purine metabolism and uric acid synthesis [91].

Notably, a recent work has elegantly demonstrated the existence of a tight link between miR-376a editing and human brain tumors [108]. Choudhury *et al.* found that RNA editing of miR-376 cluster is extremely reduced in human gliomas, with glioblastoma cells accumulating almost exclusively the unedited form of miR-376a*. The unedited miRNA promotes glioma cell migration and invasion, whilst the edited form inhibits these capacities *in vitro*. These effects are the consequence of a different mRNA target specificity of the edited and unedited form of the miRNA [108]. As ADAR2 is responsible for miR-376a* editing, these findings strengthen the notion that this enzyme plays a crucial role in glioma progression, as previously shown [45,109,110].

5.5. ADAR-Dependent Effects on Target 3' UTRs

As most of the editing sites are also located in 3' UTRs of human mRNAs [111], an additional interplay between ADAR activity and miRNAs is possible. Computational screening showed that RNA editing tends to avoid miRNA binding sites, with less than 10% of editing events occurring in 3' UTR regions recognised by miRNAs [111]. However, it was also found that editing can create new miRNA target sites [111].

More recent analyses indicate that up to 20% of the editing sites in the 3' UTR of human mRNAs may alter miRNA target sites [112], making the mRNA resistant to miRNA activity. In addition, in mouse tissues, A-to-I changes seem to be highly frequent in 3' UTR regions, including miRNA target sites [113]. Wang *et al.* provided novel insights into the mechanism by which ADAR1 and its activity regulate miRNA-mediated modulation of target gene expression [114]. Indeed, multiple A-to-I RNA editing events (mediated by ADAR1) were found within the 3' UTR of *ARHGAP26*, encoding the Rho GTPase activating protein 26. Furthermore, the authors revealed that both miR-30b* and miR-573 are able to target ARHGAP26, but that editing make this transcript resistant to repression mediated by these two miRNAs.

5.6. ADAR-Mediated Editing-Independent Effects on miRNAs

In addition to A-to-I sequence changes on miRNAs, ADARs can also act through an editing-independent mechanism by binding dsRNAs [97]. Heale *et al.* found that ADAR1 and ADAR2 editing activity can result in retargeting of human miR-376a2, as shown previously for mouse miR-376 [91]. By performing *in vitro* pri-miRNA processing assays, they also pointed out that, even in the absence of editing, ADAR2 can inhibit the processing of pri-miR-376a2 at the Drosha cleavage step [97]. Therefore, the simple binding of ADAR proteins to dsRNAs may have a range of biological roles that are still to be fully discovered.

6. Large-Scale Surveys

Initial low-throughput experiments followed by NGS approaches have been performed by several groups, adding new insights on the role of ADARs in the miRNA pathway.

One of the original systematic survey proposed that 6% of all human pri-miRNAs are edited [90]. The author determined that six out of 99 pri-miRNAs undergo editing (*i.e.*, pri-miR-151, -197, -223, -376a, -379, -99a) in humans. The extent of editing ranged from ~10% to 70%, depending on sites and different tissues analyzed. Most of the editing events were located in the mature miRNA seed sequence, suggesting that RNA editing may contribute to increase miRNA diversity. This paper established that ADARs edit miRNAs but did not elucidate the functional consequences of these events.

A couple of years later, a larger scale survey of 209 human pri-miRNAs showed that ~16% of them undergo A-to-I editing in human brain, with editing levels ranging from ~10% to 100% [92]. Then, for six randomly chosen edited pri-miRNAs (*i.e.*, pri-let-7g, pri-miR-33, -133a2, -197, -203, -379) it was discovered that editing alters either the Drosha or the Dicer cleavage step. It is worth noting that the processing of two pri-miRNAs (*i.e.*, pri-miR-197 and -203) was enhanced by editing. The authors also showed that some pri-miRs are preferentially edited by ADAR1 (*i.e.*, pri-miR-99b, -151, -376b, -411, -423), while others by ADAR2 (*i.e.*, pri-let-7g, pri-miR-27a, -99a, -203, -376a, -379) [92].

Recent advances in high-throughput small RNA sequencing (smRNA-Seq) have reshaped the miRNA research landscape, including RNA editing analysis. Using a novel strategy to avoid cross-mapping artefacts, de Hoon *et al.* found that editing prevalence in human mature miRNAs is extremely low in a human monocytic leukemia cell line (THP-1) [115]. Ten potential miRNA editing sites were found. However, eight of these were due to cross-mapping, one was due to a single nucleotide polymorphism, and the remaining editing site (in the mature miR-376c) was already identified [91]. Similar results were obtained by sequencing small RNAs from mouse brain [116].

Recently, Vesely and co workers analyzed the frequency and sequence composition of miRNA pools from transgenic *Adar* null mouse embryos by NGS [93]. *Adar2* deficiency leads to a change in the expression level of specific target mRNAs when compared to wild-type embryos. In particular, the authors detected 10 edited miRNAs, four of which had been identified previously (*i.e.*, mmu-miR-378, -376b, -381, -3099) and six were novel edited miRNAs (*i.e.*, mmu-miR-1957, -467d*, -706, -1186, -3102-5p.2, -703). Some editing events were located in the seed region, opening the possibility that editing could lead to their retargeting. However, the biological consequences of the observed editing events are difficult to interpret, especially because of the low levels detected.

Using NGS followed by bioinformatics analysis, Eisenberg and co-workers found a clear A-to-I signal in mature miRNAs of human brain [94]. Overall, 19 statistically significant modification sites (mainly due to ADAR2 activity) were detected in 18 different miRNAs, confirming previously detected editing sites as well as revealing several novel ones. Most of the detected A-to-G modifications were within the miRNA seed sequence, with editing significantly changing their binding specificity. As previously reported, a relatively low editing level was found, with few exceptions (editing percentage ranging from 0.2% to 70%) [94].

7. Stimulative Role of ADAR1

ADAR1 has been emerging as a promoter for small non-coding RNAs. Indeed, a recent study has highlighted the important role of ADAR1 in interacting with Dicer to form heterodimers [72]. Notably, the authors established that ADAR1 uses its second dsRBD to form ADAR1/Dicer heterodimers (acting as modulator of RNAi machinery) and its third dsRBD to form ADAR1/ADAR1 homodimers (acting as an RNA editing enzyme). The ADAR1/Dicer interaction increases the rate of processing from pre- to mature miRNAs, promotes the RISC loading and, consequently, the mRNA silencing efficacy [72]. It seems that neither dsRNA-binding nor deaminase activity of ADAR1 is required for these effects. As expected, the authors found that the miRNA expression is inhibited in *Adar1* null mouse embryos, as a consequence of the lack of formation of the Dicer/ADAR1 complex with a final alteration of the target genes [72].

8. Conclusions

A-to-I editing is believed to be an important way of generating protein diversity by codon alteration in mRNAs. However, editing sites in some coding targets make up only a tiny fraction of all editing events, most of which are actually located in non-coding sequences such as introns, UTRs or regulatory RNAs (miRNAs and their precursors). The biological function of editing in non-coding RNA sequences remains not completely disclosed. As far as miRNAs are concerned, the general feeling about A-to-I changes is that they regulate the levels of cellular dsRNAs, which, if not kept under control, are potent triggers of gene silencing and signaling pathway. Despite this, important questions still stand. At which extent and how diffuse is the RNA editing on mature miRNAs and their precursors? Is it a developmentally regulated or a tissue specific phenomenon? In principle, editing at any level of miRNA biogenesis may have a broad influence on expression patterns. Although the evidence is still limited, a critical examination of data reported in the literature does offer some examples of miRNA down-stream activity misregulation. One more question is whether there is any correlation between edited miRNAs and human diseases. While alterations in both substrate editing and ADAR expression/activity are often reported in different pathologies [4,5,117], the effects of edited miRNA pathways on disease onset/progression still deserves further investigation. In this context, it is worth noting that Choudhury *et al.* demonstrated that a single editing event in the miR-376a* seed sequence dramatically alters the selection of its target genes and redirects its function from inhibiting to promoting glioma cell invasion [108]. Overall, these pieces of information set the stage for further investigations, either to address the aforementioned questions and, possibly, to score against ADAR/miRNA editing-linked human diseases.

Acknowledgments

This work was supported by the PRIN 2012NA9E9Y_004 and the IG grant of AIRC (Milan, Italy) and to A.G., my First AIRC Grant to A.A., and the special project 5 × 1000 AIRC to F.L. We thank Marion Huth for the English revision.

References

1.   Lee, J.T. Epigenetic regulation by long noncoding RNAs. *Science* **2012**, *338*, 1435–1439.

2.   Baltimore, D. Our genome unveiled. *Nature* **2001**, *409*, 814–816.

3.   Gott, J.M.; Emeson, R.B. Functions and mechanisms of RNA editing. *Annu. Rev. Genet.* **2000**, *34*, 499–531.

4.   Gallo, A.; Locatelli, F. ADARs: Allies or enemies? The importance of A-to-I RNA editing in human disease: From cancer to HIV-1. *Biol. Rev. Camb. Philos. Soc.* **2011**, *87*, 95–110.

5.   Keegan, L.P.; Gallo, A.; O'Connell, M.A. The many roles of an RNA editor. *Nat. Rev. Genet.* **2001**, *2*, 869–878.

6.   Nishikura, K. Functions and regulation of RNA editing by ADAR deaminases. *Annu. Rev. Biochem.* **2010**, *79*, 321–349.

7.   Blow, M.; Futreal, P.A.; Wooster, R.; Stratton, M.R. A survey of RNA editing in human brain. *Genome Res.* **2004**, *14*, 2379–2387.

8.   Kim, D.D.; Kim, T.T.; Walsh, T.; Kobayashi, Y.; Matise, T.C.; Buyske, S.; Gabriel, A. Widespread RNA editing of embedded alu elements in the human transcriptome. *Genome Res.* **2004**, *14*, 1719–1725.

9.   Levanon, E.Y.; Eisenberg, E.; Yelin, R.; Nemzer, S.; Hallegger, M.; Shemesh, R.; Fligelman, Z.Y.; Shoshan, A.; Pollock, S.R.; Sztybel, D.; *et al.* Systematic identification of abundant A-to-I editing sites in the human transcriptome. *Nat. Biotechnol.* **2004**, *22*, 1001–1005.

10.   Athanasiadis, A.; Rich, A.; Maas, S. Widespread A-to-I RNA editing of Alu-containing mRNAs in the human transcriptome. *PLoS Biol.* **2004**, *2*, 2144–2158.

11.   Morse, D.P.; Aruscavage, P.J.; Bass, B.L. RNA hairpins in noncoding regions of human brain and Caenorhabditis elegans mRNA are edited by adenosine deaminases that act on RNA. *Proc. Natl. Acad. Sci. USA* **2002**, *99*, 7906–7911.

12.   Maas, S.; Rich, A.; Nishikura, K. A-to-I RNA editing: Recent news and residual mysteries. *J. Biol. Chem.* **2003**, *278*, 1391–1394.

13.   Nishikura, K. Editor meets silencer: Crosstalk between RNA editing and RNA interference. *Nat. Rev. Mol. Cell Biol.* **2006**, *7*, 919–931.

14.   Macbeth, M.R.; Schubert, H.L.; Vandemark, A.P.; Lingam, A.T.; Hill, C.P.; Bass, B.L. Inositol hexakisphosphate is bound in the ADAR2 core and required for RNA editing. *Science* **2005**, *309*, 1534–1539.

15.   Kuttan, A.; Bass, B.L. Mechanistic insights into editing-site specificity of ADARs. *Proc. Natl. Acad. Sci. USA* **2012**, *109*, E3295–E3304.

16.   Ryter, J.M.; Schultz, S.C. Molecular basis of double-stranded RNA-protein interactions: Structure of a dsRNA-binding domain complexed with dsRNA. *EMBO J.* **1998**, *17*, 7505–7513.

17.   Herbert, A.; Alfken, J.; Kim, Y.G.; Mian, I.S.; Nishikura, K.; Rich, A. A Z-DNA binding domain present in the human editing enzyme, double-stranded RNA adenosine deaminase. *Proc. Natl. Acad. Sci. USA* **1997**, *94*, 8421–8426.

18.   Brown, B.A., 2nd; Lowenhaupt, K.; Wilbert, C.M.; Hanlon, E.B.; Rich, A. The zalpha domain of the editing enzyme dsRNA adenosine deaminase binds left-handed Z-RNA as well as Z-DNA. *Proc. Natl. Acad. Sci. USA* **2000**, *97*, 13532–13536.

19. Weier, H.U.; George, C.X.; Greulich, K.M.; Samuel, C.E. The interferon-inducible, double-stranded RNA-specific adenosine deaminase gene (DSRAD) maps to human chromosome 1q21.1-21.2. *Genomics* **1995**, *30*, 372–375.

20. Patterson, J.B.; Samuel, C.E. Expression and regulation by interferon of a double-stranded-RNA-specific adenosine deaminase from human cells: Evidence for two forms of the deaminase. *Mol. Cell. Biol.* **1995**, *15*, 5376–5388.

21. Patterson, J.B.; Thomis, D.C.; Hans, S.L.; Samuel, C.E. Mechanism of interferon action: Double-stranded RNA-specific adenosine deaminase from human cells is inducible by alpha and gamma interferons. *Virology* **1995**, *210*, 508–511.

22. George, C.X.; Das, S.; Samuel, C.E. Organization of the mouse RNA-specific adenosine deaminase Adar1 gene 5'-region and demonstration of STAT1-independent, STAT2-dependent transcriptional activation by interferon. *Virology* **2008**, *380*, 338–343.

23. Poulsen, H.; Nilsson, J.; Damgaard, C.K.; Egebjerg, J.; Kjems, J. CRM1 mediates the export of ADAR1 through a nuclear export signal within the Z-DNA binding domain. *Mol. Cell. Biol.* **2001**, *21*, 7862–7871.

24. Fritz, J.; Strehblow, A.; Taschner, A.; Schopoff, S.; Pasierbek, P.; Jantsch, M.F. RNA-regulated interaction of transportin-1 and exportin-5 with the double-stranded RNA-binding domain regulates nucleocytoplasmic shuttling of ADAR1. *Mol. Cell. Biol.* **2009**, *29*, 1487–1497.

25. Desterro, J.M.; Keegan, L.P.; Jaffray, E.; Hay, R.T.; O'Connell, M.A.; Carmo-Fonseca, M. SUMO-1 modification alters ADAR1 editing activity. *Mol. Biol. Cell* **2005**, *16*, 5115–5126.

26. Gallo, A.; Keegan, L.P.; Ring, G.M.; O'Connell, M.A. An ADAR that edits transcripts encoding ion channel subunits functions as a dimer. *EMBO J.* **2003**, *22*, 3421–3430.

27. Valente, L.; Nishikura, K. RNA binding-independent dimerization of adenosine deaminases acting on RNA and dominant negative effects of nonfunctional subunits on dimer functions. *J. Biol. Chem.* **2007**, *282*, 16054–16061.

28. Cho, D.S.; Yang, W.; Lee, J.T.; Shiekhattar, R.; Murray, J.M.; Nishikura, K. Requirement of dimerization for RNA editing activity of adenosine deaminases acting on RNA. *J. Biol. Chem.* **2003**, *278*, 17093–17102.

29. Chilibeck, K.A.; Wu, T.; Liang, C.; Schellenberg, M.J.; Gesner, E.M.; Lynch, J.M.; MacMillan, A.M. FRET analysis of *in vivo* dimerization by RNA-editing enzymes. *J. Biol. Chem.* **2006**, *281*, 16530–16535.

30. Nie, Y.; Ding, L.; Kao, P.N.; Braun, R.; Yang, J.H. ADAR1 interacts with NF90 through double-stranded RNA and regulates NF90-mediated gene expression independently of RNA editing. *Mol. Cell. Biol.* **2005**, *25*, 6956–6963.

31. Clerzius, G.; Gelinas, J.F.; Daher, A.; Bonnet, M.; Meurs, E.F.; Gatignol, A. ADAR1 interacts with PKR during human immunodeficiency virus infection of lymphocytes and contributes to viral replication. *J. Virol.* **2009**, *83*, 10119–10128.

32. Lei, M.; Liu, Y.; Samuel, C.E. Adenovirus VAI RNA antagonizes the RNA-editing activity of the ADAR adenosine deaminase. *Virology* **1998**, *245*, 188–196.

33. Liu, Y.; Wolff, K.C.; Jacobs, B.L.; Samuel, C.E. Vaccinia virus E3L interferon resistance protein inhibits the interferon-induced adenosine deaminase A-to-I editing activity. *Virology* **2001**, *289*, 378–387.

34. Mittaz, L.; Scott, H.S.; Rossier, C.; Seeburg, P.H.; Higuchi, M.; Antonarakis, S.E. Cloning of a human RNA editing deaminase (ADARB1) of glutamate receptors that maps to chromosome 21q22.3. *Genomics* **1997**, *41*, 210–217.

35. Maas, S.; Gommans, W.M. Novel exon of mammalian ADAR2 extends open reading frame. *PLoS One* **2009**, *4*, e4225.

36. Peng, P.L.; Zhong, X.; Tu, W.; Soundarapandian, M.M.; Molner, P.; Zhu, D.; Lau, L.; Liu, S.; Liu, F.; Lu, Y. ADAR2-dependent RNA editing of AMPA receptor subunit GluR2 determines vulnerability of neurons in forebrain ischemia. *Neuron* **2006**, *49*, 719–733.

37. Yang, L.; Huang, P.; Li, F.; Zhao, L.; Zhang, Y.; Li, S.; Gan, Z.; Lin, A.; Li, W.; Liu, Y. c-Jun amino-terminal kinase-1 mediates glucose-responsive upregulation of the RNA editing enzyme ADAR2 in pancreatic beta-cells. *PLoS One* **2012**, *7*, e48611.

38. Desterro, J.M.; Keegan, L.P.; Lafarga, M.; Berciano, M.T.; O'Connell, M.; Carmo-Fonseca, M. Dynamic association of RNA-editing enzymes with the nucleolus. *J. Cell Sci.* **2003**, *116*, 1805–1818.

39. Maas, S.; Gommans, W.M. Identification of a selective nuclear import signal in adenosine deaminases acting on RNA. *Nucleic Acids Res.* **2009**, *37*, 5822–5829.

40. Jaikaran, D.C.; Collins, C.H.; MacMillan, A.M. Adenosine to inosine editing by ADAR2 requires formation of a ternary complex on the GluR-B R/G site. *J. Biol. Chem.* **2002**, *277*, 37624–37629.

41. Poulsen, H.; Jorgensen, R.; Heding, A.; Nielsen, F.C.; Bonven, B.; Egebjerg, J. Dimerization of ADAR2 is mediated by the double-stranded RNA binding domain. *RNA* **2006**, *12*, 1350–1360.

42. Mittaz, L.; Antonarakis, S.E.; Higuchi, M.; Scott, H.S. Localization of a novel human RNA-editing deaminase (hRED2 or ADARB2) to chromosome 10p15. *Hum. Genet.* **1997**, *100*, 398–400.

43. Chen, C.X.; Cho, D.S.; Wang, Q.; Lai, F.; Carter, K.C.; Nishikura, K. A third member of the RNA-specific adenosine deaminase gene family, ADAR3, contains both single- and double-stranded RNA binding domains. *RNA* **2000**, *6*, 755–767.

44. Melcher, T.; Maas, S.; Herb, A.; Sprengel, R.; Higuchi, M.; Seeburg, P.H. RED2, a brain-specific member of the RNA-specific adenosine deaminase family. *J. Biol. Chem.* **1996**, *271*, 31795–31798.

45. Cenci, C.; Barzotti, R.; Galeano, F.; Corbelli, S.; Rota, R.; Massimi, L.; Di Rocco, C.; O'Connell, M.A.; Gallo, A. Down-regulation of RNA editing in pediatric astrocytomas: ADAR2 editing activity inhibits cell migration and proliferation. *J. Biol. Chem.* **2008**, *283*, 7251–7160.

46. Bass, B.L. RNA editing by adenosine deaminases that act on RNA. *Annu. Rev. Biochem.* **2002**, *71*, 817–846.

47. Kiran, A.; Baranov, P.V. DARNED: A DAtabase of RNa EDiting in humans. *Bioinformatics* **2010**, *26*, 1772–1776.

48. Galeano, F.; Leroy, A.; Rossetti, C.; Gromova, I.; Gautier, P.; Keegan, L.P.; Massimi, L.; Di Rocco, C.; O'Connell, M.A.; Gallo, A. Human BLCAP transcript: New editing events in normal and cancerous tissues. *Int. J. Cancer* **2010**, *127*, 127–137.

49. Wahlstedt, H.; Daniel, C.; Enstero, M.; Ohman, M. Large-scale mRNA sequencing determines global regulation of RNA editing during brain development. *Genome Res.* **2009**, *19*, 978–986.

50. Lehmann, K.A.; Bass, B.L. Double-stranded RNA adenosine deaminases ADAR1 and ADAR2 have overlapping specificities. *Biochemistry* **2000**, *39*, 12875–12884.

51. Nishikura, K.; Yoo, C.; Kim, U.; Murray, J.M.; Estes, P.A.; Cash, F.E.; Liebhaber, S.A. Substrate specificity of the dsRNA unwinding/modifying activity. *EMBO J.* **1991**, *10*, 3523–3532.

52. Lehmann, K.A.; Bass, B.L. The importance of internal loops within RNA substrates of ADAR1. *J. Mol. Biol.* **1999**, *291*, 1–13.

53. Vitali, P.; Basyuk, E.; Le Meur, E.; Bertrand, E.; Muscatelli, F.; Cavaille, J.; Huttenhofer, A. ADAR2-mediated editing of RNA substrates in the nucleolus is inhibited by C/D small nucleolar RNAs. *J. Cell Biol.* **2005**, *169*, 745–753.

54. Lev-Maor, G.; Sorek, R.; Levanon, E.Y.; Paz, N.; Eisenberg, E.; Ast, G. RNA-editing-mediated exon evolution. *Genome Biol.* **2007**, *8*, R29.

55. Lev-Maor, G.; Sorek, R.; Shomron, N.; Ast, G. The birth of an alternatively spliced exon: 3' splice-site selection in Alu exons. *Science* **2003**, *300*, 1288–1291.

56. Lev-Maor, G.; Ram, O.; Kim, E.; Sela, N.; Goren, A.; Levanon, E.Y.; Ast, G. Intronic Alus influence alternative splicing. *PLoS Genet.* **2008**, *4*, e1000204.

57. Sakurai, M.; Yano, T.; Kawabata, H.; Ueda, H.; Suzuki, T. Inosine cyanoethylation identifies A-to-I RNA editing sites in the human transcriptome. *Nat. Chem. Biol.* **2010**, *6*, 733–740.

58. Zhang, Z.; Carmichael, G.G. The fate of dsRNA in the nucleus: A p54(nrb)-containing complex mediates the nuclear retention of promiscuously A-to-I edited RNAs. *Cell* **2001**, *106*, 465–475.

59. Scadden, A.D. The RISC subunit Tudor-SN binds to hyper-edited double-stranded RNA and promotes its cleavage. *Nat. Struct. Mol. Biol.* **2005**, *12*, 489–496.

60. He, L.; He, X.; Lim, L.P.; de Stanchina, E.; Xuan, Z.; Liang, Y.; Xue, W.; Zender, L.; Magnus, J.; Ridzon, D.; *et al.* A microRNA component of the p53 tumour suppressor network. *Nature* **2007**, *447*, 1130–1134.

61. Lagos-Quintana, M.; Rauhut, R.; Lendeckel, W.; Tuschl, T. Identification of novel genes coding for small expressed RNAs. *Science* **2001**, *294*, 853–858.

62. Lau, N.C.; Lim, L.P.; Weinstein, E.G.; Bartel, D.P. An abundant class of tiny RNAs with probable regulatory roles in *Caenorhabditis elegans*. *Science* **2001**, *294*, 858–862.

63. Lee, R.C.; Ambros, V. An extensive class of small RNAs in *Caenorhabditis elegans*. *Science* **2001**, *294*, 862–864.

64. Rodriguez, A.; Griffiths-Jones, S.; Ashurst, J.L.; Bradley, A. Identification of mammalian microRNA host genes and transcription units. *Genome Res.* **2004**, *14*, 1902–1910.

65. Morlando, M.; Ballarino, M.; Gromak, N.; Pagano, F.; Bozzoni, I.; Proudfoot, N.J. Primary microRNA transcripts are processed co-transcriptionally. *Nat. Struct. Mol. Biol.* **2008**, *15*, 902–909.

66. Kim, V.N.; Han, J.; Siomi, M.C. Biogenesis of small RNAs in animals. *Nat. Rev. Mol. Cell. Biol.* **2009**, *10*, 126–139.

67. Lewis, B.P.; Burge, C.B.; Bartel, D.P. Conserved seed pairing, often flanked by adenosines, indicates that thousands of human genes are microRNA targets. *Cell* **2005**, *120*, 15–20.

68. Filipowicz, W.; Bhattacharyya, S.N.; Sonenberg, N. Mechanisms of post-transcriptional regulation by microRNAs: Are the answers in sight? *Nat. Rev. Genet.* **2008**, *9*, 102–114.

69. Bartel, D.P. MicroRNAs: Genomics, biogenesis, mechanism, and function. *Cell* **2004**, *116*, 281–297.

70. Chen, X.; Liang, H.; Zhang, C.Y.; Zen, K. miRNA regulates noncoding RNA: A noncanonical function model. *Trends Biochem. Sci.* **2011**, *37*, 457–459.

71. Fabbri, M.; Paone, A.; Calore, F.; Galli, R.; Gaudio, E.; Santhanam, R.; Lovat, F.; Fadda, P.; Mao, C.; Nuovo, G.J.; *et al.* MicroRNAs bind to Toll-like receptors to induce prometastatic inflammatory response. *Proc. Natl. Acad. Sci. USA* **2012**, *109*, E2110–E2116.

72. Ota, H.; Sakurai, M.; Gupta, R.; Valente, L.; Wulff, B.E.; Ariyoshi, K.; Iizasa, H.; Davuluri, R.V.; Nishikura, K. ADAR1 forms a complex with Dicer to promote microRNA processing and RNA-induced gene silencing. *Cell* **2013**, *153*, 575–589.

73. Lee, Y.; Ahn, C.; Han, J.; Choi, H.; Kim, J.; Yim, J.; Lee, J.; Provost, P.; Radmark, O.; Kim, S.; Kim, V.N. The nuclear RNase III Drosha initiates microRNA processing. *Nature* **2003**, *425*, 415–419.

74. Han, J.; Lee, Y.; Yeom, K.H.; Nam, J.W.; Heo, I.; Rhee, J.K.; Sohn, S.Y.; Cho, Y.; Zhang, B.T.; Kim, V.N. Molecular basis for the recognition of primary microRNAs by the Drosha-DGCR8 complex. *Cell* **2006**, *125*, 887–901.

75. Winter, J.; Jung, S.; Keller, S.; Gregory, R.I.; Diederichs, S. Many roads to maturity: MicroRNA biogenesis pathways and their regulation. *Nat. Cell Biol.* **2009**, *11*, 228–234.

76. Tang, R.; Li, L.; Zhu, D.; Hou, D.; Cao, T.; Gu, H.; Zhang, J.; Chen, J.; Zhang, C.Y.; Zen, K. Mouse miRNA-709 directly regulates miRNA-15a/16-1 biogenesis at the posttranscriptional level in the nucleus: Evidence for a microRNA hierarchy system. *Cell Res.* **2012**, *22*, 504–515.

77. Yi, R.; Qin, Y.; Macara, I.G.; Cullen, B.R. Exportin-5 mediates the nuclear export of pre-microRNAs and short hairpin RNAs. *Genes Dev.* **2003**, *17*, 3011–3016.

78. Lund, E.; Guttinger, S.; Calado, A.; Dahlberg, J.E.; Kutay, U. Nuclear export of microRNA precursors. *Science* **2004**, *303*, 95–98.

79. Weinmann, L.; Hock, J.; Ivacevic, T.; Ohrt, T.; Mutze, J.; Schwille, P.; Kremmer, E.; Benes, V.; Urlaub, H.; Meister, G. Importin 8 is a gene silencing factor that targets argonaute proteins to distinct mRNAs. *Cell* **2009**, *136*, 496–507.

80. Gregory, R.I.; Chendrimada, T.P.; Cooch, N.; Shiekhattar, R. Human RISC couples microRNA biogenesis and posttranscriptional gene silencing. *Cell* **2005**, *123*, 631–640.

81. Chendrimada, T.P.; Gregory, R.I.; Kumaraswamy, E.; Norman, J.; Cooch, N.; Nishikura, K.; Shiekhattar, R. TRBP recruits the Dicer complex to Ago2 for microRNA processing and gene silencing. *Nature* **2005**, *436*, 740–744.

82. Lee, Y.; Hur, I.; Park, S.Y.; Kim, Y.K.; Suh, M.R.; Kim, V.N. The role of PACT in the RNA silencing pathway. *EMBO J.* **2006**, *25*, 522–532.

83. MacRae, I.J.; Ma, E.; Zhou, M.; Robinson, C.V.; Doudna, J.A. *In vitro* reconstitution of the human RISC-loading complex. *Proc. Natl. Acad. Sci. USA* **2008**, *105*, 512–517.

84. Diederichs, S.; Haber, D.A. Dual role for argonautes in microRNA processing and posttranscriptional regulation of microRNA expression. *Cell* **2007**, *131*, 1097–1108.

85. Nilsen, T.W. Mechanisms of microRNA-mediated gene regulation in animal cells. *Trends Genet.* **2007**, *23*, 243–249.

86. Luciano, D.J.; Mirsky, H.; Vendetti, N.J.; Maas, S. RNA editing of a miRNA precursor. *RNA* **2004**, *10*, 1174–1177.

87. Takeda, A.; Iwasaki, S.; Watanabe, T.; Utsumi, M.; Watanabe, Y. The mechanism selecting the guide strand from small RNA duplexes is different among argonaute proteins. *Plant Cell Physiol.* **2008**, *49*, 493–500.

88. Valencia-Sanchez, M.A.; Liu, J.; Hannon, G.J.; Parker, R. Control of translation and mRNA degradation by miRNAs and siRNAs. *Genes Dev.* **2006**, *20*, 515–524.

89. Carthew, R.W.; Sontheimer, E.J. Origins and Mechanisms of miRNAs and siRNAs. *Cell* **2009**, *136*, 642–655.

90. Blow, M.J.; Grocock, R.J.; van Dongen, S.; Enright, A.J.; Dicks, E.; Futreal, P.A.; Wooster, R.; Stratton, M.R. RNA editing of human microRNAs. *Genome Biol.* **2006**, *7*, R27.

91. Kawahara, Y.; Zinshteyn, B.; Sethupathy, P.; Iizasa, H.; Hatzigeorgiou, A.G.; Nishikura, K. Redirection of silencing targets by adenosine-to-inosine editing of miRNAs. *Science* **2007**, *315*, 1137–1140.

92. Kawahara, Y.; Megraw, M.; Kreider, E.; Iizasa, H.; Valente, L.; Hatzigeorgiou, A.G.; Nishikura, K. Frequency and fate of microRNA editing in human brain. *Nucleic Acids Res.* **2008**, *36*, 5270–5280.

93. Vesely, C.; Tauber, S.; Sedlazeck, F.J.; von Haeseler, A.; Jantsch, M.F. Adenosine deaminases that act on RNA induce reproducible changes in abundance and sequence of embryonic miRNAs. *Genome Res.* **2012**, *22*, 1468–1476.

94. Alon, S.; Mor, E.; Vigneault, F.; Church, G.M.; Locatelli, F.; Galeano, F.; Gallo, A.; Shomron, N.; Eisenberg, E. Systematic identification of edited microRNAs in the human brain. *Genome Res.* **2012**, *22*, 1533–1540.

95. Warf, M.B.; Shepherd, B.A.; Johnson, W.E.; Bass, B.L. Effects of ADARs on small RNA processing pathways in *C. elegans*. *Genome Res.* **2012**, *22*, 1488–1498.

96. Ekdahl, Y.; Farahani, H.S.; Behm, M.; Lagergren, J.; Ohman, M. A-to-I editing of microRNAs in the mammalian brain increases during development. *Genome Res.* **2012**, *22*, 1477–1487.

97. Heale, B.S.; Keegan, L.P.; McGurk, L.; Michlewski, G.; Brindle, J.; Stanton, C.M.; Caceres, J.F.; O'Connell, M.A. Editing independent effects of ADARs on the miRNA/siRNA pathways. *EMBO J.* **2009**, *28*, 3145–3156.

98. Yang, W.; Chendrimada, T.P.; Wang, Q.; Higuchi, M.; Seeburg, P.H.; Shiekhattar, R.; Nishikura, K. Modulation of microRNA processing and expression through RNA editing by ADAR deaminases. *Nat. Struct. Mol. Biol.* **2006**, *13*, 13–21.

99. Caudy, A.A.; Ketting, R.F.; Hammond, S.M.; Denli, A.M.; Bathoorn, A.M.; Tops, B.B.; Silva, J.M.; Myers, M.M.; Hannon, G.J.; Plasterk, R.H. A micrococcal nuclease homologue in RNAi effector complexes. *Nature* **2003**, *425*, 411–414.

100. Weissbach, R.; Scadden, A.D. Tudor-SN and ADAR1 are components of cytoplasmic stress granules. *RNA* **2012**, *18*, 462–471.

101. Kawahara, Y.; Zinshteyn, B.; Chendrimada, T.P.; Shiekhattar, R.; Nishikura, K. RNA editing of the microRNA-151 precursor blocks cleavage by the Dicer-TRBP complex. *EMBO Rep.* **2007**, *8*, 763–769.

102. Cullen, B.R. Viral and cellular messenger RNA targets of viral microRNAs. *Nature* **2009**, *457*, 421–425.

103. Skalsky, R.L.; Cullen, B.R. Viruses, microRNAs, and host interactions. *Annu. Rev. Microbiol.* **2010**, *64*, 123–141.

104. Iizasa, H.; Wulff, B.E.; Alla, N.R.; Maragkakis, M.; Megraw, M.; Hatzigeorgiou, A.; Iwakiri, D.; Takada, K.; Wiedmer, A.; Showe, L.; Lieberman, P.; Nishikura, K. Editing of Epstein-Barr virus-encoded BART6 microRNAs controls their dicer targeting and consequently affects viral latency. *J. Biol. Chem.* **2010**, *285*, 33358–33370.

105. Lei, T.; Yuen, K.S.; Tsao, S.W.; Chen, H.; Kok, K.H.; Jin, D.Y. Perturbation of biogenesis and targeting of Epstein-Barr virus-encoded miR-BART3 microRNA by A-to-I editing. *J. Gen. Virol.* **2013**, doi:10.1099/vir.0.056226-0.

106. Gandy, S.Z.; Linnstaedt, S.D.; Muralidhar, S.; Cashman, K.A.; Rosenthal, L.J.; Casey, J.L. RNA editing of the human herpesvirus 8 kaposin transcript eliminates its transforming activity and is induced during lytic replication. *J. Virol.* **2007**, *81*, 13544–13551.

107. Pfeffer, S.; Sewer, A.; Lagos-Quintana, M.; Sheridan, R.; Sander, C.; Grasser, F.A.; van Dyk, L.F.; Ho, C.K.; Shuman, S.; Chien, M.; *et al.* Identification of microRNAs of the herpesvirus family. *Nat. Methods* **2005**, *2*, 269–276.

108. Choudhury, Y.; Tay, F.C.; Lam, D.H.; Sandanaraj, E.; Tang, C.; Ang, B.T.; Wang, S. Attenuated adenosine-to-inosine editing of microRNA-376a* promotes invasiveness of glioblastoma cells. *J. Clin. Investig.* **2012**, *122*, 4059–4076.

109. Maas, S.; Patt, S.; Schrey, M.; Rich, A. Underediting of glutamate receptor GluR-B mRNA in malignant gliomas. *Proc. Natl. Acad. Sci. USA* **2001**, *98*, 14687–14692.

110. Galeano, F.; Rossetti, C.; Tomaselli, S.; Cifaldi, L.; Lezzerini, M.; Pezzullo, M.; Boldrini, R.; Massimi, L.; Di Rocco, C.M.; Locatelli, F.; Gallo, A. ADAR2-editing activity inhibits glioblastoma growth through the modulation of the CDC14B/Skp2/p21/p27 axis. *Oncogene* **2012**, *32*, 998–1009.

111. Liang, H.; Landweber, L.F. Hypothesis: RNA editing of microRNA target sites in humans? *RNA* **2007**, *13*, 463–467.

112. Peng, Z.; Cheng, Y.; Tan, B.C.; Kang, L.; Tian, Z.; Zhu, Y.; Zhang, W.; Liang, Y.; Hu, X.; Tan, X.; *et al.* Comprehensive analysis of RNA-Seq data reveals extensive RNA editing in a human transcriptome. *Nat. Biotechnol.* **2012**, *30*, 253–260.

113. Gu, T.; Buaas, F.W.; Simons, A.K.; Ackert-Bicknell, C.L.; Braun, R.E.; Hibbs, M.A. Canonical A-to-I and C-to-U RNA editing is enriched at 3'UTRs and microRNA target sites in multiple mouse tissues. *PLoS One* **2012**, *7*, e33720.

114. Wang, Q.; Hui, H.; Guo, Z.; Zhang, W.; Hu, Y.; He, T.; Tai, Y.; Peng, P.; Wang, L. ADAR1 regulates ARHGAP26 gene expression through RNA editing by disrupting miR-30b-3p and miR-573 binding. *RNA* **2013**, *19*, 1525–1536.

115. De Hoon, M.J.; Taft, R.J.; Hashimoto, T.; Kanamori-Katayama, M.; Kawaji, H.; Kawano, M.; Kishima, M.; Lassmann, T.; Faulkner, G.J.; Mattick, J.S.; *et al.* Cross-mapping and the identification of editing sites in mature microRNAs in high-throughput sequencing libraries. *Genome Res.* **2010**, *20*, 257–264.

116. Chiang, H.R.; Schoenfeld, L.W.; Ruby, J.G.; Auyeung, V.C.; Spies, N.; Baek, D.; Johnston, W.K.; Russ, C.; Luo, S.; Babiarz, J.E.; *et al.* Mammalian microRNAs: Experimental evaluation of novel and previously annotated genes. *Genes Dev.* **2010**, *24*, 992–1009.

117. Galeano, F.; Tomaselli, S.; Locatelli, F.; Gallo, A. A-to-I RNA editing: The "ADAR" side of human cancer. *Semin. Cell Dev. Biol.* **2012**, *23*, 244–250.

10

The Role of MicroRNAs in Breast Cancer Stem Cells

Daniela Schwarzenbacher, Marija Balic and Martin Pichler *

Division of Clinical Oncology, Department of Medicine, Medical University of Graz, Auenbruggerplatz 15, 8036 Graz, Austria; E-Mails: daniela.schwarzenbacher@medunigraz.at (D.S.); marija.balic@medunigraz.at (M.B.)

* Author to whom correspondence should be addressed; E-Mail: martin.pichler@medunigraz.at;

Abstract: The concept of the existence of a subset of cancer cells with stem cell-like properties, which are thought to play a significant role in tumor formation, metastasis, resistance to anticancer therapies and cancer recurrence, has gained tremendous attraction within the last decade. These cancer stem cells (CSCs) are relatively rare and have been described by different molecular markers and cellular features in different types of cancers. Ten years ago, a novel class of molecules, small non-protein-coding RNAs, was found to be involved in carcinogenesis. These small RNAs, which are called microRNAs (miRNAs), act as endogenous suppressors of gene expression that exert their effect by binding to the 3'-untranslated region (UTR) of large target messenger RNAs (mRNAs). MicroRNAs trigger either translational repression or mRNA cleavage of target mRNAs. Some studies have shown that putative breast cancer stem cells (BCSCs) exhibit a distinct miRNA expression profile compared to non-tumorigenic breast cancer cells. The deregulated miRNAs may contribute to carcinogenesis and self-renewal of BCSCs via several different pathways and can act either as oncomirs or as tumor suppressive miRNAs. It has also been demonstrated that certain miRNAs play an essential role in regulating the stem cell-like phenotype of BCSCs. Some miRNAs control clonal expansion or maintain the self-renewal and anti-apoptotic features of BCSCs. Others are targeting the specific mRNA of their target genes and thereby contribute to the formation and self-renewal process of BCSCs. Several miRNAs are involved in epithelial to mesenchymal transition, which is often implicated in the process of formation of CSCs. Other miRNAs were shown to be involved in the increased chemotherapeutic resistance of BCSCs. This review highlights the recent findings and crucial role of miRNAs in the maintenance, growth and behavior of BCSCs, thus indicating the potential for novel diagnostic, prognostic and therapeutic miRNA-based strategies.

Keywords: microRNAs; breast cancer; tumor stem cells

---

1. Introduction

1.1. Breast Cancer and Breast Cancer Stem Cells

Breast cancer is the most frequently diagnosed cancer and the leading cause of cancer-related death among women worldwide [1]. According to the American Cancer Society, an estimated number of 232,340 new cases of breast cancer will be diagnosed in women and approximately 39,620 female breast cancer deaths are estimated in the United States for 2013. Thus, it is expected that breast cancer will account for 29% of all new cases of cancer in 2013 among women [2]. Hence, it is essential to gain a better understanding of the molecular mechanisms of breast cancer formation to ensure more efficient cancer treatments [3]. Since human breast tumors are very heterogeneous regarding time since diagnosis, histological pattern and clinical course, breast cancer can be classified into several subtypes based on distinct gene expression profiles [4]. In general, heterogeneity within and among several subtypes of cancers can arise in various ways. One common model to explain the usually observed heterogeneity of tumors is the cancer stem cell model [5]. According to the cancer stem cell hypothesis, tumors are hierarchically organized with cancer stem cells (CSCs) at the top [6] and the non-tumorigenic cell population forming the bulk of the tumor [7]. The term CSC indicates that only a subset of cancer cells in the tumor has self-renewal (asymmetric and symmetric division) capacity and the ability to produce all types of cancer cells within the tumor [6,8]. Targeting CSCs is of great interest as CSCs are considered to be more resistant to radiotherapy and chemotherapy, and are also thought to be responsible for the dissemination and growth of metastases [6].

Breast cancer stem cells (BCSCs) were originally described by Al-Hajj *et al.* in 2003. They isolated a tumorigenic subset of cancer cells from human breast tumors based on the expression of the surface markers CD44$^+$, CD24$^{-/low}$ and ESA$^+$ (CD is short for cluster of differentiation, ESA is short for epithelial specific antigen).This was the first evidence for the existence of CSCs in breast cancer and they were the first to show that only the minority of breast cancer cells with a CD44$^+$, CD24$^{-/low}$ and ESA$^+$ phenotype have the ability to form new tumors in NOD/SCID mice [9]. In 2007, Ginestier *et al.* indicated that high aldehyde dehydrogenase 1 (ALDH1) expression is also characteristic for BCSCs and therefore extended the BCSC phenotype on CD44$^+$, CD24$^{-/low}$, ESA$^+$ and alternatively ALDH$^+$ [10]. Dontu and colleagues developed an *in vitro* cell culture system under non-adherent conditions for human mammary epithelial cells. Under these conditions only cells with stem cell-like properties are able to survive. These cells can proliferate and build so called mammospheres, which are multicellular formations and are thought to contain high numbers of mammary stem cells as well as progenitor cells [11]. These mammosphere cultures are commonly used in experimental studies to enrich BCSCs. However, other studies indicate that the currently used markers for BCSCs remain controversial. Lehmann *et al.* discovered that some markers used to identify putative breast tumor initiating cells do not correlate with *in vivo* tumorigenicity. Therefore it may be essential to determine other markers and/or factors that affect the increased tumorigenicity of BCSCs [12]. Another recently published study also revealed that these markers alone might not be sufficient to distinguish tumorigenic from non-tumorigenic cells. They demonstrated that tumor cells which are negative to the common CSC markers are also capable of inducing tumor growth *in vivo* [13].

1.2. Epithelial to Mesenchymal Transition in Cancer

Epithelial to mesenchymal transition (EMT) is an essential process during embryonic development in many species of mammals [14]. The transformation of epithelial to mesenchymal cells has also been associated with cancer progression, because the EMT program often becomes activated during cancer invasion and metastasis. This process is characterized by the loss of the epithelial marker E-cadherin, loss of cell–cell contact and cell polarity as well as an increased cell motility [15]. EMT has also been directly linked with the CSC phenotype. Induction of EMT in breast cancer cells leads to generation of cells with stem cell like properties [16].

1.3. Function and Biogenesis of miRNAs

It is now clear, that miRNAs together with other non-coding RNAs (long non-coding RNAs, small nucleolar RNAs and ultraconserved regions) contribute to carcinogenesis. Aberrantly expressed miRNAs are involved in initiation and progression of cancer. MiRNAs are small non-coding RNAs with a length of approximately 22 nucleotides (nt), which act as endogenous inhibitors of gene function. They modulate the expression of their target genes by either degrading their target mRNA or inhibiting their translation [17] through pairing of miRNA sequences to complementary bases on the target mRNA [18]. MiRNAs can function both as oncogenes and as tumor suppressors [19,20] and are considered as emerging potential candidates for improved cancer diagnosis, prognosis and therapy [21–24].

Biogenesis of miRNAs is a complex process. Most miRNA genes are transcribed by RNA polymerase II as long primary transcripts containing a stemloop structure. This pri-miRNA is cleaved by the RNase III endonuclease Drosha and the double-stranded RNA-binding domain (dsRBD) protein DGCR8/Pasha in the nucleus. The cleavage produces a ~70 nt hairpin precursor miRNA (pre-miRNA) with a 2-nt 3' overhang. The 3' overhang is recognized by Exportin-5, which transports the pre-miRNA into the cytoplasm. There, the pre-miRNA is cleaved by another RNase III endonuclease, Dicer. Dicer interacts with the dsRBD proteins TRBP/Loquacious and cleaving produces the mature ~22 nt miRNA: miRNA* duplex. The miRNA strand is usually incorporated to a RNA-induced silencing complex (RISC), a ribonuclein complex, while the miRNA* strand is typically degraded. When the miRNA is bound to RISC, the miRNA and its target mRNA can interact by base-pairing. The target mRNAs can then be cleaved and degraded or repressed in their translation [25].

In different breast cancer subtypes (basal, luminal cancers) miRNAs are differentially expressed and some miRNAs are associated with a specific ER, PR and Her2/neu status in human breast cancers [26]. MiRNAs can also function as potential targets of anticancer therapies. In breast cancer several miRNAs may possibly play a key role in cancer progression. Different studies have shown that silencing or overexpression of particular miRNAs can have an effect on the process of invasion and development of metastases in human breast cancers [27], showing a potential therapeutic application of miRNAs in breast cancer. The aim of this review is to summarize the involvement of different miRNAs in the formation and regulation of human BCSCs. Table 1 gives an overview of the roles of different miRNAs in BCSCs which are described in this review.

Table 1. Roles of different miRNAs in BCSCs.

miRNA	Roles in BCSCs
let-7 family	downregulated in BCSCs targets RAS and HMGA2 acts as tumor suppressor Lin28 blocks *let-7* biogenesis and promotes tumorigenic activity in breast cancer influences mammosphere formation and proliferation *in vitro*
	affects tumor formation ability and metastatic potential *in vivo*
	reduced *let-7* expression inhibits differentiation, maintains proliferation and promotes EMT
miR-200 family	downregulated in BCSCs
	targets Bmi-1 and Suz12
	regulation of EMT
	relevant for stem cell functions in cancer cells (self-renewal, clonal expansion, differentiation) *in vitro*
	induces stem-like properties
miR-30 family	downregulated in BCSCs
	targets Ubc9, ITGB3 and AVEN
	influences self-renewal capacity and anti-apoptotic features
	important for modulation of the stem-like properties of BCSCs
	regulates non-attachment growth of mammospheres and mammosphere formation ability
	controls genes involved in apoptosis and proliferation in BCSCs
miR-128	downregulated in BCSCs
	targets Bmi-1 and ABCC5
	link to chemotherapeutical resistance and survival rates of breast cancer patients
	influences number and size of mammospheres *in vitro*
	reduced tumor growth and induced apoptosis *in vivo*
miR-34c	downregulated in BCSCs
	targets Notch4
	influences self-renewal and EMT
	acts on mammosphere formation *in vitro*
	is epigenetically regulated via methylation
	controls migration of tumor cells
miR-16	downregulated in BCSCs
	targets Wip1
	influences number and size of mammospheres and cell proliferation
	responsible for sensitivity to chemotherapeutic drug doxorubicin
miR-181	upregulated in BCSCs
	targets ATM
	TGF-β induces mammosphere formation by upregulation of *miR-181*
miR-495	upregulated in BCSCs
	targets REDD1
	leads to downregulation of E-cadherin
	promotes colony formation
	leads to increased tumor formation *in vivo*
	responsible for maintaining a stem-cell line phenotype

2. Particular miRNAs and Their Role in Tumor-Initiating BCSCs

2.1. miRNAs Down-Regulated in BCSCs

2.1.1. let-7 Family

Yu and colleagues were the first to investigate the expression of miRNAs in BCSCs in 2007. They compared the miRNA expression profile in self-renewing BCSCs and differentiated cells from breast cancer cell lines as well as in samples from primary breast tumors. They enriched BCSCs of a human breast cancer cell line (SKBR3) by passaging them in NOD/SCID mice treated with chemotherapeutical agents. The tumors contained a high percentage of CD44$^+$CD24$^{-/\text{low}}$ cells and showed high mammosphere formation ability *in vitro*. The miRNA *let-7* was found to be the most consistently down-regulated miRNA in tumor-initiating cells (SK-3rd) compared to the non-self-renewing population of cancer cells. *let-7* expression increased when the cells differentiated to non-tumorigenic cancer cells. The *let-7* family functions as a well-known tumor suppressor and targets the oncogenes rat sarcoma (RAS) and high mobility group AT-hook 2 (HMGA2). The HMGA2 gene is involved in mesenchymal cell differentiation and tumor formation. Lentiviral-mediated re-expression of *let-7* resulted in reduced mammospheres formation, proliferation and a reduced number of undifferentiated stem-like cells *in vitro*. *let-7* expression also inhibited the tumor formation ability in NOD/SCID mice *in vivo* and *let-7* expressing tumors had less metastatic potential. Therefore, *let-7* is apparently responsible for the regulation of multiple stem cell-like properties of BCSCs, because reduced *let-7* expression inhibits differentiation and maintains proliferation [28]. As *let-7* is a common tumor suppressor and has anti-proliferative properties, it can regulate cell differentiation and apoptotic pathways. Its down-regulation has been reported in several cancers and reconstitution of regular *let-7* expression has been shown to inhibit cancer growth [29–31]. These findings suggest *let-7* as a potential molecular marker for BCSCs with a potential as therapeutical target in anti-cancer therapy [32]. In this context, the Lin28 protein is a RNA-binding protein which regulates *let-7* family members and expression of Lin28 blocks the biogenesis of *let-7* [33]. One recently published study indicates that suppression of *let-7* through Lin28 promotes tumorigenicity in breast cancer cells [34]. Inflammatory cytokines can lead to the induction of EMT. Guo and colleagues showed that inflammatory cytokines can trigger signal transducer and activator of transcription factor 3 (Stat3) which promotes Lin28 transcription. As a consequence, this process results in repression of *let-7* expression and up-regulation of the *let-7* target HMGA2. As HMGA2 is involved in EMT, this event leads to increased levels of mesenchymal markers. These findings suggest that the inflammation-induced and Stat3 mediated Lin28-let-7-HMGA2 signaling pathway might be involved in regulation of self-renewal and differentiation in CSCs [35].

2.1.2. miR-200 Family

The *miR-200* family consists of five members of miRNAs: *miR-200a*, *miR-200b*, *miR-200c*, *miR-141* and *miR-429*. The family can be divided into genetically different subfamilies (gene clusters) according to their location at two different chromosomes: the *miR-200b/miR-200a/miR-429* gene cluster on chromosome 1 and the *miR-200c/miR-141* gene cluster on chromosome 12 [36]. Several recent studies have associated *miR-200* family members and their target mRNAs with establishment, maintenance and regulation of the BCSC phenotype. One of the first studies that showed an involvement of the *miR-200* family in BCSCs came from Shimono *et al.* in 2009. Comparing the miRNA expression profile between fluorescence-activated cell sorted CD44$^+$CD24$^{-/low}$ lineage BCSCs and the remaining non-tumorigenic human breast cancer cells, they found 37 miRNAs differentially expressed between non-tumorigenic cancer cells and BCSCs in eleven human breast cancer samples. They showed that three clusters of miRNAs (*miRNA-200c-141*, *miR-200b-200a-429* and *miR-183-96-182*) were consistently down-regulated in BCSCs, in normal mammary stem cells and in embryonal carcinoma cells. This finding suggests that the down-regulation of these miRNAs may be relevant for stem cell functions in cancer cells, such as self-renewal or differentiation. Downregulation of *miR-200c* in BCSCs suppressed the expression of polycomb ring finger oncogene (B lymphoma Mo-MLV insertion region 1 homolog, Bmi-1), which is a regulator of stem cell self-renewal. *miR-200c* inhibited the clonal expansion of BCSCs *in vitro*. Interestingly, *miR-200c* repressed the ability of normal mammary stem cells to generate mammary ducts and also inhibits the tumor-formation capacity of BCSCs *in vivo*. These results indicate that down-regulation of *miR-200c* might be a molecular link between CSCs and normal stem cells [37]. Consistent to that the *miR-200* family was shown to be inhibited during BCSC formation in an inducible CSC model. One of the down-regulated *miR-200* family members in BCSCs was *miR-200b* and inhibition of *miR-200b* increased the formation of BCSCs. Down-regulation of *miR-200b* resulted in increased Suz12 expression (a subunit of a polycomb repressor complex, PRC2), which led to repression of E-cadherin [38]. This inhibition of E-cadherin through miRNA is sufficient to cause EMT [39]. Overexpression of *miR-200b* or inhibition of Suz12 significantly reduced BCSC growth. In tumors of breast cancer patients, *miR-200b* and Suz12 expression were inversely correlated. Apparently the *miR-200b*-Suz12-cadherin pathway is an important pathway to induce and sustain growth of BCSCs and the invasion and migration abilities of BCSCs [38]. Another recently published study confirmed the role of the *miR-200* family in BCSCs by showing that the spontaneous conversion of immortalized human mammary epithelial cells to a stem-like phenotype with mesenchymal and less differentiated properties was accompanied by loss of *miR-200* expression. In mammospheres, *miR-200a*, *miR-200b* and *miR-200c* were described as down-regulated. Expression of *miR-200* was shown to be epigenetically regulated by histone-modifications and DNA promoter methylation. Also, in samples of pleural or ascites effusions of breast cancer patients the *miR-200* family members were consistently down-regulated in CD44$^+$CD24$^{-/low}$ putative BCSCs. Re-expression of *miR-200* in these stem-like cells led to a partial reprogramming to a non-stem like phenotype, and the cells also did undergo mesenchymal to epithelial transition (MET). These data indicate that the *miR-200* family is functionally inducing the stem-like properties [40].

2.1.3. *miR-30* Family

Similar to the *let-7* family, Yu and colleagues also demonstrated the down-regulation of *miR-30*, particularly *miR-30e*, in tumor initiating BCSCs (in mammospheres SK-3rd as well as in primary BCSCs obtained from breast cancer patients). In accordance to the down-regulation of *miR-30e*, the protein levels of two direct target genes of *miR-30e*, ubiquitin-conjugating enzyme 9 (Ubc9) and integrin b3 (ITGB3), were significantly up-regulated. When *miR-30e* was constitutively expressed in

BCSCs, their self-renewal capacity was impaired. This inhibition occurred through decreased Ubc9 levels and induction of apoptosis via silencing of ITGB3. Blocking of *miR-30e* in differentiated breast cancer cells on the other hand led to regeneration of their self-renewal capacity. Overexpression of *miR-30e* in NOD/SCID mice reduced tumorigenesis and lung metastases, while blocking of *miR-30e* expression enhanced tumor formation and metastases. These results indicate that reduction of *miR-30* expression is responsible for maintaining the self-renewal and anti-apoptotic features of BCSCs. *miR-30* can therefore be considered as an important miRNA for modulation of the stem-like properties of BCSCs [41]. Down-regulation of *miR-30* family members in non-adherent mammospheres compared to breast cancer cells under adherent conditions was recently confirmed in an independent study. BCSCs growth under non-attachment conditions displayed a different miRNA expression pattern compared to adherent parental cells and members of the *miR-30* family were found to be the most consistently down-regulated miRNAs in putative BCSCs. Especially *miR-30a* was found to regulate the non-attachment growth of mammospheres. Overexpression of *miR-30a* significantly reduced the mammosphere formation ability, while inhibition of *miR-30a* dramatically increased the number of mammospheres in the human breast cancer cell line MCF-7.These results confirm the relevance of this miRNA in sustaining the growth of BCSCs under non-attachment conditions. Also down-regulation of potential *miR-30a* targets after overexpressing *miR-30a* was shown. Among the potential targets, the anti-apoptotic protein AVEN was one of the most significantly down-regulated genes after overexpression of *miR-30*.This study confirms that *miR-30* family members can control expression of genes involved in apoptosis and proliferation in BCSCs [42].

2.1.4. *miR-128*

The level of *miR-128* was shown to be significantly reduced in mammospheric BCSCs in two breast cancer cell lines (SK-3rd and MCF-7) and in BCSCs isolated from primary breast cancer patients. This reduction increased the protein levels of the polycomb ring finger oncogene Bmi-1 and ATP-binding cassette sub-family C member 5 (ABCC5), which are targets of *miR-128* [43]. Tumor initiating cells with stem cell-like features were shown to be more resistant to chemotherapeutical agents and radiotherapy than more differentiated tumor cells [9]. Ectopic expression of *miR-128* decreased Bmi-1 and ABCC5 levels in BCSCs and led to an enhanced pro-apoptotic and DNA-damaging effect when treated with the chemotherapeutical agent doxorubicin. This observation indicates a possible therapeutic potential of this miRNA. Furthermore, a reduction of *miR-128* in breast tumor tissues was linked with chemotherapeutic resistance and poor survival rates of breast cancer patients. Consequently the reduced levels of *miR-128* in BCSCs are likely to induce increased chemotherapeutic resistance [43]. In another study ectopic expression of *miR-128* led to a decreased number and size of mammospheres in an *in vitro* cell culture model, whereas *miR-128* depletion caused an increase of mammosphere growth. The *in vivo* tumor-initiating potential was also evaluated and it has been shown that overexpression of *miR-128-2* repressed the ability to form tumors in mice. Forced expression of *miR-128* reduced tumor growth *in vivo* and induced apoptosis [44].

2.1.5. *miR-34c*

MiR-34c has been identified as a putative tumor suppressor and has been reported to inhibit invasion, proliferation and to promote apoptosis. Reduced expression of *miR-34c* was revealed in two human breast cancer cell lines (MCF-7 and SK-3rd) enriched for BCSCs. Down-regulation of *miR-34c* apparently occurred via hypermethylation of the promoter region of BCSCs and resulted in increased self-renewal and epithelial-mesenchymal transition of these cells. Ectopic expression of this miRNA inhibited EMT and reduced mammosphere formation and self-renewal potential. It also led to silencing of its target gene Neurogenic Locus Notch Homolog Protein 4 (Notch4) and suppressed the migration of tumor cells. This study proposed *miR-34c* as a possible target for BCSCs, as this epigenetically regulated miRNA apparently controls self-renewal and EMT in these tumor initiating cells [45].

2.1.6. *miR-16*

A decreased level of *miR-16* has been shown in mammospheres derived from mammary tumors in mice compared to the whole tumor cell population. The oncogene wild-type p53-induced phosphatase 1 (Wip1) is apparently regulated by *miR-16* and protein levels of Wip1 were consequently increased in these mammospheres. Overexpression of *miR-16* in the human breast cancer cell line MCF-7 as well as inhibition of Wip1 decreased the number and size of mammospheres. When *miR-16* was overexpressed in the human breast cancer cell line MCF-7 it suppressed cell proliferation and led to an increased sensitivity to the chemotherapeutic drug doxorubicin. These findings indicate that *miR-16* is another miRNA that might be responsible for regulation of the proliferation and differentiation of mammary CSCs [46].

2.2. miRNAs Up-Regulated in BCSCs

2.2.1. *miR-181* Family

In three human breast cancer cell lines (BT474, MDA361 and MCF7) levels of *miR-181* family members were reported to be increased in tumor initiating mammospheres compared to non-tumorigenic parental cells. Transforming growth factor-β (TGF-β) seemed to induce sphere formation by up-regulation of *miR-181* at the post-transcriptional level. A potential target of *miR-181* is the serine/threonine kinase Ataxia telangiectasia mutated (ATM) which acts as a tumor suppressor. ATM was reduced in mammospheres and after treatment with TGF-β. This study suggests that the TGF-β pathway and the *miR-181* family interacts and plays a role in regulating the BCSC phenotype [47].

2.2.2. *miR-495*

Hwang-Verslues and colleagues isolated a novel highly tumorigenic subpopulation of BCSCs based on the surface markers PROCR$^+$/ESA$^+$ (PROCR is short for protein C receptor). In this BCSC subpopulation and also in the more commonly used CD44$^+$CD24$^{-/low}$ subpopulation, *miR-495* was highly up-regulated. As *miR-495* was found to be up-regulated in two distinct BCSC subpopulations, this mechanism might be important for maintaining stem cell-like features. This up-regulation of *miR-495* is regulated by the transcription factors E12 and E47. Overexpression of *miR-495 in vitro* promoted colony formation. *miR-495* overexpressing cells in mice led to significantly higher tumor formation *in vivo*. These results indicate that ectopic expression of *miR-495* in human breast cancer cells increases tumorigenesis *in vivo* and enhances colony formation *in vitro*. E-cadherin expression, a marker considered surrogate for EMT, was down-regulated by overexpression of *miR-495*. Decreased E-cadherin expression was responsible for promoting cell invasion. *miR-495* also targets REDD1 (short for regulated in development and DNA damage responses) which is a factor for enhanced hypoxia resistant cell proliferation. Summarizing, these findings suggest an up-regulation of *miR-495* by E12 and E47 which in turn contributes to down-regulation of E-cadherin and REDD1, finally resulting in maintaining a stem cell-like phenotype in breast cancer [48].

3. Conclusions

Understanding the role of miRNAs in the biology of CSCs can provide promising advances for cancer treatment and might be helpful to improve cancer diagnosis. As miRNAs are post-transcriptional regulators of gene expression, they also play important roles in carcinogenesis. Several independent studies that are reviewed here have shown a dysregulation of several different miRNAs in BCSCs. Anticancer-therapy with miRNAs could eliminate the CSC self-renewal capacity and their anti-apoptotic features which can improve the development of resistance against current cancer treatment. For this reason, future research should address the therapeutic potential of miRNAs to prevent cancer progression, relapse and formation of metastases by eliminating CSCs.

Acknowledgements

This work was supported by funds of the Oesterreichische Nationalbank (Anniversary Fund, project number: 14869) and by the Start foundation of the Medical University of Graz (both to Martin Pichler).

References

1. Jemal, A.; Bray, F.; Center, M.; Ferlay, J.; Ward, E.; Forman, D. Global cancer statistics. *CA: Cancer J. Clin.* **2011**, *61*, 69–90.

2. Siegel, R.; Naishadham, D.; Jemal, A. Cancer statistics, 2013. *CA: Cancer J. Clin.* **2013**, *63*, 11–30.

3. Liu, H. MicroRNAs in breast cancer initiation and progression. *Cell. Mol. Life Sci.* **2012**, *69*, 3587–3599.

4. Perou, C.M.; Sørlie, T.; Eisen, M.B.; van de Rijn, M.; Jeffrey, S.S.; Rees, C.A.; Pollack, J.R.; Ross, D.T.; Johnsen, H.; Akslen, L.A.; *et al.* Molecular portraits of human breast tumours. *Nature* **2000**, *406*, 747–752.

5. Magee, J.; Piskounova, E.; Morrison, S.J. Cancer stem cells: Impact, heterogeneity, and uncertainty. *Cancer Cell* **2012**, *21*, 283–296.

6. Visvader, J.E.; Lindeman, G. Cancer stem cells in solid tumours: accumulating evidence and unresolved questions. *Nat. Rev. Cancer.* **2008**, *8*, 755–768.

7. Vincent, A.; van Seuningen, I. On the epigenetic origin of cancer stem cells. *Biochim. Biophys. Acta* **2012**, *1826*, 83–88.

8. Wicha, M.; Liu, S.; Dontu, G. Cancer stem cells: An old idea—A paradigm shift. *Cancer Res.* **2006**, *66*, 1883–1890.

9.   Al-Hajj, M. Prospective identification of tumorigenic breast cancer cells. *Proc. Natl. Acad. Sci. USA* **2003**, *100*, 3983–3988.

10. Ginestier, C.; Hur, M.; Charafe-Jauffret, E. ALDH1 is a marker of normal and malignant human mammary stem cells and a predictor of poor clinical outcome. *Cell Stem Cell* **2007**, *1*, 555–567.

11. Dontu, G.; Abdallah, W. *In vitro* propagation and transcriptional profiling of human mammary stem/progenitor cells. *Genes Dev.* **2003**, *17*, 1253–1270.

12. Lehmann, C.; Jobs, G.; Thomas, M.; Burtscher, H.; Kubbies, M. Established breast cancer stem cell markers do not correlate with *in vivo* tumorigenicity of tumor-initiating cells. *Int. J. Oncol.* **2012**, *41*, 1932–1942.

13. Huang, S.-D.; Yuan, Y.; Tang, H.; Liu, X.-H.; Fu, C.-G.; Cheng, H.-Z.; Bi, J.-W.; Yu, Y.-W.; Gong, D.-J.; Zhang, W.; *et al*. Tumor cells positive and negative for the common cancer stem cell markers are capable of initiating tumor growth and generating both progenies. *PLoS One* **2013**, doi:10.1371/journal.pone.0054579.

14. Kiesslich, T.; Berr, F.; Alinger, B.; Kemmerling, R.; Pichler, M.; Ocker, M.; Neureiter, D. Current status of therapeutic targeting of developmental signalling pathways in oncology. *Curr. Pharm. Biotechnol.* **2012**, *13*, 2184–2220.

15. Thiery, J.P. Epithelial-mesenchymal transitions in tumour progression. *Nat. Rev. Cancer* **2002**, *2*, 442–454.

16. Mani, S.A.; Guo, W.; Liao, M.-J.; Eaton, E.N.; Ayyanan, A.; Zhou, A.Y.; Brooks, M.; Reinhard, F.; Zhang, C.C.; Shipitsin, M.; *et al*. The epithelial-mesenchymal transition generates cells with properties of stem cells. *Cell* **2008**, *133*, 704–715.

17. Nana-Sinkam, S.P.; Croce, C.M. Clinical applications for microRNAs in cancer. *Clin. Pharmacol. Ther.* **2013**, *93*, 98–104.

18. Munker, R.; Calin, G.A. MicroRNA profiling in cancer. *Clin. Sci.* **2011**, *121*, 141–158.

19. Calin, G.A.; Croce, C.M. MicroRNA signatures in human cancers. *Nat. Rev. Cancer* **2006**, *6*, 857–866.

20. Al-Ali, B.M.; Ress, A.L.; Gerger, A.; Pichler, M. MicroRNAs in renal cell carcinoma: implications for pathogenesis, diagnosis, prognosis and therapy. *Anticancer Res.* **2012**, *32*, 3727–3732.

21. Pichler, M.; Winter, E.; Stotz, M.; Eberhard, K.; Samonigg, H.; Lax, S.; Hoefler, G. Down-regulation of KRAS-interacting miRNA-143 predicts poor prognosis but not response to EGFR-targeted agents in colorectal cancer. *Br. J. Cancer* **2012**, *106*, 1826–1832.

22. Bach, D.; Fuereder, J.; Karbiener, M.; Scheideler, M.; Ress, A.L.; Neureiter, D.; Kemmerling, R.; Dietze, O.; Wiederstein, M.; Berr, F.; *et al*. Comprehensive analysis of alterations in the miRNome in response to photodynamic treatment. *J. Photochem. Photobiol. B* **2013**, *120*, 74–81.

23. Van Roosbroeck, K.; Pollet, J.; Calin, G. miRNAs and long noncoding RNAs as biomarkers in human diseases. *Expert Rev. Mol. Diagn.* **2013**, *13*, 183–204.

24. Calin, G.A.; Konopleva, M. Small gene, big number, many effects. *Blood* **2012**, *120*, 240–241.

25. Bushati, N.; Cohen, S.M. microRNA functions. *Annu. Rev. Cell Dev. Biol.* **2007**, *23*, 175–205.

26. Iorio, M.V.; Croce, C.M. MicroRNA dysregulation in cancer: diagnostics, monitoring and therapeutics. A comprehensive review. *EMBO Mol. Med.* **2012**, *4*, 143–159.

27. Ahmad, A. Pathways to breast cancer recurrence. *ISRN Oncol.* **2013**, *2013*, doi:10.1155/2013/290568.

28. Yu, F.; Yao, H.; Zhu, P.; Zhang, X.; Pan, Q.; Gong, C.; Huang, Y.; Hu, X.; Su, F.; Lieberman, J.; et al. Let-7 regulates self renewal and tumorigenicity of breast cancer cells. *Cell* **2007**, *131*, 1109–1123.

29. Johnson, C.D.; Esquela-Kerscher, A.; Stefani, G.; Byrom, M.; Kelnar, K.; Ovcharenko, D.; Wilson, M.; Wang, X.; Shelton, J.; Shingara, J.; et al. The let-7 microRNA represses cell proliferation pathways in human cells. *Cancer Res.* **2007**, *67*, 7713–7722.

30. Kumar, M.S.; Erkeland, S.J.; Pester, R.E.; Chen, C.Y.; Ebert, M.S.; Sharp, P.A.; Jacks, T. Suppression of non-small cell lung tumor development by the let-7 microRNA family. *Proc. Natl. Acad. Sci. USA* **2008**, *105*, 3903–3908.

31. Wang, Y.; Hu, X.; Greshock, J.; Shen, L.; Yang, X.; Shao, Z.; Liang, S.; Tanyi, J.L.; Sood, A.K.; Zhang, L. Genomic DNA copy-number alterations of the let-7 family in human cancers. *PLoS One* **2012**, doi:10.1371/journal.pone.0044399.

32. Barh, D.; Malhotra, R.; Ravi, B.; Sindhurani, P. Micro rna let-7: An emerging next-generation cancer therapeutic. *Curr. Oncol.* **2010**, *17*, 70–80.

33. Newman, M.; Thomson, J.; Hammond, S. Lin-28 interaction with the Let-7 precursor loop mediates regulated microRNA processing. *RNA* **2008**, *14*, 1539–1549.

34. Sakurai, M.; Miki, Y.; Masuda, M.; Hata, S.; Shibahara, Y.; Hirakawa, H.; Suzuki, T.; Sasano, H. LIN28: A regulator of tumor-suppressing activity of let-7 microRNA in human breast cancer. *J. Steroid Biochem. Mol. Biol.* **2012**, *131*, 101–106.

35. Guo, L.; Chen, C.; Shi, M.; Wang, F.; Chen, X.; Diao, D.; Hu, M.; Yu, M.; Qian, L.; Guo, N. Stat3-coordinated Lin-28-let-7-HMGA2 and miR-200-ZEB1 circuits initiate and maintain oncostatin M-driven epithelial-mesenchymal transition. *Oncogene* **2013**, doi:10.1038/onc.2012.573.

36. Park, S.-M.; Gaur, A.B.; Lengyel, E.; Peter, M.E. The miR-200 family determines the epithelial phenotype of cancer cells by targeting the E-cadherin repressors ZEB1 and ZEB2. *Genes Dev.* **2008**, *22*, 894–907.

37. Shimono, Y.; Ugalde, M.Z.; Cho, R.W.; Lobo, N.; Dalerba, P.; Qian, D.; Diehn, M.; Liu, H.; Panula, S.P.; Chiao, E.; et al. Down-regulation of miRNA-200c links breast cancer stem cells with normal stem cells. *Cell* **2009**, *138*, 592–603.

38. Iliopoulos, D.; Lindahl-Allen, M. Loss of miR-200 inhibition of Suz12 leads to polycomb-mediated repression required for the formation and maintenance of cancer stem cells. *Mol. Cell* **2010**, *39*, 761–772.

39. Onder, T.T.; Gupta, P.B.; Mani, S.A.; Yang, J.; Lander, E.S.; Weinberg, R.A. Loss of E-cadherin promotes metastasis via multiple downstream transcriptional pathways. *Cancer Res.* **2008**, *68*, 3645–3654.

40. Lim, Y.; Wright, J.; Attema, J.; Gregory, P.; Bert, A.; Smith, E.; Thomas, D.; Drew, P.; Khew-Goodall, Y.; Goodall, G. Epigenetic modulation of the miR-200 family is associated with transition to a breast cancer stem cell-like state. *J. Cell Sci.* **2013**, doi:10.1242/jcs.122275.

41. Yu, F.; Deng, H.; Yao, H.; Liu, Q.; Su, F.; Song, E. MiR-30 reduction maintains self-renewal and inhibits apoptosis in breast tumor-initiating cells. *Oncogene* **2010**, *29*, 4194–4204.

42. Ouzounova, M.; Vuong, T.; Ancey, P.-B.; Ferrand, M.; Durand, G.; Le-Calvez Kelm, F.; Croce, C.; Matar, C.; Herceg, Z.; Hernandez-Vargas, H. MicroRNA miR-30 family regulates non-attachment growth of breast cancer cells. *BMC Genomics* **2013**, doi:10.1186/1471-2164-14-139.

43. Zhu, Y.; Yu, F.; Jiao, Y.; Feng, J.; Tang, W.; Yao, H.; Gong, C.; Chen, J.; Su, F.; Zhang, Y.; *et al.* Reduced *miR-128* in breast tumor-initiating cells induces chemotherapeutic resistance via Bmi-1 and ABCC5. *Clin. Cancer Res.* **2011**, *17*, 7105–7115.

44. Qian, P.; Banerjee, A.; Wu, Z.-S.; Zhang, X.; Wang, H.; Pandey, V.; Zhang, W.-J.; Lv, X.-F.; Tan, S.; Lobie, P.E.; *et al.* Loss of SNAIL regulated *miR-128*-2 on chromosome 3p22.3 targets multiple stem cell factors to promote transformation of mammary epithelial cells. *Cancer Res.* **2012**, *72*, 6036–6050.

45. Yu, F.; Jiao, Y.; Zhu, Y.; Wang, Y.; Zhu, J.; Cui, X.; Liu, Y.; He, Y.; Park, E.-Y.; Zhang, H.; *et al.* MicroRNA 34c gene down-regulation via DNA methylation promotes self-renewal and epithelial-mesenchymal transition in breast tumor-initiating cells. *J. Biol. Chem.* **2012**, *287*, 465–473.

46. Zhang, X.; Wan, G.; Mlotshwa, S.; Vance, V.; Berger, F.; Chen, H.; Lu, X. Oncogenic Wip1 phosphatase is inhibited by *miR-16* in the DNA damage signaling pathway. *Cancer Res.* **2010**, *70*, 7176–7186.

47. Wang, Y.; Yu, Y.; Tsuyada, A.; Ren, X.; Wu, X.; Stubblefield, K.; Rankin-Gee, E.; Wang, S. Transforming growth factor β regulates the sphere-initiating stem cell-like feature in breast cancer through miRNA-181 and ATM. *Oncogene* **2011**, *30*, 1470–1480.

48. Hwang-Verslues, W.W.; Chang, P.-H.; Wei, P.-C.; Yang, C.-Y.; Huang, C.-K.; Kuo, W.-H.; Shew, J.-Y.; Chang, K.-J.; Lee, E.Y.-H.P.; Lee, W.-H. *miR-495* is upregulated by E12/E47 in breast cancer stem cells, and promotes oncogenesis and hypoxia resistance via downregulation of E-cadherin and REDD1. *Oncogene* **2011**, *30*, 2463–2474.

11

Regulation of Huntingtin Gene Expression by miRNA-137, -214, -148a and their Respective isomiRs

Emilia Kozlowska, Wlodzimierz J. Krzyzosiak * and Edyta Koscianska *

Department of Molecular Biomedicine, Institute of Bioorganic Chemistry, Polish Academy of Sciences, Noskowskiego 12/14 Str., 61-704 Poznan, Poland; E-Mail: emiliak@ibch.poznan.pl

* Authors to whom correspondence should be addressed;
  E-Mails: wlodkrzy@ibch.poznan.pl (W.J.K.); edytak@ibch.poznan.pl (E.K.);

Abstract: With the advent of deep sequencing technology, a variety of miRNA length and sequence variants, termed isomiRNAs (isomiRs), have been discovered. However, the functional roles of these commonly detected isomiRs remain unknown. In this paper, we demonstrated that miRNAs regulate the expression of the *HTT* gene, whose mutation leads to Huntington's disease (HD), a hereditary degenerative disorder. Specifically, we validated the interactions of canonical miRNAs, miR-137, miR-214, and miR-148a, with the HTT 3'UTR using a luciferase assay. Moreover, we applied synthetic miRNA mimics to examine whether a slight shifting of miRNA seed regions might alter the regulation of the *HTT* transcript. We also examined miR-137, miR-214, and miR-148a isomiRs and showed the activity of these isoforms on reporter constructs bearing appropriate sequences from the HTT 3'UTR. Hence, we demonstrated that certain 5'-end variants of miRNAs might be functional for the regulation of the same targets as canonical miRNAs.

Keywords: miRNA; isomiR; target validation; luciferase assay; huntingtin; Huntington's disease

1. Introduction

MicroRNAs (miRNAs) are 21- to 24-nucleotide noncoding RNAs that fine-tune gene expression. These molecules act at the posttranscriptional level through modulation of translational efficiency and/or destabilization of target transcripts (reviewed in [1]). miRNAs exert their functions through imperfect pairing with the 3' untranslated region (UTR) of target mRNAs. Nucleotides 2 through 8 of the miRNA, termed the "seed" sequence, are essential for target recognition and binding [2].

The canonical pathway of animal miRNA biogenesis includes two subsequent cleavages (reviewed in [3–6]). Briefly, precursor miRNAs (~60-nt pre-miRNAs) are generated from primary transcripts (pri-miRNAs) through cleavage with the ribonuclease Drosha and exported to the cytoplasm by Exportin-5. Then, ~22-nt miRNA duplexes are generated through cleavage with the ribonuclease Dicer. Only one miRNA strand (the guide strand) of the duplex induces Argonaute proteins (AGO) to form the programmed RNA-induced silencing complex (RISC); the other strand (the passenger strand, or miRNA*) is released and degraded. The thermodynamic stability of the ends of the miRNA duplexes plays a crucial role in miRNA strand selection.

Currently, more than 2000 mature human miRNAs have been deposited in the miRNA repository (miRBase, Release 19) [7]. The deep sequencing of short RNAs has not only enabled the identification of novel miRNAs but also revealed that miRNAs are heterogeneous and differ in length. Heterogeneous miRNA variants are referred to as isomiRNAs (isomiRs) [8]. The primary source of the heterogeneity of miRNA length is imprecise cleavage by the ribonucleases Drosha and Dicer [8–11], which can be further biased at the AGO2 binding step [12]. However, miRNA length variation might also reflect various downstream effects, such as limited miRNA degradation by exonucleases, the addition of extra nucleotides [13–15], and miRNA sequencing artifacts [16,17]. It has recently been shown that the human trans-activation response (TAR) RNA-binding protein (TRBP), a molecular partner of Dicer, might also contribute to miRNA length heterogeneity. Specifically, TRBP triggered the production of isomiRs that were longer at the 5' strand than the canonical miRNAs by a single nucleotide. As a result, different mRNAs were targeted due to changes in guide-strand selection [18]. It has also been reported in *Drosophila* that the Nibbler (Nbr) 3'–5' exonuclease trims the 3'ends of miR-34 generating isomiRs shorter than the canonical sequence [19]; however, there is no evidence for similar exonuclease activity in vertebrates.

miRNAs control the expression of the majority of human genes [20], and these molecules are involved in many physiological and pathological processes. The alteration of miRNA expression has been associated with numerous diseases, including neurodegenerative disorders, such as Huntington's disease (HD). HD is the most common fatal polyglutamine (polyQ) disorder and results from the expansion of a CAG repeat in exon 1 of the huntingtin (*HTT*) gene. The precise mechanism of HD pathogenesis is not fully understood, but both the mutant protein (reviewed in [21]) and mutant transcript might be toxic to cells (reviewed in [22]). Of particular interest is the potential involvement of miRNA in the regulation of the *HTT* gene. The global deregulation of miRNAs in samples obtained from HD patients was demonstrated using Illumina massively parallel sequencing [23]. Most importantly, miRNA of varying lengths and/or sequences (isomiRs) were observed for the vast majority of miRNAs detected in two forebrain areas, the frontal cortex (FC) and striatum (ST), of both healthy individuals and HD patients [23].

In general, the miRNA heterogeneity observed in deep sequencing might have important functional implications. Most importantly, miRNAs with shifted 5'-ends have different seed sequences responsible for the recognition of a complementary sequence and the binding to mRNA. Therefore, it is assumed that heterogeneous 5' isomiRs might regulate different targets [10,15,24,25]. Moreover, both 5' and 3' isomiRs might exhibit modified turnover properties [24,26] and altered strand selection within the RISC because strand selection is influenced by the extent of the 3' overhang and the degree of pairing for any miRNA-miRNA* duplex [27,28].

An early evidence supporting the hypothesis of isomiRs functionality comes from an experiment that showed a difference in target cleavage between miR-142-5p and its variant, which contained two extra nucleotides at the 5'-end [29]. A putative functional role for isomiRs has been suggested in many reports because isomiRs actively associate with the RISC and translational machinery [24,30–32] (reviewed in [33]). This assumption was further supported by the observation that isomiRs exhibit differential expression across tissues and developmental stages [26,34,35]. Nevertheless, the real biological significance of isomiRs is not fully understood because few studies concerning isomiR regulation at the cellular level have been reported, and thus far, only variants of miR-133, miR-101, and miR-31 have been experimentally examined. Specifically, it was shown that 5'-isomiR-101, which is highly expressed in the brain, associates with AGO2 immunocomplexes and decreases the expression of five validated miR-101 targets but to a lesser degree than the canonical miR-101 [35]. Differential mRNA targeting was demonstrated in the case of two prevalent 5' isomiRs of the key cardiac regulator miR-133a [31]. Three miR-31 isoforms that differed only slightly in their 5'- and/or 3'-end sequences were compared (namely, hsa-miR-31, ptr-miR-31, and mmu-miR-31), implicating isomiR-31s in the concordant and discordant regulation of six known target genes [36].

In this paper, we validated miRNA-mRNA interactions that might be involved in the regulation of the HTT transcript. Specifically, we experimentally assessed the validity of three predicted interactions and demonstrated that the canonical miR-137, miR-214, and miR-148a bind to the 3'UTR of the *HTT* gene. These results provide the first evidence that miR-137 and miR-148a regulate the expression of huntingtin and confirm that this regulation is also mediated by miR-214, as previously reported [37]. Moreover, using luciferase reporter assays, we investigated the regulation of huntingtin using select miRNA isoforms. We focused on 5'-end isomiRs with the shifted seed sequence that is the primary determinant of mRNA target recognition. Here, we showed that certain 5'-end isomiRs of miR-214 are functional for the downregulation of huntingtin expression.

2. Results and Discussion

2.1. Prediction of miRNA Interactions with the HTT Transcript

In a previous study, we predicted potential miRNA interactions with mRNAs derived from genes triggering hereditary neurological disorders known as trinucleotide repeat expansion diseases (TREDs), including Huntington's disease (HD) [38]. The results of this in-depth *in silico* analysis prompted further research on the potential miRNA-mediated regulation of the HTT transcript in the context of the pathogenesis and therapy of HD. We compared different target prediction algorithms and verified our predictions using the available data gathered in various databases dedicated to miRNA target prediction (e.g., miRWalk database [39] and miRTarBase [40]). We selected interactions with miR-137, miR-214, and miR-148a for experimental verification. The deregulation of the expression of these miRNAs in HD patients or in cellular models of HD has been reported. Specifically, miR-137 was downregulated in the striatum of HD patients [23], while both miR-214 and miR-148a were upregulated in ST*Hdh*^{Q111}/*Hdh*^{Q111} cells [41]. Moreover, miR-137 is highly expressed in the nervous system, suggesting the involvement/potential role of this miRNA in the pathogenesis of HD. miR-137 has also been recently identified as a direct target of the repressor element-1 (RE-1) silencing transcription factor (REST) [42]. The second candidate, miR-214, has been positively verified in previous studies; miR-214, along with three other miRNAs (miRs 150, 146a, and 125b), downregulated the expression of huntingtin [37]. The same study also showed that these miRNAs affect the formation of mutHTT aggregates, the toxicity induced by mutHTT, and the expression of brain-derived neurotrophic factor (BDNF), thereby collectively contributing to HD pathogenesis.

The candidate miRNAs (miRs 137, 214, and 148a) ranked high in the results generated by either algorithm based on conservation criteria, *i.e.*, Diana-micro T [43], miRanda [44], or PicTar [45]. However, our prediction was primarily based on the use of the TargetScanHuman algorithm (Release 6.2) [46]. According to TargetScan, a site for miR-137 is highly conserved among vertebrates, and sites for miRs 214 and 148a are poorly conserved among mammals or vertebrates. In addition, the miR-137 and miR-148a sites were 8mers (defined as exact matches to positions 2–8 of the mature miRNA, followed by an adenine), while the selected miR-214 site was a 7mer-m8 (an exact match to positions 2–8 of the mature miRNA). The positions of the miR-137, miR-214, and miR-148a binding sites in the 3'UTR of the huntingtin transcript and the base pairing of these miRNAs with target sequences are presented in Figure 1. The binding parameters of these miRNAs met the recommended bioinformatics criteria, and their experimental validation was of particular interest in the light of current knowledge of potential involvement of miRNAs in neurodegeneration and the entire competing endogenous RNA (ceRNA) activity network [47], which recently has been shown to be implicated in neurodegenerative diseases including HD [48,49].

Figure 1. Graphical presentation of selected miRNA target site distribution in the 3' untranslated region (3'UTR) of the huntingtin transcript. To predict miRNAs that potentially target the HTT 3'UTR, the TargetScanHuman algorithm (Release 6.2) [46] was used. (**A**) Regions of interaction for the miRNAs selected for experimental validation; (**B**) miRNA base pairing with an appropriate target sequence is schematically presented.

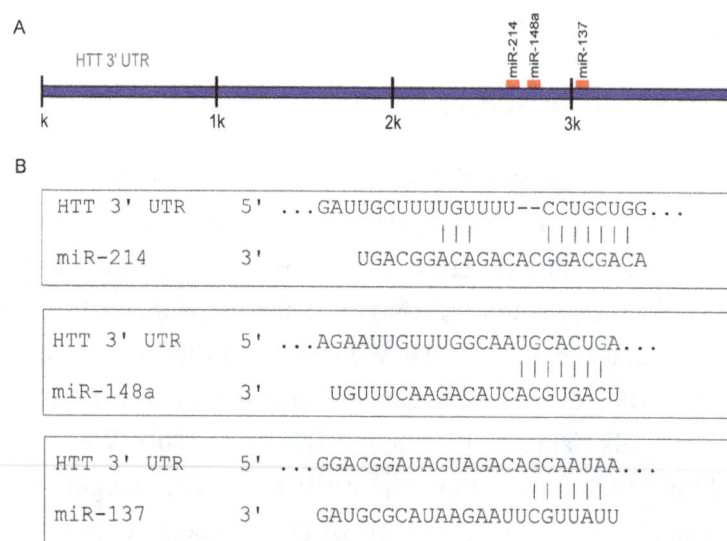

2.2. Canonical miR-137, miR-214, and miR-148a Regulate the Expression of the HTT Gene

For the experimental validation of the predicted binding of the selected canonical miRNAs (miRs 137, 214, and 148a) to their target sites in the HTT 3'UTR, experiments using reporter constructs and luciferase assays were performed as described previously [50]. However, sequences carrying binding sites for the appropriate miRNAs were cloned into pmirGLO vector (Promega), which is considered optimal for miRNA-mRNA interaction studies. Constructs bearing single miRNA binding sites were generated and defined as wild-type reporters (WT). Constructs with mutations that disrupted native pairing within the binding region (5' seed site) of the candidate miRNAs (MUT) and constructs that showed perfect complementarity (PM) to these sites were also generated to provide negative and positive controls, respectively (details in the Experimental section).

We transfected HEK293T cells with either reporter carrying potential miRNA binding sites. Four constructs were transfected into cells and tested in parallel. To determine whether the miRNAs of interest were expressed in the HEK293T cells, we performed northern blot analysis. The expression of miR-137 was not detected in the HEK293T cells. miR-214 and miR-148a were expressed at low and moderate levels (Figure 2), respectively, consistent with the available deep sequencing results. Therefore, our experimental system required miRNA overexpression, and we used miRNA-coding plasmid vectors (System Biosciences, Open Biosystems) for this purpose (compare endogenous miRNA levels and those expressed from vectors in Figure 2).

Figure 2. Endogenous expression and overexpression of miR-137, miR-214, and miR-148a in HEK293T cells. Northern blot detection of miRs 137, 214, and 148a in non-treated HEK293T cells and cells transfected with miRNA-coding plasmids (System Biosciences, Open Biosystems). M denotes the size marker, end-labeled 17, 19, 21, 23, and 25-nt oligoribonucleotides. En and Ex indicate the miRNA levels, endogenous and expressed from appropriate vectors, respectively. Hybridization to U6 RNA provides a loading control.

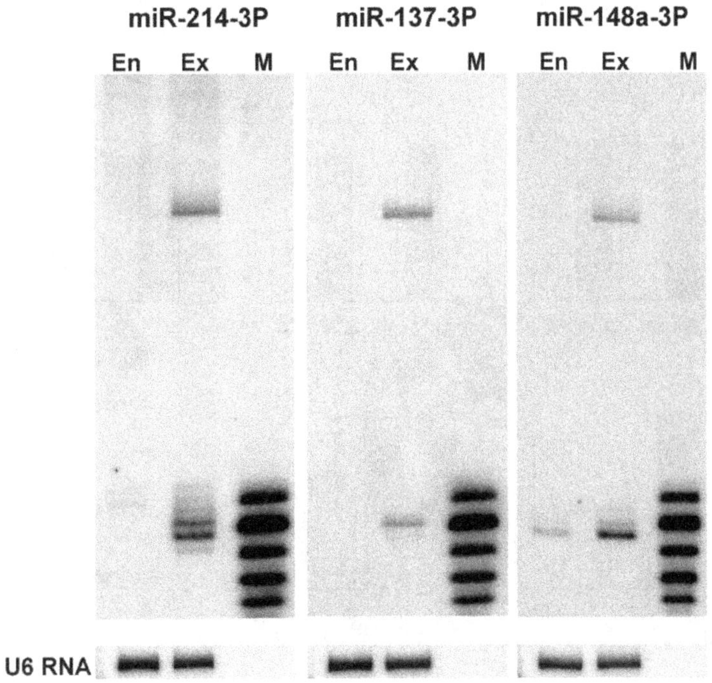

In the luciferase assays, we obtained considerable repression of the luciferase expression after the transfection of reporter constructs for the three miRNAs tested (Figure 3A). Specifically, we observed a significant reduction in luciferase activity when reporter constructs bearing binding sites for miRs 137 and 214 were used (reductions to 83% and 79%, respectively) and a slightly weaker but reproducible and statistically significant suppression of the luciferase activity in the case of miR-148a (suppression to 87%). The luciferase activity for all of the MUT constructs showed efficient de-repression nearly equal to that in the control experiment; the positive controls (PMs) repressed luciferase at low levels, ranging from 17% to 33% (for miR-148a PM and miR-214 PM, respectively) of the empty reporter construct. These results verify the reliability of the experimental system used.

Figure 3. Regulation of the huntingtin (HTT) expression by canonical miRNAs. (**A**) Relative repression of the luciferase expression. Reporter constructs carrying a single binding site for miR-137, miR-214, and miR-148a were tested. For each luciferase. experiment, the miRNA activity on four constructs was measured in parallel: an empty pmirGlO vector (Control), a wild-type potential binding site for the appropriate miRNA (WT), a mutated binding site (MUT), and a site with full complementarity (PM). The firefly luciferase activity was normalized against *Renilla* luciferase activity. An average result from at least three independent experiments is shown (details in the text); (**B**) Relative HTT mRNA levels. Real-time PCR performed 48 h after transfection of HEK293T cells with miR-137, miR-214, and miR-148a. The bar graphs show the quantification of the HTT mRNA levels normalized to actin mRNAs based on data collected from three independent experiments; (**C**) Relative HTT protein levels. Western blot analysis of the cellular levels of HTT protein 72 h after transfection of HEK293T cells with miR-137, miR-214, and miR-148a. The bar graphs show the quantification of the protein levels detected in three western blot experiments. A representative blot is shown. The asterisks indicate statistical significance; a single asterisk at p-value < 0.05 and a double asterisk at p-value < 0.01.

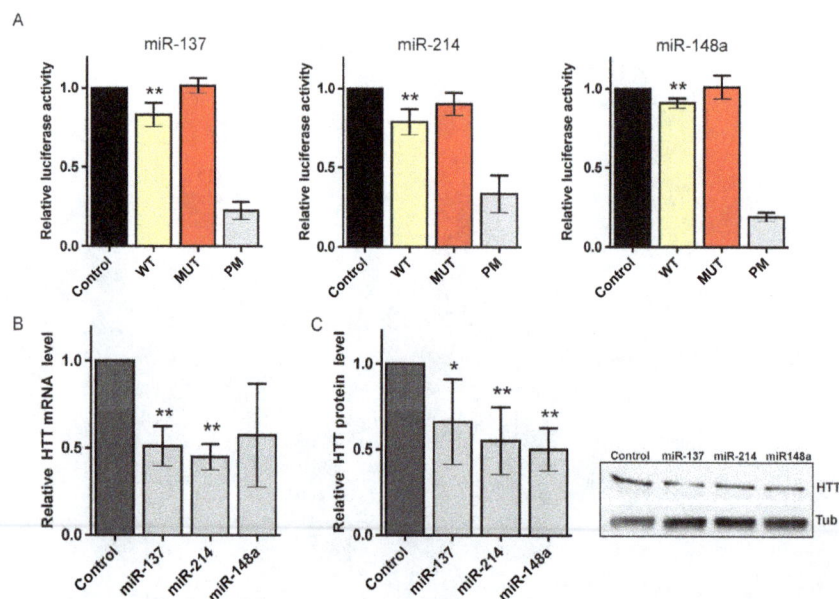

We also monitored huntingtin expression at the mRNA and protein levels following the transfection of HEK293T cells with miRNA-coding plasmids. Real-time PCR performed 48 h after transfection with miR-137, miR-214, or miR-148a showed a strong decrease in the HTT mRNA level (Figure 3B). Similarly, the HTT protein level was significantly reduced 72 h posttransfection in cells overexpressing any of the miRNAs (Figure 3C, Figure S1). This observation is consistent with the finding that miRNA binding reduces the cellular levels of targeted transcripts [51,52]. However, other studies have reported that no or minimal changes in the respective mRNA levels were observed or that these changes were only reported for certain targets [35]. Overall, miR-137, -214 and -148a were positively verified as negative regulators of the *HTT* gene. The lack of regulation of the huntingtin expression, demonstrated in both luciferase assays and western blotting, was observed for the other miRNA (miR-107) and shown for comparison as supplementary data (Figure S2). The strongest reduction in the luciferase activity and the greatest and second-greatest repression at the mRNA and protein levels were observed with miR-214. Thus, this study provides further support for the regulatory potential of miR-214, which was previously validated in a different experimental system [37]. Moreover, this study provides the first evidence of HTT regulation by miR-137 and miR-148a.

2.3. 5'-End Variants of miRNAs Are Functional and Might Regulate the Same Targets as Canonical miRNAs

Although many reports suggest isomiR functionality [24,30–32], there is still little research that address this issue experimentally. Specifically, one variant of miR-101 [35] and two isomiRs of miR-133 [31] and miR-31 [36] have been investigated. In these reports, the isomiRs were less effective than their canonical analogs [35] or exhibited differences in effectiveness depending on the regulated target [31].

Here, we determined whether the 5'-end variants of three miRNAs (5'-end isomiRs), namely miRs 137, 214 and 148a, might function in the same experimental system (*i.e.*, whether these miRs reduce the luciferase activity when appropriate reporter constructs are used). We designed and synthesized miRNA variants with seed sequences shifted by −1, +1, or +2 nt (Integrated DNA Technologies) (Figure 4A). We selected miRNA 5' isoforms that are relatively highly represented in deep sequencing data because we considered sequence abundance a prerequisite for the functionality of these molecules. We based this selection on the sequencing data gathered in the YM 500 database [53] but we also evaluated the expression levels of isomiRs in other sources [32]. The only exception was isomiR-137+1, whose sequence is barely detectable using deep sequencing. This isomiR variant was added to the analysis to examine the same miRNA seed shifts for all isomiRs tested. Moreover, trimming variants that affect the 5' end of miRNAs were reported to be abundant species, and the vast majority of these 5' isomiRs affected a single nucleotide upstream of the reference miRNA [35]. A strong correlation between the expression of miRNAs and isomiRs was also observed [30].

According to the TargetScanHuman Custom (Release 5.2) [46] prediction, none of the selected isomiRs targeted HTT (Figure 4A); thus, we verified the targeting of these molecules experimentally. In addition, we assessed *in silico* how the overall number of genes targeted by the analyzed miRNAs and isomiRs might vary due to the change introduced into their seed regions. Potential targets for the 5'-end variants of miR-137, miR-214, and miR-148a were predicted using the TargetScan Custom 5.2

algorithm [46] and are shown in the Venn diagrams by overlaps (Figure 4B). Specifically, targets for the canonical miRNAs were compared with the targets of the miRNAs with seed regions shifted by −1, +1, and +2 nt. This analysis revealed that the number of predicted targets changed, but apart from unique targets, many genes were still predicted as targets for both miRNAs and isomiRs, confirming that isomiRs might share certain common mRNA targets but not all mRNA targets [36]. These results are also consistent with the suggestion that isomiRs function cooperatively to target common biological pathways [30]. However, distinct functions for miRs and isomiRs have also been suggested [31,35].

Figure 4. Graphical presentation of selected isomiR variants and their potential to target different genes. (**A**) Nucleotide sequences of miR-137, miR-214, miR-148a, and their isoforms. miRNA sequences are marked in red, and isomiR sequences are shown in blue, green, and violet for −1-, +1-, and +2-nt seed shifting, respectively. The miRNA seed sequences are labeled with black rectangles. Information on the miRNA lengths, as well as their potential for targeting the *HTT* gene and isomiR expression levels, is also provided. (*) Ability to interact with the HTT 3′UTR, as predicted by the TargetScanHuman algorithm (Release 6.2) for miRNAs and the TargetScan Custom (Release 5.2) for isomiRs [46], (**) isomiR read number according to the YM500 database [53]; (**B**) Venn diagrams showing the predicted miRNA targets for selected isomiRs. Potential targets for the 5′-end variants of miR-137, miR-214, and miR-148a were predicted using the TargetScan Custom algorithm (Release 5.2) and are shown as overlaps in the Venn diagrams. Targets for the canonical miRNAs are compared with the targets for the miRNAs with the shifted seed regions and are depicted in the same colors as in panel A. The numbers inside the circles denote the numbers of potential targets predicted for the appropriate miRNA variants.

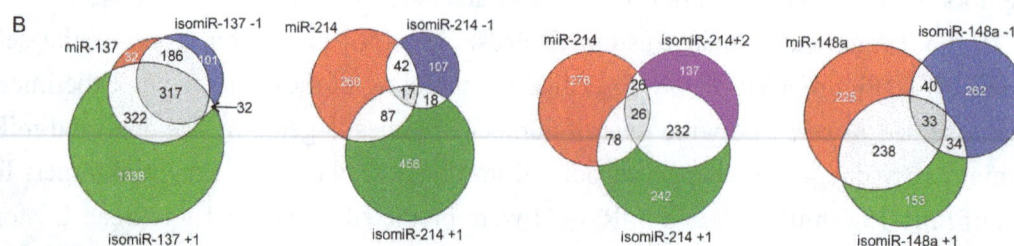

To validate the regulation of the HTT transcript by canonical miRNAs in a luciferase assay, we overexpressed the desired miRNAs from plasmid vectors. To study the interactions of the HTT transcript with isomiRs, appropriate isomiR sequences had to be introduced into cells as synthetic oligonucleotides. Thus, we transfected HEK293T cells with both the miR-137 mimic and miR-137 vector (System Biosciences) to determine whether these two experimental systems generate the same results (Figure 5A). Moreover, we examined miRNA mimic activities at different final concentrations (10, 30, and 50 nM) to determine the optimal concentration for these experiments (Figure 5B). A clear correlation between the results of the luciferase experiments with the miR-137-coding plasmid and the synthetic miR-137 mimic was observed; thus, we further investigated the functionality of our 5'-end isomiRs using appropriate miRNA mimics. In the luciferase assays, we obtained considerable and significant repression of the luciferase expression after the transfection of the reporter constructs and all three miR-214 isomiR mimics, namely, isomiR-214+1, isomiR-214+2, and isomiR-214-1 (luciferase repression equal to 71%, 80%, and 79%, respectively). Moreover, this reduction in the luciferase activity was comparable to the reduction induced by the canonical miR-214 mimic (71%) (Figure 6A). In contrast, the luciferase activity was not reduced when miR-137 isomiRs were used, in the case of neither isomiR-137+1 nor isomiR-137-1, compared with the considerable repression observed using the canonical miR-137 mimic (79%) (Figure 6B). Similarly, in the case of isomiR-148a+1 and isomiR-148a-1, the activity of luciferase was slightly reduced (9% and 6%, respectively), while the reduction obtained for the canonical miR-148a mimic was much stronger (80%) (Figure 6C). The observed difference in the functionality of the analyzed isomiRs raises the question when miRNA-mRNA pairing conforms to strict rules and when some flexibility in the miRNA seed region is permitted, and which additional mechanisms other than the base paring of the seed region might affect target genes repression by isomiRs.

Several factors influence the recognition of a target site by miRNA, e.g., the sequence composition of the 3'-UTR [54], the immediate environment of the putative target site [55], and the structural accessibility of the target site [2,56]. Moreover, endogenous natural antisense transcripts transcribed from the opposite strand of a protein-coding gene or a non-protein coding gene [34] and the RNA-binding proteins [57] could directly bind to mRNA, thereby masking the miRNA binding site of a target gene and preventing the inhibitory effects of the miRNA on target gene translation. These factors, however, are of importance to canonical miRNA binding. Here, we examined several 5' isomiRs of slightly different lengths that previously demonstrated canonical miRNA targeting. Therefore, the structural features and genomic context of these molecules did not significantly differ between the canonical miRNAs and their isomiRs or between the isomiRs themselves.

A distinct feature of the functional isomiR-214 variants and the two other isomiRs examined in this study was the fact that miR-214 is a 7mer with compensatory base pairing at the 3' end (see Figure 1). Although canonical miRNA-target specificity is primarily triggered by complementarity within the seed region, non-canonical interactions depend also on 3' compensatory sites [2,58], which might be important for miR-214 and its variants. The miRNA/isomiR length was also suggested as a factor that might affect functionality. In a study of isomiRs, the analysis of two miR-133a mimics (22/23 nt) was performed, followed by the analysis of two other variants that represented the respective other length for each miR-133a variant. However, the luciferase repression did not depend on mimic length within this range [31]. Therefore, alterations to the 3' end of the miR-133a mimic did not affect the level of mRNA repression, suggesting that the 3' end is not essential for efficient target binding in this case. Another important factor that might account for the disparate functioning of isomiRs is differential binding capacity with the Argonaute complex (affinity of a given miRNA to AGO). Previous studies have shown that some miRNA variants were differentially loaded onto AGOs, and the 5'-end nucleotide of small RNA was critical for its interaction with AGO proteins [12,59–61]. However, miR-101 was more efficiently loaded into the RISC than its isomiR [35], and the 5'-end nucleotide of isomiR-31s was not a rigorous criterion for AGO complex loading [36]. In this study, in the case of the most effective miRNA, namely miR-214, all variants were functional regardless of the different nucleotides at their 5' end (Figure 2). Small changes in the miRNA sequence profoundly affected the functional asymmetry of the miRNA duplex, altering which strand of a miRNA duplex functions in mRNA silencing [18]. Therefore, it cannot be ruled out that, in the case of the nonfunctional isomiRs of miR-137 and miR-148a, the passenger strands were incorporated into the RISC and did not target their binding sites.

Figure 5. Correlation between the results of the luciferase experiments conducted with miR-137-coding plasmid and synthetic miR-137 mimics. (**A**) Relative repression of the luciferase expression. Reporter constructs carrying a single binding site for miR-137 were tested; miRNA activity on four constructs was measured in parallel (Control, WT, MUT, and PM), as described in Figure 3. Left—miRNA expression from the synthetic oligonucleotide (miR-137 mimic), right—miRNA overexpressed from the miR-137 vector. The firefly luciferase activity was normalized against *Renilla* luciferase activity. The standard errors are calculated from three independent experiments; (**B**) The relative repression of the luciferase expression resulted from the miRNA mimic activity. Four reporter constructs were tested (Control, WT, MUT, and PM) but with the addition of miR-137 mimic at different final concentrations, specifically 10, 30, and 50 nM, as denoted in the figure. The standard errors were calculated from one experiment performed in triplicate.

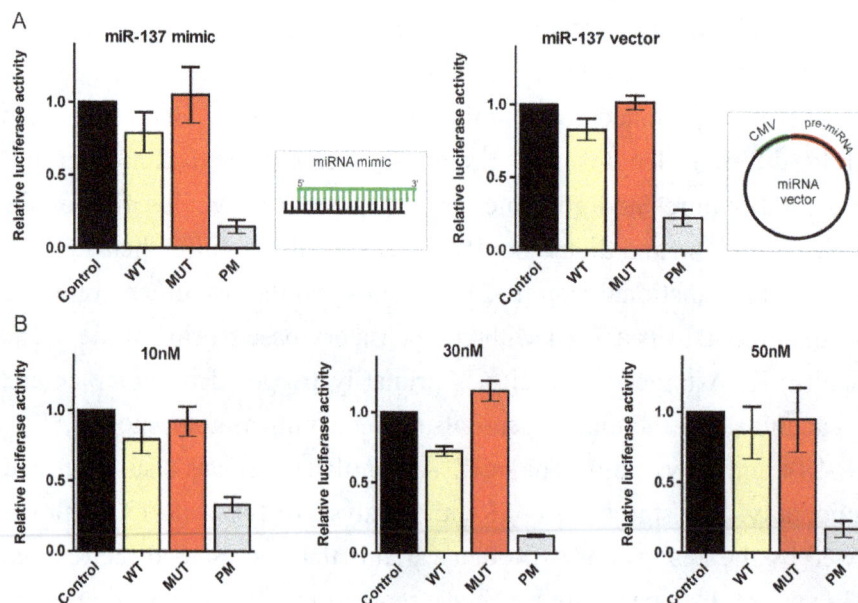

Figure 6. Regulation of the huntingtin expression by isomiRs. Relative repression of the luciferase expression for miR-214, miR-137, miR-148a, and their isomiRs (+1, +2, or −1). Reporter constructs carrying single binding sites for the appropriate miRNAs were tested, namely miR-137 (**A**), miR-214 (**B**), and miR-148a (**C**), as depicted in the figure. For each luciferase experiment, the miRNA activity on four constructs (Control, WT, MUT and PM) was measured in parallel, as described in Figures 3 and 5. The firefly luciferase activity was normalized against *Renilla* luciferase activity. The standard errors were calculated from three independent experiments. The asterisks indicate statistical significance; a single asterisk at p-value < 0.05 and a double asterisk at p-value < 0.01.

A

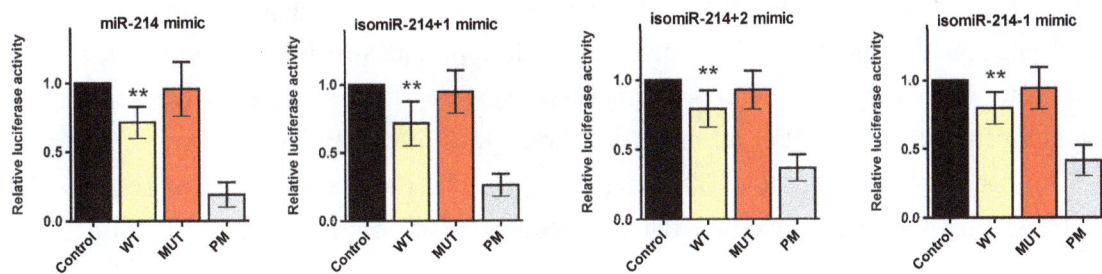

B

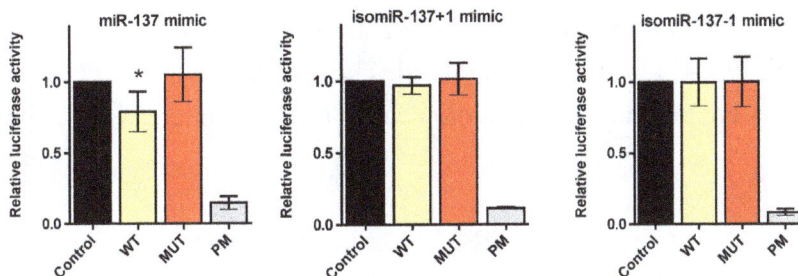

C

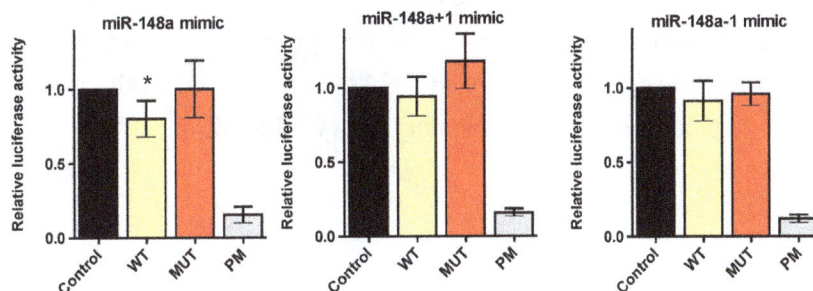

3. Experimental Section

3.1. Cell Culture

HEK293T cells were obtained from the American Type Culture Collection (ATCC) and grown in Dulbecco's Modified Eagle's Medium (DMEM, Lonza, Wakersville, MD, USA) supplemented with 8% fetal bovine serum (FBS) (Sigma-Aldrich, St. Louis, MO, USA), 2 mM L-glutamine, and an antibiotic-antimycotic solution (Sigma-Aldrich, St. Louis, MO, USA) at 37 °C in a humidified atmosphere of 5% CO_2. At 24 h prior to transfection, the HEK293T cells were plated in 12-well or 6-well dishes in DMEM growing medium and harvested 24, 48, and 72 h post-transfection for the luciferase assay, real-time PCR, and western blot analyses, respectively.

3.2. Plasmid Constructs and Synthetic miRNA Oligonucleotides

To generate reporter constructs bearing miRNA-binding sites, the pmirGLO Dual-Luciferase miRNA Target Expression Vector was used (Promega, Madison, WI, USA). This vector is based on Promega dual-luciferase technology, with firefly luciferase (*luc2*) as the primary reporter for monitoring mRNA regulation and *Renilla* luciferase (*hRluc-neo*) as a control reporter for normalization and selection. Specific oligonucleotides with *Dra*I and *Xba*I ends containing single binding sites for the analyzed miRNA (HTT b.s. for miRs 214, 137, and 148a) were synthesized (IBB Warsaw). The appropriate oligos were annealed by boiling and gradual cooling and subsequently phosphorylated and cloned into the pmirGLO vector, previously digested with *Dra*I (Fermentas, St.-Leon-Rot, Germany) and *Xba*I (Fermentas, St.-Leon-Rot, Germany) restriction enzymes, downstream of the *luc2* gene. For all miRNAs, three types of constructs were prepared, namely wild type (WT), carrying mutations (MUT) and perfect match (PM) constructs (for sequences refer to Table S1), which all have 10-nucleotide flanking sequences, as described previously [50].

For miRNA overexpression, commercial plasmid constructs expressing miRNA precursors (pri-miR-148a (Open Biosystems, Huntsville, AL, USA), pri-miR-137, or pri-miR-214 (System Biosciences, Mountain View, CA, USA)) were used. These plasmids contain pri-miRNA sequences in their natural genome context to ensure biologically relevant interactions with the endogenous processing machinery.

Synthetic miRNA mimics (miR-137, miR-214, and miR-148a mimics) and their length variants were chemically synthesized (Integrated DNA Technologies). The following modifications were introduced: (1) 2'-*O*-methyl modification on positions 1 and 2 and a two-nucleotide UU overhang on the 3' end of the miRNA mimic sense strand, (2) 5' phosphorylation and a two-nucleotide overhang based on nucleotide types found in natural pre-miRNAs on the 3' end of the miRNA mimic antisense strand. All sequences are presented as supplementary data (Table S2).

3.3. Cell Transfection

HEK293T cells were transfected using Lipofectamine 2000 (Invitrogen, Carlsbad, CA, USA) according to the manufacturer's protocols. For luciferase assays, the cells were transfected in 12-well plates at ~80% confluence. For each transfection experiment, 200 ng of the appropriate reporter construct and either 250 ng of the appropriate miRNA-coding vector or 30 nM of miRNA mimic were used. The cells were harvested 24 h after transfection and assayed for luciferase activity. For miRNA overexpression required for real-time PCR and western blot analyses, the cells were grown to 80% and 60% confluence, respectively, transfected in 6-well plates with 1 μg/mL pri-miRNA plasmid vectors, and harvested at 48 and 72 h, respectively.

3.4. Luciferase Reporter Assay

After harvesting, the cells were lysed in a passive lysis buffer (Promega, Madison, WI, USA). The luciferase activity was measured using a Dual-Luciferase Reporter Assay System (Promega, Madison, WI, USA) according to the manufacturer's instructions with a Centro LB 960 luminometer (Berthold Technologies, Oak Ridge, TN, USA).

3.5. RNA Isolation and Real-Time PCR

Total RNA from HEK293T cells was isolated using TRI Reagent (MRC, Inc., BioShop, Cincinnati, OH, USA) according to the manufacturer's instructions. The RNA concentration was estimated using a NanoDrop spectrophotometer. cDNA was obtained from 500 ng of total RNA using Superscript III (Life Technologies, Carlsbad, CA, USA) and random hexamer primers (Promega, Madison, WI, USA). For subsequent quantitative real-time analyses, 50 ng of cDNA was used. Real Time PCR was performed on a LightCycler 480 II system (Roche Diagnostics, Mannheim, Germany) using TaqMan Gene Expression Assays and TaqMan Universal Master Mix II (Applied Biosystems, Foster City, CA, USA). The results obtained for the assessment of huntingtin mRNA levels were normalized to the levels of actin mRNA.

3.6. Northern Blotting

High-resolution northern blotting was performed as previously described [62,63]. Briefly, 25 µg of total RNA was extracted from HEK293T cells and resolved on a 12% denaturing polyacrylamide gel in 0.5× TBE. The RNA was transferred to a GeneScreen Plus hybridization membrane (PerkinElmer, Spokane, WA, USA) using semi-dry electroblotting (Sigma-Aldrich, St. Louis, MO, USA), immobilized by subsequent UV irradiation (120 mJ/cm^2) (UVP), and baked in an oven at 80 °C for 30 min. The membranes were probed with specific DNA oligonucleotides (Table S3) complementary to the annotated human miRNAs miR-137-3P, miR-214-3P, and miR-148a-3P (miRBase). The probes were labeled with [γ^{32}P] ATP (5000 Ci/mmol; Hartmann Analytics, Braunschweig, Germany) using USB OptiKinase (Affymetrix, Cleveland, OH, USA). The hybridizations were performed at 37 °C overnight in a PerfectHyb buffer (Sigma-Aldrich, St. Louis, MO, USA). The marker lanes contained a mixture of radiolabeled RNA oligonucleotides (17-, 19-, 21-, 23-, and 25-nt in length). Hybridizations to U6 RNA provided loading controls. Radioactive signals were quantified by phosphorimaging (Multi Gauge v3.0; Fujifilm).

3.7. Western Blotting

A total of 15 µg of protein was diluted in sample buffer containing 2-mercaptoethanol, denatured for 5 min, and separated on 3%–8% gradient Tris-Acetate gels (Invitrogen, Carlsbad, CA, USA) in XT Tricine Buffer (BioRad, Hercules, CA, USA). After electrophoresis, the proteins were electrotransferred onto a nitrocellulose membrane (Sigma, St. Louis, MO, USA). All immunodetection steps were performed on a SNAPid (Millipore, Billerica, MA, USA) in PBS buffer containing 0.25% nonfat milk and 0.1% Tween 20, and the membranes were washed in PBS/Tween. For huntingtin and tubulin detection, the blots were probed with the primary anti-huntingtin (1:500, Millipore, Billerica,

MA, USA) and anti-alpha-tubulin (1:5000, Covance, Emeryville, CA, USA) antibodies, respectively, and subsequently probed with HRP-conjugated secondary antibodies (1:500, Sigma, St. Louis, MO, USA). The immunoreaction was detected using Western Bright Quantum (Advansta, CA, USA). The protein amounts were quantified using GelPro 3.1 software (Media Cybernetics, Bethesda, MD, USA).

3.8. Statistical Analysis

All experiments were repeated at least three times. Graphs were generated using GraphPad Prism 5 (GraphPad Software). The figures for the luciferase assays were generated after averaging the results from the repeat experiments for a particular construct. The values for error bars (mean with SD) and the statistical significance were calculated using GraphPad Prism 5. The statistical significance of the luciferase reduction in the case of transfection with constructs carrying miRNA-binding sites was assessed using a one-sample t-test with a hypothetical value of 1 assigned to cells transfected with a control empty vector. p-values < 0.05 (two-tailed) were considered significant.

4. Conclusions

This study presents new evidence that *HTT* gene expression is regulated by miRNAs and, most importantly, demonstrates that certain isomiRs are functional and regulate the same target as canonical miRNAs.

IsomiRs are commonly reported in deep-sequencing studies and have been described in all studied organisms and tissues. The existence of miRNA variants might contribute considerably to the complexity of target regulation by miRNAs and strongly increase the regulatory potential of these molecules. The presence of isomiRs could have far-reaching implications for miRNA therapeutic applications; it must be taken into account in various diagnostic tests as well as in the design of miRNA mimics or anti-miRs as therapeutic agents. Therefore, of particular importance is to identify factors that determine the biological relevance of isomiRs.

Acknowledgments

This work was supported by funding from the Polish Ministry of Science and Higher Education (N N301 523038), the National Science Centre (2011/03/B/NZ1/03259), and the European Regional Development Fund within the Innovative Economy Programme (POIG.01.03.01-00-098/08). The real-time PCR analyses were performed on a LightCycler 480 II system (Roche) in the Laboratory of Subcellular Structures Analysis at the Institute of Bioorganic Chemistry, PAS, in Poznan.

References

1. Chekulaeva, M.; Filipowicz, W. Mechanisms of miRNA-mediated post-transcriptional regulation in animal cells. *Curr. Opin. Cell Biol.* **2009**, *21*, 452–460.

2. Bartel, D.P. MicroRNAs: Target recognition and regulatory functions. *Cell* **2009**, *136*, 215–233.

3. Kim, V.N.; Han, J.; Siomi, M.C. Biogenesis of small RNAs in animals. *Nat. Rev. Mol. Cell Biol.* **2009**, *10*, 126–139.

4. Krol, J.; Loedige, I.; Filipowicz, W. The widespread regulation of microRNA biogenesis, function and decay. *Nat. Rev. Genet.* **2010**, *11*, 597–610.

5. Starega-Roslan, J.; Koscianska, E.; Kozlowski, P.; Krzyzosiak, W.J. The role of the precursor structure in the biogenesis of microRNA. *Cell. Mol. Life Sci.* **2011**, *68*, 2859–2871.

6. Winter, J.; Jung, S.; Keller, S.; Gregory, R.I.; Diederichs, S. Many roads to maturity: microRNA biogenesis pathways and their regulation. *Nat. Cell Biol.* **2009**, *11*, 228–234.

7. Griffiths-Jones, S.; Saini, H.K.; van Dongen, S.; Enright, A.J. miRBase: Tools for microRNA genomics. *Nucleic Acids Res.* **2008**, *36*, D154–D158.

8. Morin, R.D.; O'Connor, M.D.; Griffith, M.; Kuchenbauer, F.; Delaney, A.; Prabhu, A.L.; Zhao, Y.; McDonald, H.; Zeng, T.; Hirst, M.; *et al.* Application of massively parallel sequencing to microRNA profiling and discovery in human embryonic stem cells. *Genome Res.* **2008**, *18*, 610–621.

9. Starega-Roslan, J.; Krol, J.; Koscianska, E.; Kozlowski, P.; Szlachcic, W.J.; Sobczak, K.; Krzyzosiak, W.J. Structural basis of microRNA length variety. *Nucleic Acids Res.* **2011**, *39*, 257–268.

10. Seitz, H.; Ghildiyal, M.; Zamore, P.D. Argonaute loading improves the 5' precision of both MicroRNAs and their miRNA* strands in flies. *Curr. Biol.* **2008**, *18*, 147–151.

11. Wu, H.; Ye, C.; Ramirez, D.; Manjunath, N. Alternative processing of primary microRNA transcripts by Drosha generates 5' end variation of mature microRNA. *PLoS One* **2009**, *4*, e7566.

12. Frank, F.; Sonenberg, N.; Nagar, B. Structural basis for 5'-nucleotide base-specific recognition of guide RNA by human AGO2. *Nature* **2010**, *465*, 818–822.

13. Landgraf, P.; Rusu, M.; Sheridan, R.; Sewer, A.; Iovino, N.; Aravin, A.; Pfeffer, S.; Rice, A.; Kamphorst, A.O.; Landthaler, M.; *et al.* A mammalian microRNA expression atlas based on small RNA library sequencing. *Cell* **2007**, *129*, 1401–1414.

14. Ruby, J.G.; Jan, C.; Player, C.; Axtell, M.J.; Lee, W.; Nusbaum, C.; Ge, H.; Bartel, D.P. Large-scale sequencing reveals 21U-RNAs and additional microRNAs and endogenous siRNAs in *C. elegans*. *Cell* **2006**, *127*, 1193–1207.

15. Wu, H.; Neilson, J.R.; Kumar, P.; Manocha, M.; Shankar, P.; Sharp, P.A.; Manjunath, N. miRNA profiling of naive, effector and memory CD8 T cells. *PLoS One* **2007**, *2*, e1020.

16. Huse, S.M.; Huber, J.A.; Morrison, H.G.; Sogin, M.L.; Welch, D.M. Accuracy and quality of massively parallel DNA pyrosequencing. *Genome Biol.* **2007**, *8*, R143.

17. Tian, G.; Yin, X.; Luo, H.; Xu, X.; Bolund, L.; Zhang, X.; Gan, S.Q.; Li, N. Sequencing bias: Comparison of different protocols of microRNA library construction. *BMC Biotechnol.* **2010**, *10*, 64.

18. Lee, H.Y.; Doudna, J.A. TRBP alters human precursor microRNA processing *in vitro*. *RNA* **2012**, *18*, 2012–2019.

19. Liu, N.; Abe, M.; Sabin, L.R.; Hendriks, G.J.; Naqvi, A.S.; Yu, Z.; Cherry, S.; Bonini, N.M. The exoribonuclease Nibbler controls 3' end processing of microRNAs in *Drosophila*. *Curr. Biol.* **2011**, *21*, 1888–1893.

20. Friedman, R.C.; Farh, K.K.; Burge, C.B.; Bartel, D.P. Most mammalian mRNAs are conserved targets of microRNAs. *Genome Res.* **2009**, *19*, 92–105.

21. Ross, C.A.; Tabrizi, S.J. Huntington's disease: From molecular pathogenesis to clinical treatment. *Lancet Neurol.* **2011**, *10*, 83–98.

22. Fiszer, A.; Krzyzosiak, W.J. RNA toxicity in polyglutamine disorders: Concepts, models, and progress of research. *J. Mol. Med.* **2013**, *91*, 683–691.

23. Marti, E.; Pantano, L.; Banez-Coronel, M.; Llorens, F.; Minones-Moyano, E.; Porta, S.; Sumoy, L.; Ferrer, I.; Estivill, X. A myriad of miRNA variants in control and Huntington's disease brain regions detected by massively parallel sequencing. *Nucleic Acids Res.* **2010**, *38*, 7219–7235.

24. Chiang, H.R.; Schoenfeld, L.W.; Ruby, J.G.; Auyeung, V.C.; Spies, N.; Baek, D.; Johnston, W.K.; Russ, C.; Luo, S.; Babiarz, J.E.; *et al.* Mammalian microRNAs: Experimental evaluation of novel and previously annotated genes. *Genes Dev.* **2010**, *24*, 992–1009.

25. Kawahara, Y.; Megraw, M.; Kreider, E.; Iizasa, H.; Valente, L.; Hatzigeorgiou, A.G.; Nishikura, K. Frequency and fate of microRNA editing in human brain. *Nucleic Acids Res.* **2008**, *36*, 5270–5280.

26. Fernandez-Valverde, S.L.; Taft, R.J.; Mattick, J.S. Dynamic isomiR regulation in *Drosophila* development. *RNA* **2010**, *16*, 1881–1888.

27. Khvorova, A.; Reynolds, A.; Jayasena, S.D. Functional siRNAs and miRNAs exhibit strand bias. *Cell* **2003**, *115*, 209–216.

28. Schwarz, D.S.; Hutvagner, G.; Du, T.; Xu, Z.; Aronin, N.; Zamore, P.D. Asymmetry in the assembly of the RNAi enzyme complex. *Cell* **2003**, *115*, 199–208.

29. Azuma-Mukai, A.; Oguri, H.; Mituyama, T.; Qian, Z.R.; Asai, K.; Siomi, H.; Siomi, M.C. Characterization of endogenous human Argonautes and their miRNA partners in RNA silencing. *Proc. Natl. Acad. Sci. USA* **2008**, *105*, 7964–7969.

30. Cloonan, N.; Wani, S.; Xu, Q.; Gu, J.; Lea, K.; Heater, S.; Barbacioru, C.; Steptoe, A.L.; Martin, H.C.; Nourbakhsh, E.; *et al.* MicroRNAs and their isomiRs function cooperatively to target common biological pathways. *Genome Biol.* **2011**, *12*, R126.

31. Humphreys, D.T.; Hynes, C.J.; Patel, H.R.; Wei, G.H.; Cannon, L.; Fatkin, D.; Suter, C.M.; Clancy, J.L.; Preiss, T. Complexity of murine cardiomyocyte miRNA biogenesis, sequence variant expression and function. *PLoS One* **2012**, *7*, e30933.

32. Lee, L.W.; Zhang, S.; Etheridge, A.; Ma, L.; Martin, D.; Galas, D.; Wang, K. Complexity of the microRNA repertoire revealed by next-generation sequencing. *RNA* **2010**, *16*, 2170–2180.

33. Neilsen, C.T.; Goodall, G.J.; Bracken, C.P. IsomiRs—The overlooked repertoire in the dynamic microRNAome. *Trends Genet.* **2012**, *28*, 544–549.

34. Faghihi, M.A.; Zhang, M.; Huang, J.; Modarresi, F.; van der Brug, M.P.; Nalls, M.A.; Cookson, M.R.; St-Laurent, G., 3rd; Wahlestedt, C. Evidence for natural antisense transcript-mediated inhibition of microRNA function. *Genome Biol.* **2010**, *11*, R56.

35. Llorens, F.; Banez-Coronel, M.; Pantano, L.; Del Rio, J.A.; Ferrer, I.; Estivill, X.; Marti, E. A highly expressed miR-101 isomiR is a functional silencing small RNA. *BMC Genomics* **2013**, *14*, 104.

36. Chan, Y.T.; Lin, Y.C.; Lin, R.J.; Kuo, H.H.; Thang, W.C.; Chiu, K.P.; Yu, A.L. Concordant and discordant regulation of target genes by miR-31 and its isoforms. *PLoS One* **2013**, *8*, e58169.

37. Sinha, M.; Ghose, J.; Bhattarcharyya, N.P. Micro RNA-214,-150,-146a and-125b target Huntingtin gene. *RNA Biol.* **2011**, *8*, 1005–1021.

38. Witkos, T.M.; Koscianska, E.; Krzyzosiak, W.J. Practical aspects of microRNA target prediction. *Curr. Mol. Med.* **2011**, *11*, 93–109.

39. Dweep, H.; Sticht, C.; Pandey, P.; Gretz, N. miRWalk—Database: Prediction of possible miRNA binding sites by "walking" the genes of three genomes. *J. Biomed. Inform.* **2011**, *44*, 839–847.

40. Hsu, S.D.; Lin, F.M.; Wu, W.Y.; Liang, C.; Huang, W.C.; Chan, W.L.; Tsai, W.T.; Chen, G.Z.; Lee, C.J.; Chiu, C.M.; *et al.* miRTarBase: A database curates experimentally validated microRNA-target interactions. *Nucleic Acids Res.* **2011**, *39*, D163–D169.

41. Sinha, M.; Ghose, J.; Das, E.; Bhattarcharyya, N.P. Altered microRNAs in STHdh(Q111)/Hdh(Q111) cells: miR-146a targets TBP. *Biochem. Biophys. Res. Commun.* **2010**, *396*, 742–747.

42. Soldati, C.; Bithell, A.; Johnston, C.; Wong, K.Y.; Stanton, L.W.; Buckley, N.J. Dysregulation of REST-regulated coding and non-coding RNAs in a cellular model of Huntington's disease. *J. Neurochem.* **2013**, *124*, 418–430.

43. Kiriakidou, M.; Nelson, P.T.; Kouranov, A.; Fitziev, P.; Bouyioukos, C.; Mourelatos, Z.; Hatzigeorgiou, A. A combined computational-experimental approach predicts human microRNA targets. *Genes Dev.* **2004**, *18*, 1165–1178.

44. John, B.; Enright, A.J.; Aravin, A.; Tuschl, T.; Sander, C.; Marks, D.S. Human microRNA targets. *PLoS Biol.* **2004**, *2*, e363.

45. Krek, A.; Grun, D.; Poy, M.N.; Wolf, R.; Rosenberg, L.; Epstein, E.J.; MacMenamin, P.; da Piedade, I.; Gunsalus, K.C.; Stoffel, M.; *et al.* Combinatorial microRNA target predictions. *Nat. Genet.* **2005**, *37*, 495–500.

46. Lewis, B.P.; Burge, C.B.; Bartel, D.P. Conserved seed pairing, often flanked by adenosines, indicates that thousands of human genes are microRNA targets. *Cell* **2005**, *120*, 15–20.

47. Salmena, L.; Poliseno, L.; Tay, Y.; Kats, L.; Pandolfi, P.P. A ceRNA hypothesis: The Rosetta Stone of a hidden RNA language? *Cell* **2011**, *146*, 353–358.

48. Bicchi, I.; Morena, F.; Montesano, S.; Polidoro, M.; Martino, S. MicroRNAs and molecular mechanisms of neurodegeneration. *Genes* **2013**, *4*, 244–263.

49. Costa, V.; Esposito, R.; Aprile, M.; Ciccodicola, A. Non-coding RNA and pseudogenes in neurodegenerative diseases: "The (un)Usual Suspects". *Front. Genet.* **2012**, *3*, 231.

50. Koscianska, E.; Baev, V.; Skreka, K.; Oikonomaki, K.; Rusinov, V.; Tabler, M.; Kalantidis, K. Prediction and preliminary validation of oncogene regulation by miRNAs. *BMC Mol. Biol.* **2007**, *8*, 79.

51. Baek, D.; Villen, J.; Shin, C.; Camargo, F.D.; Gygi, S.P.; Bartel, D.P. The impact of microRNAs on protein output. *Nature* **2008**, *455*, 64–71.

52. Guo, H.; Ingolia, N.T.; Weissman, J.S.; Bartel, D.P. Mammalian microRNAs predominantly act to decrease target mRNA levels. *Nature* **2010**, *466*, 835–840.

53. Cheng, W.C.; Chung, I.F.; Huang, T.S.; Chang, S.T.; Sun, H.J.; Tsai, C.F.; Liang, M.L.; Wong, T.T.; Wang, H.W. YM500: A small RNA sequencing (smRNA-seq) database for microRNA research. *Nucleic Acids Res.* **2013**, *41*, D285–D294.

54. Robins, H.; Press, W.H. Human microRNAs target a functionally distinct population of genes with AT-rich 3' UTRs. *Proc. Natl. Acad. Sci. USA* **2005**, *102*, 15557–15562.

55. Grimson, A.; Farh, K.K.; Johnston, W.K.; Garrett-Engele, P.; Lim, L.P.; Bartel, D.P. MicroRNA targeting specificity in mammals: Determinants beyond seed pairing. *Mol. Cell* **2007**, *27*, 91–105.

56. Kertesz, M.; Iovino, N.; Unnerstall, U.; Gaul, U.; Segal, E. The role of site accessibility in microRNA target recognition. *Nat. Genet.* **2007**, *39*, 1278–1284.

57. Goswami, S.; Tarapore, R.S.; Teslaa, J.J.; Grinblat, Y.; Setaluri, V.; Spiegelman, V.S. MicroRNA-340-mediated degradation of microphthalmia-associated transcription factor mRNA is inhibited by the coding region determinant-binding protein. *J. Biol. Chem.* **2010**, *285*, 20532–20540.

58. Brennecke, J.; Stark, A.; Russell, R.B.; Cohen, S.M. Principles of microRNA-target recognition. *PLoS Biol.* **2005**, *3*, e85.

59. Ebhardt, H.A.; Tsang, H.H.; Dai, D.C.; Liu, Y.; Bostan, B.; Fahlman, R.P. Meta-analysis of small RNA-sequencing errors reveals ubiquitous post-transcriptional RNA modifications. *Nucleic Acids Res.* **2009**, *37*, 2461–2470.

60. Felice, K.M.; Salzman, D.W.; Shubert-Coleman, J.; Jensen, K.P.; Furneaux, H.M. The 5' terminal uracil of let-7a is critical for the recruitment of mRNA to Argonaute2. *Biochem. J.* **2009**, *422*, 329–341.

61. Mi, S.; Cai, T.; Hu, Y.; Chen, Y.; Hodges, E.; Ni, F.; Wu, L.; Li, S.; Zhou, H.; Long, C.; *et al.* Sorting of small RNAs into Arabidopsis argonaute complexes is directed by the 5' terminal nucleotide. *Cell* **2008**, *133*, 116–127.

62. Koscianska, E.; Starega-Roslan, J.; Czubala, K.; Krzyzosiak, W.J. High-resolution northern blot for a reliable analysis of microRNAs and their precursors. *ScientificWorldJournal* **2011**, *11*, 102–117.

63. Koscianska, E.; Starega-Roslan, J.; Sznajder, L.J.; Olejniczak, M.; Galka-Marciniak, P.; Krzyzosiak, W.J. Northern blotting analysis of microRNAs, their precursors and RNA interference triggers. *BMC Mol. Biol.* **2011**, *12*, 14.

12

Non-Coding RNAs in Muscle Dystrophies

Daniela Erriquez [1], **Giovanni Perini** [1,2,]*** and Alessandra Ferlini** [3,]***

[1]   Department of Pharmacy and Biotechnology, University of Bologna, Bologna 40126, Italy;
      E-Mail: daniela.erriquez@hotmail.it
[2]   Health Sciences and Technologies–Interdepartmental Center for Industrial Research,
      University of Bologna, Bologna 40064, Italy
[3]   Section of Microbiology and Medical Genetics, Department of Medical Sciences,
      University of Ferrara, Ferrara 44100, Italy

*   Authors to whom correspondence should be addressed; E-Mails: giovanni.perini@unibo.it (G.P.);
    fla@unife.it (A.F.);

Abstract: ncRNAs are the most recently identified class of regulatory RNAs with vital functions in gene expression regulation and cell development. Among the variety of roles they play, their involvement in human diseases has opened new avenues of research towards the discovery and development of novel therapeutic approaches. Important data come from the field of hereditary muscle dystrophies, like Duchenne muscle dystrophy and Myotonic dystrophies, rare diseases affecting 1 in 7000–15,000 newborns and is characterized by severe to mild muscle weakness associated with cardiac involvement. Novel therapeutic approaches are now ongoing for these diseases, also based on splicing modulation. In this review we provide an overview about ncRNAs and their behavior in muscular dystrophy and explore their links with diagnosis, prognosis and treatments, highlighting the role of regulatory RNAs in these pathologies.

Keywords: microRNAs (miRNAs); long non-coding RNAs (lncRNAs); Duchenne muscular dystrophy (DMD); Becker muscular dystrophy (BMD); Myotonic dystrophies (DM1 and DM2); Facioscapulohumeral dystrophy (FSHD)

1. Introduction

Transcription of the eukaryotic genome yields only 1%–2% of protein coding transcripts and the remainder is classified as non-coding RNAs (ncRNAs). In other words, non-coding RNAs are the main output of the global transcription process, highlighting the idea that such an intense cellular effort cannot be just simple noise. Rather, it is reasonable to speculate that this underscored transcriptome possesses specific vital functions [1–3].

In general, non-coding RNAs are divided into structural and regulatory RNAs. The first ones include ribosomal, transfer, small nuclear and small nucleolar RNAs (rRNAs, tRNAs, snRNAs and snoRNAs respectively), which have been deeply characterized at the functional level. The second ones are a very broad class of RNAs whose main categorization essentially relies on their length.

Small ncRNAs are defined as transcripts shorter than 200 nucleotides. The most functionally characterized are microRNAs (miRNAs), piwi-interacting RNAs (piRNAs) and small interfering RNAs (siRNAs), which are critical for the assembly and the activity of the RNA interference machinery.

RNAs longer than 200 nucleotides are named long non-coding RNAs (lncRNAs) and are a very heterogeneous group of molecules. Because there is not an official way to classify them, they can be placed in one or more categories depending on their genome localization and/or on their orientation (sense, antisense, bidirectional, intronic or intergenic lncRNAs) [4,5].

In the past years, several reports have increased our knowledge about additional levels of regulation of many physiological processes that are mediated by ncRNAs. Even more interesting, these flexible molecules have been found to be dysregulated in many pathological human disorders.

In this review, we will focus on RNAs involved in human skeletal muscle dystrophies. There is a continuous flow of new scientific reports that underpin functional links between ncRNAs and skeletal muscle biology, suggesting that these molecules can play a crucial function both in physiological muscle development and in pathological muscle disorders.

Muscular dystrophies (MDs) are strictly inherited conditions recognized as a common pathogenic mechanism of disruption/impairment of the muscle cell membrane (sarcolemma) which causes a cascade of pathogenic events, including: inflammation, cell necrosis and cell death with progressive fibrosis replacing the muscle mass. MDs represent diseases of extraordinary interest both in medical genetics and biology. The high translational value of research about MDs has recently driven scientific findings toward precise genetic diagnoses as well as novel therapies [6–9].

There are more than 30 different types of inherited dystrophies that are characterized by muscle wasting and weakness of variable distribution and severity, manifesting at any age from birth to middle years, resulting in mild to severe disability and even short life expectancy in the worse cases. Clinical and pathological features are generally the parameters to classify the most common type of MDs. The broad spectrum of MDs arises from many different genetic mutations that reflect defects not only in structural proteins, but also in signaling molecules and enzymes. Dystrophin was the first mutant structural protein shown to cause MD. Mutations in the dystrophin gene lead to two more common type of dystrophy: the severe Duchenne muscular dystrophy (DMD OMIM 300677) due to out-of-frame mutations, and the milder Becker muscular dystrophy (BMD OMIM 300376) associated with in-frame mutations. Some "exceptions to the reading frame rule" are associated with intermediate phenotypes. The genetic causes of the highly heterogeneous Limb Girdle Muscular Dystrophies (LGMDs) reside in many genes (such as α, β, γ, δ, and ε sarcoglycans) encoding for structural proteins that are part of the complex sarcolemma network and deeply involved, together with dystrophin, in force transduction [10,11].

There are other muscular dystrophies, such as the Facioscapulohumeral muscular dystrophy (FSHD OMIM 158900) and Myotonic dystrophies (DM1 and DM2, see below), that are due to mutations in genes with a main regulatory function. FSHD is due to deletions in non-coding RNA which cause modification of the chromatin assembly in the 4q34 chromosomal region; Myotonic dystrophies (DMs) are related to trinucleotide (DM1) and tetranucleotide (DM2) repeat expansions that produce toxic mutant mRNA with subsequent interference of RNA-splicing mechanisms [12,13].

Many lines of evidence reveal that aberrant expression levels of non-coding RNAs can result in novel types of defects that cause remarkable changes in processes such as mRNA maturation, translation, signaling pathways or gene regulation. To date, it is clear that there is involvement of several miRNAs in the muscular dystrophies, on the contrary, very little is known about the role of long ncRNAs [14].

In this review we try to recapitulate the emerging studies about this intriguing category of molecules, summarizing what is known in muscle, both in physiological and in pathological contexts; new insights are revealing that they are important players in processes such as cellular lineage commitment, growth and differentiation of skeletal muscle. Since muscle differentiation and regeneration are key features that require to be considered when designing novel therapies, addressing the role of ncRNAs in MDs is of high clinical relevance.

2. Muscle-Specific and Ubiquitously Expressed miRNAs in Skeletal Muscle

microRNAs control the stability and/or the translational efficiency of target messenger RNAs, thus causing post-transcriptional gene silencing. Mammalian miRNAs are transcribed as long primary transcripts (pri-miRNAs) and encode one or more miRNAs. Pri-miRNAs are processed by RNase III Drosha in the nucleus to generate stem-loop structures of ~70 nucleotides (pre-miRNAs) and then exported to the cytoplasm where they are further processed by RNase Dicer to yield ~22 bp mature miRNAs. A mature miRNA, incorporated into the RNA-induced silencing complex (RISC), anneals to the 3' UTRs of its target mRNAs by its complementary strand, thus causing post-transcriptional gene silencing via translational repression or mRNA degradation. In the last years new paradigms of miRNA biogenesis are also emerging in which the processing of miRNA does not require all steps mentioned above [15].

Vertebrate skeletal muscle is derived from the somites, the first metameric structures in mammalian embryos, that progressively subdivide into embryonic compartments, thus giving rise to dermomyotome and subsequently to myotome to produce differentiated muscular tissue. The process of generating muscle—myogenesis—is highly complex and requires a broad spectrum of signaling molecules, either during embryonic development and in postnatal life, that converges on specific transcription and chromatin-remodeling factors, as well as on regulatory RNAs, to activate gene and microRNA expression program [16,17].

The fate of myogenic precursor cellsfirst determined by paired -homeodomain transcription factors, Pax3/Pax7, followed by regulation of highly conserved MyoD (also named MyoD1, myogenic differentiation 1), Myf5 (myogenic factor 5), MyoG (myogenin), and MRF4 factors, expressed in the skeletal muscle lineage and therefore referred as myogenic regulatory factors (MRFs). The MRFs differ in the timing and the stages of myogenesis, reflecting their different roles during muscle cell commitment and differentiation. MyoD and Myf5 are both considered markers of terminal commitment to muscle fate. Myf5 is the first MRF expressed during the formation of the myotome, followed by expression of MyoD. Specifically, in the majority of muscle progenitors, MyoD functions downstream from Pax3 and Pax7 in the genetic hierarchy of myogenic regulators, whereas Myf5, depending on the context, can also act in parallel with the Pax transcription factors [18–20]. Instead, MyoG and MRF4 act subsequently to specify the immature muscle cells (myoblasts) for terminal differentiation. Myoblasts exit from cell cycle after a defined proliferation time, to become terminally differentiated myocytes [21,22]. Muscle-specific genes such as myosin heavy chain genes (*MyHC* genes) and muscle creatine kinase (*M-CK*) are expressed in the last phase of this multi-regulated program, where mononucleated myocytes specifically fuse to each other to form multinucleated myotubes [22–28].

Dicer loss-of function studies clarified the importance of miRNAs in normal skeletal muscle development [29]. miRNAs actively take part in the proliferation and differentiation of skeletal muscle cells as an integral component of genetic regulatory circuitries.

miR-1, miR-133a/b and miR-206 are largely studied and defined muscle-specific miRNAs (myomiRs). They are regulated in muscular transcriptional networks via MRFs and via others key-regulators of the myogenic program, MEF2 (myocyte enhancer factor 2) and SRFs (serum response factors). Recently, a new regulatory pathway, the mechanistic target of rapamycin (mTOR) signaling was seen to regulate miR-1 expression and was also found responsible for MyoD stability [30–36]. It is possible to functionally define miR-133 as enhancer of myoblast proliferation while miR-1 and miR-206 as enhancers of skeletal muscle differentiation [37–40]. An up-to-date list of the identified targets of miR-1, miR-133 and miR-206, together with a plethora of specific muscular pathways they are involved in, is reported in a recent review [40] and some of these will be also discussed in the next paragraph to highlight how these important families of miRNAs contribute to determine typical deficiencies occurring in a pathological muscular context. Intriguingly, these myomiRs have been shown to behave as serum biomarkers in DMD patients. They are released into the bloodstream as a consequence of fiber damage and their power as diagnostic tools is promising since increased miRNA levels correlate with severity of the disease, significantly better than other commonly utilized markers, such as creatine kinase (CK). Moreover, their major serum stability is another aspect that may make them useful not only for diagnosis but also for monitoring the condition of affected individuals after a therapeutic treatment [41,42].

miR-208b/miR-499, also named myomiRs because of their muscle-restricted expression, are produced from the introns of two myosin genes, *β-MHC* and *Myh7b*. They are functionally redundant and play a dominant role in the specification of muscle fiber identity by activating slow and repressing fast myofiber gene programs [43].

Interestingly, many miRNAs are defined as "non-muscle specific" (or also ubiquitously expressed), because essentially they are not exclusively expressed in muscular tissue. It has been, however, demonstrated that they play key-roles in modulating important pathways involved in the regulation of muscular metabolism and cellular commitment. Many miRNAs fall into this category and we report here a few relevant examples, providing for each miRNA the context in which they were studied and highlighting their global effects on muscular metabolism (Table 1).

Table 1. miRNAs expressed in muscular tissue (in an exclusive manner or not) and their global effect on muscle metabolism.

miRNA	Role in Muscle Metabolism [Refs.]	Tissue Expression
miR-1	enhancer of skeletal muscle differentiation [37–40]	muscle-specific
miR-133a/b	enhancer of myoblast proliferation [37–40]	muscle-specific
miR-206	enhancer of skeletal muscle differentiation [37–40]	muscle-specific
miR-208b	involved in specification of muscle fiber identity [43]	muscle-specific
miR-499	involved in specification of muscle fiber identity [43]	muscle-specific
miR-24	promotes myoblast differentiation [44]	ubiquitous
miR-26a	promotes myoblast differentiation [45,46]	ubiquitous
miR-27b	promotes entry into differentiation program [47]	ubiquitous
miR-29	enhancer of differentiation [48,49]	ubiquitous
miR-125b	negatively contributes to the myoblast differentiation and muscle regeneration [50–52]	ubiquitous
miR-155	represses myoblast differentiation [53]	ubiquitous
miR-181	regulates skeletal muscle differentiation and regeneration after injury [54]	ubiquitous
miR-146a	promotes satellite cell differentiation [55,56]	ubiquitous
miR-214	promotes cell cycle exit and differentiation [57]	ubiquitous
miR-221/222	promote cell cycle progression [58]	ubiquitous
miR-322/424; miR-503	promote myogenesis interfering with the progression through the cell cycle [59]	ubiquitous
miR-486	positively regulates myoblast differentiation [60,61]	muscle-enriched

Some of these miRNAs counteract the differentiation process since their activity is aimed to positively regulate the proliferation phase during muscular development.

miR-125b, one of the few down-regulated miRNAs during myogenesis, together with miR-221/222, negatively contributes to myoblast differentiation and muscle regeneration, taking part in the regulatory axis that includes mTOR and IGF-II [50–52]. Similarly, miR-155 mediates the repression of differentiation targeting MEF2A, a member of MEF2 family of transcription factors. By this negative regulation, miR-155 functions as an important regulator of muscle gene expression and myogenesis [53]. miR-221/222 instead are involved in maintenance of the proliferative state promoting cell cycle progression. They are under control of the Ras-MAPK axis and inhibit the cell-cycle regulator p27 (Cdkn1b/Kip1). Their ectopic expression, indeed, lead to defects in the transition from myoblasts to myocytes and in the assembly of sarcomeres in myotubes [58].

Figure 1. Overview of muscle-specific and ubiquitously expressed miRNAs that contribute to myogenesis and muscle regeneration processes and their regulatory activity on the muscular specific targets/chromatin modifying enzymes/cell cycle regulators (for details see the text). The main regulatory factors that exert a fundamental role during each step of normal muscle development are also reported as well as their eventual regulatory activity on the described miRNAs.

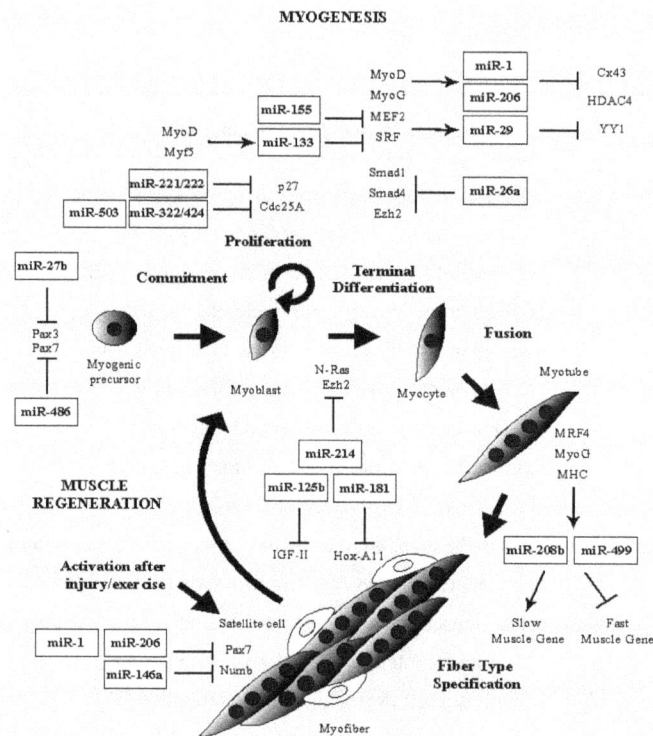

In contrast to this set of miRNAs, many other "non-muscle specific" miRNAs exert an active role in muscle differentiation through different mechanisms: miR-24, for example, has been shown to be essential for the modulation of transforming growth factor β/bone morphogenetic protein (TGF-β/BMP) pathway, a well-known inhibitor of differentiation, although its specific muscular targets are yet unknown [44]; miR-26a is involved in TGF-β/BMP pathway, where it negatively regulates the transcription factors Smad1 and Smad4, critical components of that signaling; miR26a targets the polycomb complex member Ezh2, involved in chromatin silencing of skeletal muscle genes [45,46]; miR-27b promotes entry into differentiation program both *in vitro* and *in vivo* regenerating muscles by down-regulating Pax3 [47]; miR-29 in general is defined as an enhancer of differentiation. During myogenesis it is up-regulated by SRFs and MEF2, and in a self-regulatory manner, it suppresses YY1 and HDAC4 translation by targeting their 3'-UTRs [48,49]; miR-146a is another positive regulator of myogenesis, since it modulates the activity of NUMB protein, which promotes satellite cell differentiation towards muscle cells by inhibiting Notch signaling [55,56]; miR-181 is involved in skeletal muscle differentiation and regeneration after injury and one of its targets is Hox-A11, which in turn represses transcription of MyoD [54]; miR-214 was identified in zebrafish as regulating the muscle development. Here it is expressed in skeletal muscle cell progenitors and was shown to specify muscle cell type during somitogenesis by modulating the response of muscle progenitors to Hedgehog proteins signaling [57]. Its involvement in muscle is also confirmed in C2C12 myoblasts and in skeletal myofibers of mouse where it promotes cell cycle exit and thus differentiation, targeting proto-oncogene N-*Ras* and the repressor of myogenesis Ezh2 respectively [62,63]; miR-322/424 and -503 promote myogenesis interfering with the progression through the cell cycle [59]; while miR-486 was reported to positively regulate myoblast differentiation targeting phosphatase and tensin homolog (PTEN) and Foxo1a, which negatively affect phosphoinositide-3-kinase (PI3K)/Akt signaling and down-regulate the transcription factor Pax7, required only for muscle satellite cell biogenesis and specification of the myogenic precursor lineage [60,61]. All these data clearly show the vast scenario of functions in which miRNAs are involved and their specific activities they play in the skeletal muscle physiology (Figure 1).

3. miRNAs in Muscular Dystrophies

Muscle is a dynamic tissue that goes through many recurrent phases of degeneration and regeneration throughout an individual's lifetime. During normal muscle development, specific molecular circuitries and signaling pathways control several events in different cell types such as activation of satellite cell proliferation, progenitor cell maintenance, myoblast differentiation, muscle cell homeostasis and immune cell recruitment. It is therefore not surprising that their deregulation heavily contributes to the degeneration of dystrophic muscles and is the object of intense research [64] (Table 2).

Table 2. miRNAs found deregulated in MDs and their specific activity on muscular targets or involvement in muscular processes.

miRNA/miRNAs	Deregulated in MDs [References]	Type of Deregulation	Muscular Targets/Process [References]
miR-1 (myomiR)	DMD [65,66]; DM1 [67,68]	down-regulated	HDAC4; Cx43; Pax7; c-Met; G6PD [40]
miR-133 (myomiR)	DMD [66]	down-regulated	SRF; nPTB; UCP2 [40]
miR-206 (myomiR)	DMD [65]; DM1 [69]	up-regulated	DNApolα; Fstl1; Utrn; Pax7; Cx43; HDAC4; c-Met [40]
miR-29b/c	DMD [65,66]; DM1 [67]	down-regulated	YY1; Col1a1; Eln; HDAC4 [40,62,63]
miR-135a	DMD [65]	down-regulated	muscle degeneration [65]
miR-30c	DMD [66]	down-regulated	-
miR-31	DMD [65,70]	up-regulated	DMD [70]
miR-34c; miR-449; miR-494	DMD [65]	up-regulated	muscle regeneration [65]
miR-146b; miR-155	DMD; BMD; LGMD; FSHD [71]	up-regulated	-; MEF2A [53]
miR-214	DMD; BMD; LGMD; FSHD [71]	up-regulated	Ezh2; N-Ras [40,62,63]

Table 2. *Cont.*

miRNA/miRNAs	Deregulated in MDs [References]	Type of Deregulation	Muscular Targets/Process [References]
miR-221; miR-222	DMD; BDM; LGMD; FSHD [71]	up-regulated	p27(Cdkn1b/Kip1) [58]; Sntb1 [72]
miR-223	DMD [65]	up-regulated	muscle inflammation [65]
miR-335	DMD [65]; DM1 [67]	up-regulated	muscle regeneration [65]
miR-33	DM1 [67]	down-regulated	-
miR-34a-5p; miR-34b-3p; miR-34c-5p; miR-146b-5p; miR-208a; miR-221-3p; miR-381	DM2 [73]	up-regulated	-
miR-125b-5p; miR-193a-3p; miR-193b-3p; miR-378a-3p	DM2 [73]	down-regulated	-

Eisenberg *et al.* analyzing 10 primary muscular disorders (including DMD, BDM, LGMD and FSHD samples) have identified five miRNAs (miR-146b, miR-221, miR-155, miR-214, and miR-222) consistently deregulated in almost all samples taken into consideration, suggesting their involvement in common regulatory mechanisms. Other miRNAs however showed a disease-specific profile. Functional correlation between miRNAs and mRNA targets in DMD biopsies draw a tight posttranscriptional regulation network in secondary response functions and in muscle regeneration [71].

Greco and coworkers have divided a DMD-signature of miRNAs into three main classes relative to their functional link to specific muscular pathway. Regeneration-miRNAs were up-regulated (miR-31, miR-34c, miR-206, miR-335, miR-449, and miR-494), while degenerative-miRNAs (miR-1, miR-29c, and miR-135a) were down-regulated in *mdx* mice and in DMD patients' muscles. The third class are named inflammatory-miRNAs, (miR-222 and miR-223), being expressed in damaged muscle areas only [65].

Muscle specific myomiR miR-1 and miR-133 and the ubiquitous miR-29c and miR-30c are down-regulated in *mdx* mice. It is possible to restore WT levels of these miRNAs by treating animals with an exon-skipping approach to restore a partially functional dystrophin protein, an experimental strategy that overcomes an out-frame mutation in the *DMD* locus. The same results are confirmed also in human DMD samples. These results corroborate the direct correlation between miRNAs levels and dystrophin protein levels. In contrast with the other myomiRs, miR-206 shows an increased expression in distrophic *mdx* muscle because it activates satellite cell differentiation program through Pax7 and HDAC4 repression. Another interesting target of miR-206 is Utrophin (Utrn), a dystrophin protein homolog, involved in a compensatory mechanism in DMD pathology [31,66,74].

miR-31-repressing activity seems to regulate muscle terminal differentiation directly targeting the 3'-UTR of dystrophin. Also miR-31, as miR-206, has a preferential localization in regenerating myoblasts, and is highly expressed in Duchenne muscles, probably due to an intensified activation of satellite cells. In both human and murine wild-type conditions its expression is detected in early phases of myoblast differentiation, supporting the idea that it contributes to avoid early expression of late differentiation markers. For this reason it is linked to a delay in the maturation program occurring in the pathological context [70].

Dystrophin is a structural protein that links the cytoskeleton to a large membrane-associated multiprotein complex (dystrophin-associated protein complex, DAPC) to stabilize the sarcolemma. Via

Syntrophins (SNTA1, SNTB1, SNTB2, SNTG2), members of DAPC, the enzyme neuronal Nitric Oxide Synthase (nNOS) is localized to the membrane of muscle fibers and regulates intramuscular generation of nitric oxide (NO) [75–77]. nNOS signaling determines the status of nitrosilation of Histone Deacetilases (HDACs) and thus their chromatin association to muscular specific gene-targets. Upon myoblast differentiation, HDACs are displaced from chromatin to promote muscle-specific gene transcriptional activation [78,79]. Some miRNAs involved in DMD pathology have been recently discovered to undergo this type of transcriptional regulation [66]. The absence of dystrophin in DMD patients and *mdx* mice leads to a dramatic decrease of DAPC and a consequential impairment of NO production [80,81]. The expression of a specific subset of miRNAs is modulated by HDAC2 via Dystrophin/nNOS pathway. In particular the activation of both human and murine miR-1 and miR-29 is tightly linked to HDAC2 release from their respective promoters. The functional role of these two miRNAs in muscular metabolism is also been highlighted. miR-1 controls Glucose-6-phosphate dehydrogenase (G6PD), a relevant enzyme involved in the response to oxidative stress while miR-29 controls fibrotic process since it targets the structural component of extracellular matrix, collagen (Col1a1) and elastin (Eln). Moreover, miR-222 targeting β1-Syntrophin (Sntb1) may also contribute to deregulation of the Dystophin-Syntrophins-nNOS pathway [72] (Figure 2).

Figure 2. Schematic representation of the functional/physical relationship between Dystrophin-Syntrophins-nNOS pathway and miRNAs involved in such signaling, both in a WT and a DMD context.

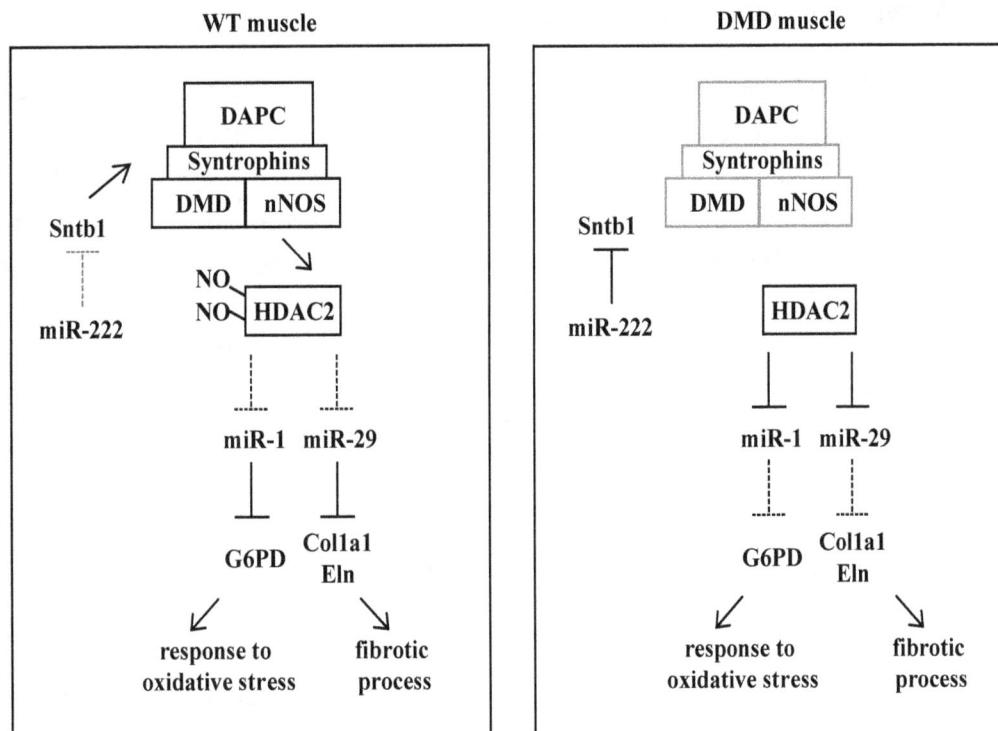

Myotonic dystrophy (DM) is the most common adult onset, progressive muscular dystrophy. DM is a multi-systemic disease and it is characterized by a generalized muscle weakness and wasting, associated with peripheral neuropathy, heart rhythm defects, and cataracts. The myotonia phenomenon is due to the peculiar muscle membrane depolarization activities. Two type of DM exist, type-1 (DM1, OMIM 160900) and type-2 (DM2, OMIM 602668). DM1 is caused by an expansion of the CTG triplet repeats in the 3'-untraslated region (UTR) of the Dystrophic Myotonic Protein Kinase (*DMPK*), while DM2 is caused by the expansion of a tetranucleotide repeat CCTG in the first intron of CCHC-type zinc finger nucleic acid binding protein (*CNPB*). These gene expansions do not disrupt the relative protein coding sequence, the repeats being in non-coding regions. However, both expanded RNAs accumulate in the nucleus and trigger a toxic gain of function that interferes with RNA splicing of other genes [82–86]. Perbellini and colleagues have performed expression analysis in DM1 biopsies obtained from 15 patients. They found specific deregulated miRNAs: miR-1 and -335 are up-regulated, whereas miR-29b, -29c and -33 are down-regulated compared to control muscles [67,68]. Gambardella and co-workers profiled a specific pattern of myomiRs involved in myogenesis of cardiac and skeletal muscle and found lines of evidence of miR-206 overexpression in five DM1 patients [69]. A similar investigation has been made in DM2 patients. Eleven miRNAs have been shown to be deregulated. Nine displayed higher levels compared to controls (miR-34a-5p, miR-34b-3p, miR-34c-5p, miR-146b-5p, miR-208a, miR-221-3p and miR-381), while four were decreased (miR-125b-5p, miR-193a-3p, miR-193b-3p and miR-378a-3p). Moreover the potential involvement of these miRNAs in relevant skeletal muscle pathways and functions has been validated by bioinformatics analyses [73]. Recently a novel therapeutic approach has been proposed to target the CTG repeat expansion on RNA using antisense oligonucleotides [8,87,88]. Therefore improving knowledge concerning the transcription regulation of the *DMPK* gene, also via ncRNAs, will greatly benefit this new therapy.

4. Long Non-Coding RNAs in Skeletal Muscle and Muscular Dystrophies

Increasing lines of evidence support the biological relevance of lncRNAs. They are regulated during development and involved in almost all levels of gene expression and cellular functions including chromosomal dosage compensation, chromatin modification, cell cycle regulation, control of imprinting, alternative splicing, intracellular trafficking, cellular differentiation, and reprogramming of stem cells [89]. Recently, lncRNAs related to muscle are emerging both in physiological and pathological context (Table 3).

Key features of dystrophic muscle include central nuclei, small regenerating fibers and accumulation of connective tissue and fatty tissue. Muscle differentiation *in vitro* is a useful system to investigate the activity of long non-coding RNAs that show muscular specific pattern of expression. Recently, a new regulatory network involving cross-talk of several ncRNAs has been identified by Cesana and colleagues. Relying on ability of myomiRs to orchestrate muscular proliferation and differentiation, the genomic region of miR-206/-133b has been analyzed in detail. Thus a novel muscle specific transcript has been identified. Because of its non-coding potential and its activated expression upon myoblast differentiation it was termed linc-MD1. More specifically linc-MD1 is expressed in newly regenerating fibers and is abundant in dystrophic condition, however no expression is detected in mature differentiated fibers. linc-MD1 is localized in the cytoplasm and is a polyadenylated transcript. Through a series of functional studies it was possible to define its competing endogenous activity (ceRNA). linc-MD1 acts as a natural decoy for miR-133 and -135, thus interfering with miRNA repressing activity on the important targets involved in myogenic differentiation MAML1 (Mastermind-like 1) and MEF2, respectively [90].

Table 3. Recently discovered lncRNAs related to muscle, both in physiological and pathological context.

lncRNA/lncRNAs [References]	Expression in Muscular Districts	Deregulated in MDs	Activity
linc-MD1 [90]	Expressed in newly regenerating fibers	DMD	natural decoy for miR-133 and -135 (ceRNA)
Malat1 [91]	up-regulated during the differentiation of myoblasts into myotubes	?	regulation of cell growth
Men ε/β lncRNAs [92–94]	up-regulated upon differentiation of C2C12 myoblats	?	critical structural/organizational components of paraspeckles
SRA ncRNA [95–97]	increased expression during myogenic differentiation	DM1	co-activator of MYOD transcription factor
NRON [98,99]	enriched also in muscle	?	regulates NFAT's subcellular localization (scaffold)
lncINT44s; lncINT44s2; lncINT55s [14]	transcribed contextually with dystrophin isoforms and upon MYOD-induced myogenic differentiation	?	negative modulation of endogenous dystrophin full-length isoforms
KUCG1 [100]	expressed at low levels in the brain	DMD with mental retardation	possible candidate gene that contribute to develop of mental retardation in the index case
DBT-E [101]	not-physiological lncRNA	FSHD	coordinates de-repression of genes located in the 4q35 region

Metastasis associated lung adenocarcinoma transcript 1 (Malat1) is a highly conserved 8.7 kb non-coding transcript that is abundantly expressed in cancer cells and a strong predictor of metastasis [102]. Malat1 has been proposed to regulate alternative splicing [103], transcriptional activation and the expression of nearby genes [104,105]. Numerous experimental examples support its functional role in the regulation of cell growth, but the exact mechanism of action of Malat1 in different physiological and pathological conditions still needs to be elucidated. By a microarray data analysis obtained using skeletal muscle of mice (gastrocnemius muscle) treated with recombinant myostatin it was observed that the Malat1 expression levels are significantly decreased. Myostatin is a potent negative regulator of myogenesis that inhibits myoblast proliferation and differentiation [106,107]. Further expression analysis confirmed a persistent up-regulation of Malat1 during the differentiation of myoblasts into myotubes in C2C12 cells as well as in primary human skeletal muscle cells. Conversely, targeted knockdown of Malat1 using siRNA suppressed myoblast proliferation by arresting cell growth in the G0/G1 phase. These results reveal Malat1 as a novel downstream target of myostatin with a considerable ability to regulate myogenesis. Although Malat1 appears largely dispensable for normal mouse development [108,109] it is plausible that Malat1 has a role in the transition from the proliferative phase to differentiation in skeletal myogenesis, as well as in the commitment to muscle differentiation [91].

Many lncRNA have been discovered but not yet fully characterized, as for example Men ε/β lncRNAs. To date it is known that two long non-coding isoforms (Men ε/β lncRNAs) which are expressed in several human tissues, including muscle, arise from the Multiple Endocrine Neoplasia I locus (*MEN1*). Experimental lines of evidence show their up-regulation upon differentiation of C2C12 myoblats, although their biological role in muscular development is not yet clear. Men ε (also known as NEAT1) and Men β are transcribed from the same RNA polymerase II promoter and are both retained in the nucleus. Suwoo and colleagues formally demonstrated that Men ε/β transcripts are critical structural/organizational components of paraspeckles, organelles localized in the nucleoplasm close to nuclear speckles, where RNA-binding proteins and *Cat2*-transcribed nuclear RNA (CTN-RNA) are stored [110]. Moreover, large-scale analysis revealed that many other lncRNAs are differentially expressed in C2C12 cells upon myoblast differentiation into myotubes, although their biological functions have not been investigated [92–94].

Between the many functions ascribed to lncRNA there are examples of lncRNAs modulating the activity of transcriptional activators or co-activators, directly or through the regulation of their sub-cellular localization [89]. Two of these have been seen also in a muscular context. The steroid receptor RNA activator (SRA) RNA is a very peculiar transcript that exists as both a non-coding and a coding RNA (yielding SRA ncRNA and protein SRAP respectively). The SRA ncRNA is highly expressed in skeletal muscle and works as a co-activator of MYOD transcription factor, a master regulator of skeletal myogenesis. To address the significance of the enigmatic bifunctional property of this transcript, Hube and colleagues performed an exhaustive analysis clarifying the opposite function of non-protein coding SRA versus ORF-containing transcripts. The balance between coding and non-coding SRA isoforms changes during myogenic differentiation in primary human cells. In particular it is shown that an increased expression of SRA ncRNA and a parallel decrease of protein SRAP occurs during myogenic differentiation in healthy muscle satellite cells. This does not happen in cells isolated from DM1 patients, probably because of a delay in differentiation program. Remarkably, only the ncRNA species enhances MYOD transcriptional activity. The protein SRAP prevents this SRA RNA-dependent co-activation through interaction with its RNA counterpart [95–97]. However how this is achieved is not known.

Non-coding repressor of NFAT (NRON) is another case of lncRNA that shows a regulatory activity on a transcription factor. NRON is not highly expressed but it has a distinct tissue specific expression. It has been found enriched in placenta, muscle, and lymphoid tissues. NFAT is a transcription factor responsive to local changes in calcium signals. It is essential for the T cell receptor–mediated immune response and plays a critical role in the development of heart and vasculature, musculature, and nervous tissue. The first study about the role of NRON showed that it regulates NFAT's subcellular localization rather than its transcriptional activity. Sharma and coworkers confirmed these data demonstrating that NRON takes part in a large cytoplasmic RNA-protein complex that acts as a scaffold for NFAT to modulate its nuclear trafficking and thus its response activity [98,99].

Little is yet known about the dystrophin gene regulation. *DMD* is the largest gene in the human genome that comprises 79 exons spanning >2500 kb on chromosome Xp21.2, which gives rise to 7 isoforms that are finely regulated in terms of tissue specificity [111]. Mutations in the *DMD* gene range from single-nucleotide changes to chromosomal abnormalities (http://www.dmd.nl/). Deletions encompassing one or more exons of the dystrophin gene are the most common cause of the severe Duchenne muscular dystrophy (DMD) resulting in an absence of dystrophin or expression of a non-functional protein. Becker muscular dystrophy (BMD) instead is a milder form of dystrophy because it is associated with reduction of wild-type dystrophin or expression of a partially functional protein. DMD is the most common inherited muscle disease affecting approximately one in

3500 males and is characterized by progressive muscle wasting during childhood. Heterozygous females for dystrophin mutations are named carriers of DMD mutations [112,113]. Many of them are asymptomatic, but a certain number, defined as "manifesting" or "symptomatic", develop symptoms of the disease, which vary from a mild muscle weakness to a DMD-like clinical course. Despite intensively explored, the pathogenic mechanism underlying clinical manifestation in DMD female carriers still remains a controversial issue [114]. For these reasons *DMD* regulation is a field of intense interest to shed light on this complex scenario.

Using a custom-made tiling array the entire *DMD* gene has been explored in the search for non-coding transcripts originating within the dystrophin locus. The major tissues of dystrophin synthesis, namely human brain, heart and skeletal muscle, were used as test tissues in array. The data analysis has highlighted a variety of novel long non-coding RNAs (lncRNAs), both sense and antisense oriented, whose expression profiles mirror that of *DMD* gene. Importantly, these transcripts are intronic in origin, specifically localized to the nucleus and are transcribed contextually with dystrophin isoforms or in fibroblast upon MYOD-induced myogenic differentiation. To characterize their possible functional role on the *DMD* locus three sense-oriented lncRNAs (lncINT44s, lncINT44s2 and lncINT55s) isolated from skeletal muscle were further investigated. Their forced ectopic expression in both human muscle and neuronal cells causes a negative regulation of endogenous full-length dystrophin isoforms, denoted B for brain (Dp427b), M for muscle (Dp427m) and P for Purkinje (Dp427p). Importantly, no variation was observed with regard to the ubiquitous Dp71 transcript, suggesting that the effect of sense lncRNAs on full-length dystrophin isoforms may be specific. In particular, reporter assay confirmed their repressive role on the minimal promoter regions of the muscle dystrophin isoform. A possible mechanism of action involves specific DMD lncRNAs that control muscle dystrophin isoforms by down-modulating dystrophin transcription levels. An inverse correlation between ncRNAs expression and muscle dystrophin has been also found *in vivo*, analyzing muscle samples of DMD female carriers, either healthy or mildly affected, reinforcing the idea that a negative relationship between lncRNAs and dystrophin mRNA levels may exist [14].

In severe DMD one third of patients display also mental retardation, but the pathogenesis is unknown. In a singular case of DMD complicated by mental retardation, an intra-chromosomal inversion (inv(X)p21.2;q28) has been identified. The genetic rearrangement has been molecularly characterized to find a possible disrupted gene because of the inversion, and that might be responsible for the neurological symptoms associated with dystrophy. A novel gene named *KUCG1* was discovered at break point on Xq28. The 658-bp transcript displays an mRNA-like structure but not having coding potential is been classified as long non-coding RNA. KUCG1 lncRNA is expressed at low levels in a tissue-specific manner, as well in the brain. It is possible that the disruption of KUCG1 transcript contributes to the development of mental retardation in the index case [115] since other experimental lines of evidence suggest that a subset of lncRNAs could contribute to neurological disorders when they become deregulated [100].

Polycomb (PcG) and Trithorax (TrxG) group proteins antagonistically act in the epigenetic regulation of gene expression. Typically, TrxG counteracts PcG-mediated epigenetic gene silencing. Among the many lncRNAs interacting with chromatin remodeling enzymes the most famous are Xist and HOTAIR, both acting as a negative regulators of gene expression by recruitment of PRC2 (Polycomb Repressive Complex 2) on PcG target genes [116,117]. Cabianca *et al*. were the first to discover an lncRNA interacting with the TrxG in the Facioscapulohumeral muscular dystrophy (FSHD). FSHD is an autosomal-dominant disease characterized by progressive wasting of facial, upper arm, and shoulder girdle muscles. In up to 95% of cases, the genetic defect is mapped to the subtelomeric region of chromosome 4q35 containing a macrosatellite tandem array of 3.3 Kb long D4Z4 repeats. FSHD is caused by deletions reducing copy number of D4Z4 below 11 units rather than a classical mutation in a coding-protein gene. D4Z4 deletion is associated to a loss of repressive epigenetic marks and thus to a switch from a heterochromatic/close state to a more euchromatic/open conformation of chromatin structure. A novel long non-coding RNA, named DBT-E is produced selectively in FSHD patients. DBT-E is transcribed from D4Z4 repeats and is a chromatin-associated lncRNA that coordinates de-repression of genes located in the 4q35 region. DBT-E recruits the Trithorax group protein Ash1L to the FSHD locus driving histone H3 lysine dimethylation and thus chromatin remodeling [101].

5. Discussion

It is surprising how ncRNAs are tightly interconnected with the main fundamental aspects of muscular tissue: development, differentiation and regeneration. At the molecular level miRNAs and lncRNAs take part in almost all levels of regulation in these key processes. Chromatin modifying enzymes, positive and negative transcription factors, cell cycle regulators, and enzymatic and structural proteins involved in signaling circuitries are under their fine-modulation. Moreover, ncRNAs are often found to be under the regulation of their own targets, thus determining feedback loops that drive developmental switches ensuring a perfect synergy between stimuli and responses. Both time- and tissue-specific gene regulation are the fulcrum on which the fine-tuning of a healthy organism is based. Disrupting the physiological pattern of expression not only in codifying genes, but also in regulatory RNAs, can heavily modify specific cell processes.

If this is true in physiological conditions, increased lines of evidence show that regulatory RNAs play a crucial role also in the etiology of many human diseases. Among these, muscular dystrophies represent a field of intense research, also because of the recent creation of novel experimental treatments. This has encouraged studies on expression regulation in diseases muscle cells, both *in vivo* (animal models) and *in vitro*. These studies have shown that mutant proteins in MDs result in perturbations of many cellular components. Indeed MDs have been associated with mutations in structural proteins, signaling molecules and enzymes as well as mutations that result in aberrant processing of mRNA or alterations in post-translational modifications of proteins. These findings have not only revealed important insights for cell biologists, but have also provided unexpected and exciting new approaches for therapy. Moreover, in muscular dystrophies as well as in other diseases, such as cancer, regulatory RNAs may serve as biomarkers, providing information on disease course, disease severity and response to therapies. miRNA dosing in serum is a very appealing field of investigation since they are easily accessible, peculiar to defined conditions and can facilitate the early identification of the muscular disease, potentially avoiding invasive techniques such as a biopsy, or in some cases to reduce the time and the costs of diagnosis. Biomarkers are particularly important in the field of personalized treatments. Pharmacogenomics aims at predicting which drug will be most effective and safe in the individuals. This can be established via genome sequence and SNP association (pharmacogenetics) and expression profiling. miRNAs have been shown to play a pivotal role in drug efficacy and toxicity, having powerful implications in personalized medicine [118]. Indeed, as we have described, miRNAs can negatively regulate gene expression and can profile the disease severity, as in the case of myomiRs and DMD [41]. miRNAs show a linear relationship with genes and drugs, since drug function can be influenced or even hampered by changes in genes expression level or in specific isoforms representation, as supported by several data on cancer [119]. Many pharmacogenomically relevant genes are regulated by miRNAs, as summarized and shown in the Pharmacogenomcis Knowledge Base (PharmGKB, www.pharmgkb.org/), a very useful resource listing genes known to be relevant for drug response.

Conversely, miRNAs can vary in their expression level following drug treatments [120]. Within the muscle field, we do have increased knowledge on the miRNAs network, especially those governing the muscle transcriptional network. It is clearly emerging how miRNAs can regulate differentiation and homeostasis of skeletal muscle progenitor cells, providing robustness to the MYOD-induced myoblast differentiation and myogenesis [121,122]. Disclosing the role miRNAs have in regulating the intermediate steps of the myogenesis cascade will be of outmost importance in identifying drugs that may act as adjuvants/enhancers of gene/protein re-synthesis in clinical trials, as for exon skipping therapies in DMD.

More complex is our current understanding of the role of lncRNAs in muscle biology and pathology. We have just started to explore the peripheral areas of this "*terra incognita*". So far lncRNAs have been involved in numerous molecular processes such as remodelling of chromatin architecture, or regulation of gene transcription. For instance, some pharmacodynamic studies on corticosteroids, which represent the gold standard in the routine therapy of DMD, revealed that the steroid receptor RNA activator (SRA) transcript functions as both a lncRNA and template for synthesis of a protein (SRAP). Interestingly, the SRA ncRNA increases the activity of nuclear receptors (not only for corticosteroids) and acts as a master regulator of MYOD expression [95]. lncRNAs can also exert their function through a more passive role. For instance they are valued for their ability to work as molecular sponges by annealing to small RNAs and thereby preventing them from their normal activity. Furthermore, in some other cases lncRNAs have been shown to provide a kind of structural backbone for the assembly of ribonucleic particles whose functions are still to be disclosed. In this respect, it has been crucial to determine in which intracellular compartments these RNA/protein particles form. Despite the fact that so far most investigated lncRNAs are confined to nuclei, a few recent studies have, indeed, shown that some lncRNAs can also abundantly localize inside the cytoplasm with functions that still remain to be determined.

Given that just a few lncRNAs have been tackled on functional levels and that the annotated ones are in the order of thousands with many more expected to be discovered, it is plausible to speculate that their involvement in novel functions and roles will be rapidly identified with repercussions on many fields of cell biology and pathology and with the possibility to potentially employ them as biological markers as well as drugs to treat major diseases such as muscular dystrophies.

6. Conclusions

Although in the majority of cases the etiology of muscular dystrophies is not ascribed to functional non-coding RNA molecules (with the exception of FSHD), they appear as powerful regulators of several key-pathways and show how actively they can contribute to the progression of disease. This reflects the strong ability of miRNAs and lncRNAs in the modulation of the phenotype of dystrophic affected individuals via fine regulatory pathways that can lead to increased transcript stability, mRNA splicing control, enhanced protein production, posttranslational protein modification and other mechanisms. These versatile roles support the idea to use regulatory RNAs as novel targeted molecules acting as enhancers or inhibitors in well-established therapeutic strategies (based both on drugs and gene therapies). After all, the general goal is to ameliorate the final output of the specific treatments.

Acknowledgments

The BIO-NMD EU project (N. 241665 to A.F.) is acknowledged.

References

1. Carninci, P.; Kasukawa, T.; Katayama, S.; Gough, J.; Frith, M.C.; Maeda, N.; Oyama, R.; Ravasi, T.; Lenhard, B.; Wells, C.; *et al.* The transcriptional landscape of the mammalian genome. *Science* **2005**, *309*, 1559–1563.

2. Mattick, J.S.; Makunin, I.V. Non-coding RNA. *Hum. Mol. Genet.* **2006**, *15*, R17–R29.

3. Kapranov, P.; Cheng, J.; Dike, S.; Nix, D.A.; Duttagupta, R.; Willingham, A.T.; Stadler, P.F.; Hertel, J.; Hackermuller, J.; Hofacker, I.L.; *et al.* RNA maps reveal new RNA classes and a possible function for pervasive transcription. *Science* **2007**, *316*, 1484–1488.

4. Wang, X.Q.; Crutchley, J.L.; Dostie, J. Shaping the genome with non-coding RNAs. *Curr. Genomics* **2011**, *12*, 307–321.

5. Ponting, C.P.; Oliver, P.L.; Reik, W. Evolution and functions of long noncoding RNAs. *Cell* **2009**, *136*, 629–641.

6. Cirak, S.; Arechavala-Gomeza, V.; Guglieri, M.; Feng, L.; Torelli, S.; Anthony, K.; Abbs, S.; Garralda, M.E.; Bourke, J.; Wells, D.J.; *et al.* Exon skipping and dystrophin restoration in patients with duchenne muscular dystrophy after systemic phosphorodiamidate morpholino oligomer treatment: An open-label, phase 2, dose-escalation study. *Lancet* **2011**, *378*, 595–605.

7. Goemans, N.M.; Tulinius, M.; van den Akker, J.T.; Burm, B.E.; Ekhart, P.F.; Heuvelmans, N.; Holling, T.; Janson, A.A.; Platenburg, G.J.; Sipkens, J.A.; *et al.* Systemic administration of pro051 in duchenne's muscular dystrophy. *N. Engl. J. Med.* **2011**, *364*, 1513–1522.

8. Evers, M.M.; Pepers, B.A.; van Deutekom, J.C.; Mulders, S.A.; den Dunnen, J.T.; Aartsma-Rus, A.; van Ommen, G.J.; van Roon-Mom, W.M. Targeting several cag expansion diseases by a single antisense oligonucleotide. *PLoS One* **2011**, *6*, e24308.

9. Ferlini, A.; Neri, M.; Gualandi, F. The medical genetics of dystrophinopathies: Molecular genetic diagnosis and its impact on clinical practice. *Neuromuscul. Disord. NMD* **2013**, *23*, 4–14.

10. Mitsuhashi, S.; Kang, P.B. Update on the genetics of limb girdle muscular dystrophy. *Semin. Pediatr. Neurol.* **2012**, *19*, 211–218.

11. Pegoraro, E.; Hoffman, E.P. Limb-girdle Muscular Dystrophy Overview. In *Genereviews*; Pagon, R.A., Adam, M.P., Bird, T.D., Dolan, C.R., Fong, C.T., Stephens, K., Eds.; University of Washington: Seattle, WA, USA, 1993.

12. Davies, K.E.; Nowak, K.J. Molecular mechanisms of muscular dystrophies: Old and new players. *Nat. Rev. Mol. Cell Biol.* **2006**, *7*, 762–773.

13. Johnson, N.E.; Heatwole, C.R. Myotonic dystrophy: From bench to bedside. *Semin. Neurol.* **2012**, *32*, 246–254.

14. Bovolenta, M.; Erriquez, D.; Valli, E.; Brioschi, S.; Scotton, C.; Neri, M.; Falzarano, M.S.; Gherardi, S.; Fabris, M.; Rimessi, P.; *et al.* The dmd locus harbours multiple long non-coding RNAs which orchestrate and control transcription of muscle dystrophin mRNA isoforms. *PLoS One* **2012**, *7*, e45328.

15. Graves, P.; Zeng, Y. Biogenesis of mammalian micro RNAs: A global view. *Genomics Proteomics Bioinforma.* **2012**, *10*, 239–245.

16. Bentzinger, C.F.; von Maltzahn, J.; Rudnicki, M.A. Extrinsic regulation of satellite cell specification. *Stem Cell Res. Ther.* **2010**, *1*, 27.

17. Bentzinger, C.F.; Wang, Y.X.; Rudnicki, M.A. Building muscle: Molecular regulation of myogenesis. *Cold Spring Harb Perspect Biol.* **2012**, *4*, doi:10.1101/cshperspect.a008342.

18. Bismuth, K.; Relaix, F. Genetic regulation of skeletal muscle development. *Exp. Cell Res.* **2010**, *316*, 3081–3086.

19. Bryson-Richardson, R.J.; Currie, P.D. The genetics of vertebrate myogenesis. *Nat. Rev. Genet.* **2008**, *9*, 632–646.

20. Punch, V.G.; Jones, A.E.; Rudnicki, M.A. Transcriptional networks that regulate muscle stem cell function. *Wiley Interdiscip Rev. Syst. Biol. Med.* **2009**, *1*, 128–140.

21. Kassar-Duchossoy, L.; Gayraud-Morel, B.; Gomes, D.; Rocancourt, D.; Buckingham, M.; Shinin, V.; Tajbakhsh, S. Mrf4 determines skeletal muscle identity in myf5:Myod double-mutant mice. *Nature* **2004**, *431*, 466–471.

22. Nabeshima, Y.; Hanaoka, K.; Hayasaka, M.; Esumi, E.; Li, S.; Nonaka, I. Myogenin gene disruption results in perinatal lethality because of severe muscle defect. *Nature* **1993**, *364*, 532–535.

23. Charge, S.B.; Rudnicki, M.A. Cellular and molecular regulation of muscle regeneration. *Physiol. Rev.* **2004**, *84*, 209–238.

24. Parker, M.H.; Seale, P.; Rudnicki, M.A. Looking back to the embryo: Defining transcriptional networks in adult myogenesis. *Nat. Rev. Genet.* **2003**, *4*, 497–507.

25. Pownall, M.E.; Gustafsson, M.K.; Emerson, C.P., Jr. Myogenic regulatory factors and the specification of muscle progenitors in vertebrate embryos. *Annu. Rev. Cell Dev. Biol.* **2002**, *18*, 747–783.

26. Buckingham, M. Skeletal muscle formation in vertebrates. *Curr. Opin. Genet. Dev.* **2001**, *11*, 440–448.

27. Hasty, P.; Bradley, A.; Morris, J.H.; Edmondson, D.G.; Venuti, J.M.; Olson, E.N.; Klein, W.H. Muscle deficiency and neonatal death in mice with a targeted mutation in the myogenin gene. *Nature* **1993**, *364*, 501–506.

28. Yokoyama, S.; Asahara, H. The myogenic transcriptional network. *Cell Mol. Life Sci.* **2011**, *68*, 1843–1849.

29. O'Rourke, J.R.; Georges, S.A.; Seay, H.R.; Tapscott, S.J.; McManus, M.T.; Goldhamer, D.J.; Swanson, M.S.; Harfe, B.D. Essential role for dicer during skeletal muscle development. *Dev. Biol.* **2007**, *311*, 359–368.

30. Zhao, Y.; Samal, E.; Srivastava, D. Serum response factor regulates a muscle-specific microRNA that targets hand2 during cardiogenesis. *Nature* **2005**, *436*, 214–220.

31. Rosenberg, M.I.; Georges, S.A.; Asawachaicharn, A.; Analau, E.; Tapscott, S.J. Myod inhibits fstl1 and utrn expression by inducing transcription of mir-206. *J. Cell Biol.* **2006**, *175*, 77–85.

32. Rao, P.K.; Kumar, R.M.; Farkhondeh, M.; Baskerville, S.; Lodish, H.F. Myogenic factors that regulate expression of muscle-specific microRNAs. *Proc. Natl. Acad. Sci. USA* **2006**, *103*, 8721–8726.

33. Liu, N.; Williams, A.H.; Kim, Y.; McAnally, J.; Bezprozvannaya, S.; Sutherland, L.B.; Richardson, J.A.; Bassel-Duby, R.; Olson, E.N. An intragenic mef2-dependent enhancer directs muscle-specific expression of microRNAs 1 and 133. *Proc. Natl. Acad. Sci. USA* **2007**, *104*, 20844–20849.

34. Sun, Y.; Ge, Y.; Drnevich, J.; Zhao, Y.; Band, M.; Chen, J. Mammalian target of rapamycin regulates miRNA-1 and follistatin in skeletal myogenesis. *J. Cell Biol.* **2010**, *189*, 1157–1169.

35. Weintraub, H. The myod family and myogenesis: Redundancy, networks, and thresholds. *Cell* **1993**, *75*, 1241–1244.

36. Naya, F.J.; Olson, E. Mef2: A transcriptional target for signaling pathways controlling skeletal muscle growth and differentiation. *Curr. Opin. Cell Biol.* **1999**, *11*, 683–688.

37. Van Rooij, E.; Liu, N.; Olson, E.N. MicroRNAs flex their muscles. *Trends Genet. TIG* **2008**, *24*, 159–166.

38. Chen, J.F.; Callis, T.E.; Wang, D.Z. MicroRNAs and muscle disorders. *J. Cell Sci.* **2009**, *122*, 13–20.

39. Eisenberg, I.; Alexander, M.S.; Kunkel, L.M. miRNAs in normal and diseased skeletal muscle. *J. Cell. Mol. Med.* **2009**, *13*, 2–11.

40. Ge, Y.; Chen, J. MicroRNAs in skeletal myogenesis. *Cell Cycle* **2011**, *10*, 441–448.

41. Cacchiarelli, D.; Legnini, I.; Martone, J.; Cazzella, V.; D'Amico, A.; Bertini, E.; Bozzoni, I. miRNAs as serum biomarkers for duchenne muscular dystrophy. *EMBO Mol. Med.* **2011**, *3*, 258–265.

42. Mitchell, P.S.; Parkin, R.K.; Kroh, E.M.; Fritz, B.R.; Wyman, S.K.; Pogosova-Agadjanyan, E.L.; Peterson, A.; Noteboom, J.; O'Briant, K.C.; Allen, A.; *et al.* Circulating microRNAs as stable blood-based markers for cancer detection. *Proc. Natl. Acad. Sci. USA* **2008**, *105*, 10513–10518.

43. Van Rooij, E.; Quiat, D.; Johnson, B.A.; Sutherland, L.B.; Qi, X.; Richardson, J.A.; Kelm, R.J., Jr.; Olson, E.N. A family of microRNAs encoded by myosin genes governs myosin expression and muscle performance. *Dev. Cell* **2009**, *17*, 662–673.

44. Sun, Q.; Zhang, Y.; Yang, G.; Chen, X.; Cao, G.; Wang, J.; Sun, Y.; Zhang, P.; Fan, M.; Shao, N.; *et al.* Transforming growth factor-beta-regulated mir-24 promotes skeletal muscle differentiation. *Nucleic Acids Res.* **2008**, *36*, 2690–2699.

45. Caretti, G.; Di Padova, M.; Micales, B.; Lyons, G.E.; Sartorelli, V. The polycomb ezh2 methyltransferase regulates muscle gene expression and skeletal muscle differentiation. *Genes Dev.* **2004**, *18*, 2627–2638.

46. Dey, B.K.; Gagan, J.; Yan, Z.; Dutta, A. Mir-26a is required for skeletal muscle differentiation and regeneration in mice. *Genes Dev.* **2012**, *26*, 2180–2191.

47. Crist, C.G.; Montarras, D.; Pallafacchina, G.; Rocancourt, D.; Cumano, A.; Conway, S.J.; Buckingham, M. Muscle stem cell behavior is modified by microRNA-27 regulation of pax3 expression. *Proc. Natl. Acad. Sci. USA.* **2009**, *106*, 13383–13387.

48. Wang, H.; Garzon, R.; Sun, H.; Ladner, K.J.; Singh, R.; Dahlman, J.; Cheng, A.; Hall, B.M.; Qualman, S.J.; Chandler, D.S.; *et al.* Nf-kappab-yy1-mir-29 regulatory circuitry in skeletal myogenesis and rhabdomyosarcoma. *Cancer Cell* **2008**, *14*, 369–381.

49. Li, Z.; Hassan, M.Q.; Jafferji, M.; Aqeilan, R.I.; Garzon, R.; Croce, C.M.; van Wijnen, A.J.; Stein, J.L.; Stein, G.S.; Lian, J.B. Biological functions of mir-29b contribute to positive regulation of osteoblast differentiation. *J. Biol. Chem.* **2009**, *284*, 15676–15684.

50. Ge, Y.; Sun, Y.; Chen, J. Igf-ii is regulated by microRNA-125b in skeletal myogenesis. *J. Cell Biol.* **2011**, *192*, 69–81.

51. Erbay, E.; Park, I.H.; Nuzzi, P.D.; Schoenherr, C.J.; Chen, J. Igf-ii transcription in skeletal myogenesis is controlled by mtor and nutrients. *J. Cell Biol.* **2003**, *163*, 931–936.

52. Ge, Y.; Wu, A.L.; Warnes, C.; Liu, J.; Zhang, C.; Kawasome, H.; Terada, N.; Boppart, M.D.; Schoenherr, C.J.; Chen, J. Mtor regulates skeletal muscle regeneration *in vivo* through kinase-dependent and kinase-independent mechanisms. *Am. J. Physiol. Cell Physiol.* **2009**, *297*, C1434–C1444.

53. Seok, H.Y.; Tatsuguchi, M.; Callis, T.E.; He, A.; Pu, W.T.; Wang, D.Z. Mir-155 inhibits expression of the mef2a protein to repress skeletal muscle differentiation. *J. Biol. Chem.* **2011**, *286*, 35339–35346.

54. Naguibneva, I.; Ameyar-Zazoua, M.; Polesskaya, A.; Ait-Si-Ali, S.; Groisman, R.; Souidi, M.; Cuvellier, S.; Harel-Bellan, A. The microRNA mir-181 targets the homeobox protein hox-a11 during mammalian myoblast differentiation. *Nat. Cell Biol.* **2006**, *8*, 278–284.

55. Kuang, W.; Tan, J.; Duan, Y.; Duan, J.; Wang, W.; Jin, F.; Jin, Z.; Yuan, X.; Liu, Y. Cyclic stretch induced mir-146a upregulation delays c2c12 myogenic differentiation through inhibition of numb. *Biochem. Biophys. Res. Commun.* **2009**, *378*, 259–263.

56. Conboy, I.M.; Rando, T.A. The regulation of notch signaling controls satellite cell activation and cell fate determination in postnatal myogenesis. *Dev. Cell* **2002**, *3*, 397–409.

57. Flynt, A.S.; Li, N.; Thatcher, E.J.; Solnica-Krezel, L.; Patton, J.G. Zebrafish mir-214 modulates hedgehog signaling to specify muscle cell fate. *Nat. Genet.* **2007**, *39*, 259–263.

58. Cardinali, B.; Castellani, L.; Fasanaro, P.; Basso, A.; Alema, S.; Martelli, F.; Falcone, G. MicroRNA-221 and microRNA-222 modulate differentiation and maturation of skeletal muscle cells. *PLoS One* **2009**, *4*, e7607.

59. Sarkar, S.; Dey, B.K.; Dutta, A. Mir-322/424 and -503 are induced during muscle differentiation and promote cell cycle quiescence and differentiation by down-regulation of cdc25a. *Mol. Biol. Cell* **2010**, *21*, 2138–2149.

60. Dey, B.K.; Gagan, J.; Dutta, A. Mir-206 and -486 induce myoblast differentiation by downregulating pax7. *Mol. Cell Biol.* **2011**, *31*, 203–214.

61. Small, E.M.; O'Rourke, J.R.; Moresi, V.; Sutherland, L.B.; McAnally, J.; Gerard, R.D.; Richardson, J.A.; Olson, E.N. Regulation of pi3-kinase/akt signaling by muscle-enriched microRNA-486. *Proc. Natl. Acad. Sci. USA* **2010**, *107*, 4218–4223.

62. Juan, A.H.; Kumar, R.M.; Marx, J.G.; Young, R.A.; Sartorelli, V. Mir-214-dependent regulation of the polycomb protein ezh2 in skeletal muscle and embryonic stem cells. *Mol. Cell* **2009**, *36*, 61–74.

63. Liu, J.; Luo, X.J.; Xiong, A.W.; Zhang, Z.D.; Yue, S.; Zhu, M.S.; Cheng, S.Y. MicroRNA-214 promotes myogenic differentiation by facilitating exit from mitosis via down-regulation of proto-oncogene n-ras. *J. Biol. Chem.* **2010**, *285*, 26599–26607.

64. Marrone, A.K.; Shcherbata, H.R. Dystrophin orchestrates the epigenetic profile of muscle cells via miRNAs. *Front. Genet.* **2011**, *2*, doi:10.3389/fgene.2011.0006.

65. Greco, S.; de Simone, M.; Colussi, C.; Zaccagnini, G.; Fasanaro, P.; Pescatori, M.; Cardani, R.; Perbellini, R.; Isaia, E.; Sale, P.; *et al.* Common micro-RNA signature in skeletal muscle damage and regeneration induced by duchenne muscular dystrophy and acute ischemia. *FASEB J.* **2009**, *23*, 3335–3346.

66. Cacchiarelli, D.; Martone, J.; Girardi, E.; Cesana, M.; Incitti, T.; Morlando, M.; Nicoletti, C.; Santini, T.; Sthandier, O.; Barberi, L.; *et al.* MicroRNAs involved in molecular circuitries relevant for the duchenne muscular dystrophy pathogenesis are controlled by the dystrophin/nnos pathway. *Cell Metab.* **2010**, *12*, 341–351.

67. Perbellini, R.; Greco, S.; Sarra-Ferraris, G.; Cardani, R.; Capogrossi, M.C.; Meola, G.; Martelli, F. Dysregulation and cellular mislocalization of specific miRNAs in myotonic dystrophy type 1. *Neuromuscul. Disord. NMD* **2011**, *21*, 81–88.

68. Rau, F.; Freyermuth, F.; Fugier, C.; Villemin, J.P.; Fischer, M.C.; Jost, B.; Dembele, D.; Gourdon, G.; Nicole, A.; Duboc, D.; *et al.* Misregulation of mir-1 processing is associated with heart defects in myotonic dystrophy. *Nat. Struct. Mol. Biol.* **2011**, *18*, 840–845.

69. Gambardella, S.; Rinaldi, F.; Lepore, S.M.; Viola, A.; Loro, E.; Angelini, C.; Vergani, L.; Novelli, G.; Botta, A. Overexpression of microRNA-206 in the skeletal muscle from myotonic dystrophy type 1 patients. *J. Transl. Med.* **2010**, *8*, doi:10.1186/1479-5876-8-48.

70. Cacchiarelli, D.; Incitti, T.; Martone, J.; Cesana, M.; Cazzella, V.; Santini, T.; Sthandier, O.; Bozzoni, I. Mir-31 modulates dystrophin expression: New implications for duchenne muscular dystrophy therapy. *EMBO Rep.* **2011**, *12*, 136–141.

71. Eisenberg, I.; Eran, A.; Nishino, I.; Moggio, M.; Lamperti, C.; Amato, A.A.; Lidov, H.G.; Kang, P.B.; North, K.N.; Mitrani-Rosenbaum, S.; *et al.* Distinctive patterns of microRNA expression in primary muscular disorders. *Proc. Natl. Acad. Sci. USA* **2007**, *104*, 17016–17021.

72. De Arcangelis, V.; Serra, F.; Cogoni, C.; Vivarelli, E.; Monaco, L.; Naro, F. Beta1-syntrophin modulation by mir-222 in mdx mice. *PLoS One* **2010**, *5*, e12098.

73. Greco, S.; Perfetti, A.; Fasanaro, P.; Cardani, R.; Capogrossi, M.C.; Meola, G.; Martelli, F. Deregulated microRNAs in myotonic dystrophy type 2. *PLoS One* **2012**, *7*, e39732.

74. Williams, A.H.; Valdez, G.; Moresi, V.; Qi, X.; McAnally, J.; Elliott, J.L.; Bassel-Duby, R.; Sanes, J.R.; Olson, E.N. MicroRNA-206 delays als progression and promotes regeneration of neuromuscular synapses in mice. *Science* **2009**, *326*, 1549–1554.

75. Durbeej, M.; Campbell, K.P. Muscular dystrophies involving the dystrophin-glycoprotein complex: An overview of current mouse models. *Curr. Opin. Genet. Dev.* **2002**, *12*, 349–361.

76. Ervasti, J.M.; Sonnemann, K.J. Biology of the striated muscle dystrophin-glycoprotein complex. *Int. Rev. Cytol.* **2008**, *265*, 191–225.

77. Matsumura, K.; Tome, F.M.; Collin, H.; Leturcq, F.; Jeanpierre, M.; Kaplan, J.C.; Fardeau, M.; Campbell, K.P. Expression of dystrophin-associated proteins in dystrophin-positive muscle fibers (revertants) in duchenne muscular dystrophy. *Neuromuscul. Disord. NMD* **1994**, *4*, 115–120.

78. McKinsey, T.A.; Zhang, C.L.; Lu, J.; Olson, E.N. Signal-dependent nuclear export of a histone deacetylase regulates muscle differentiation. *Nature* **2000**, *408*, 106–111.

79. Puri, P.L.; Iezzi, S.; Stiegler, P.; Chen, T.T.; Schiltz, R.L.; Muscat, G.E.; Giordano, A.; Kedes, L.; Wang, J.Y.; Sartorelli, V. Class i histone deacetylases sequentially interact with myod and prb during skeletal myogenesis. *Mol. Cell* **2001**, *8*, 885–897.

80. Brenman, J.E.; Chao, D.S.; Xia, H.; Aldape, K.; Bredt, D.S. Nitric oxide synthase complexed with dystrophin and absent from skeletal muscle sarcolemma in duchenne muscular dystrophy. *Cell* **1995**, *82*, 743–752.

81. Colussi, C.; Mozzetta, C.; Gurtner, A.; Illi, B.; Rosati, J.; Straino, S.; Ragone, G.; Pescatori, M.; Zaccagnini, G.; Antonini, A.;*et al.* Hdac2 blockade by nitric oxide and histone deacetylase inhibitors reveals a common target in duchenne muscular dystrophy treatment. *Proc. Natl. Acad. Sci. USA* **2008**, *105*, 19183–19187.

82. Brook, J.D.; McCurrach, M.E.; Harley, H.G.; Buckler, A.J.; Church, D.; Aburatani, H.; Hunter, K.; Stanton, V.P.; Thirion, J.P.; Hudson, T.; *et al.* Molecular basis of myotonic dystrophy: Expansion of a trinucleotide (ctg) repeat at the 3' end of a transcript encoding a protein kinase family member. *Cell* **1992**, *68*, 799–808.

83. Fu, Y.H.; Pizzuti, A.; Fenwick, R.G., Jr.; King, J.; Rajnarayan, S.; Dunne, P.W.; Dubel, J.; Nasser, G.A.; Ashizawa, T.; de Jong, P.; *et al.* An unstable triplet repeat in a gene related to myotonic muscular dystrophy. *Science* **1992**, *255*, 1256–1258.

84. Mahadevan, M.; Tsilfidis, C.; Sabourin, L.; Shutler, G.; Amemiya, C.; Jansen, G.; Neville, C.; Narang, M.; Barcelo, J.; O'Hoy, K.; *et al.* Myotonic dystrophy mutation: An unstable ctg repeat in the 3' untranslated region of the gene. *Science* **1992**, *255*, 1253–1255.

85. Liquori, C.L.; Ricker, K.; Moseley, M.L.; Jacobsen, J.F.; Kress, W.; Naylor, S.L.; Day, J.W.; Ranum, L.P. Myotonic dystrophy type 2 caused by a cctg expansion in intron 1 of znf9. *Science* **2001**, *293*, 864–867.

86. Wojciechowska, M.; Krzyzosiak, W.J. Cellular toxicity of expanded RNA repeats: Focus on RNA foci. *Hum. Mol. Genet.* **2011**, *20*, 3811–3821.

87. Sicot, G.; Gomes-Pereira, M. RNA toxicity in human disease and animal models: From the uncovering of a new mechanism to the development of promising therapies. *Biochim. Biophys. Acta* **2013**, *1832*, 1390–1409.

88. Udd, B.; Krahe, R. The myotonic dystrophies: Molecular, clinical, and therapeutic challenges. *Lancet Neurol.* **2012**, *11*, 891–905.

89. Li, X.; Wu, Z.; Fu, X.; Han, W. Long noncoding RNAs: Insights from biological features and functions to diseases. *Med. Res. Rev.* **2013**, *33*, 517–553.

90. Cesana, M.; Cacchiarelli, D.; Legnini, I.; Santini, T.; Sthandier, O.; Chinappi, M.; Tramontano, A.; Bozzoni, I. A long noncoding RNA controls muscle differentiation by functioning as a competing endogenous RNA. *Cell* **2011**, *147*, 358–369.

91. Watts, R.; Johnsen, V.L.; Shearer, J.; Hittel, D.S. Myostatin-induced inhibition of the long noncoding RNA malat1 is associated with decreased myogenesis. *Am. J. Physiol. Cell Physiol.* **2013**, *304*, C995–C1001.

92. Sunwoo, H.; Dinger, M.E.; Wilusz, J.E.; Amaral, P.P.; Mattick, J.S.; Spector, D.L. Men epsilon/beta nuclear-retained non-coding RNAs are up-regulated upon muscle differentiation and are essential components of paraspeckles. *Genome Res.* **2009**, *19*, 347–359.

93. Sasaki, Y.T.; Ideue, T.; Sano, M.; Mituyama, T.; Hirose, T. Menepsilon/beta noncoding RNAs are essential for structural integrity of nuclear paraspeckles. *Proc. Natl. Acad. Sci. USA* **2009**, *106*, 2525–2530.

94. Clemson, C.M.; Hutchinson, J.N.; Sara, S.A.; Ensminger, A.W.; Fox, A.H.; Chess, A.; Lawrence, J.B. An architectural role for a nuclear noncoding RNA: Neat1 RNA is essential for the structure of paraspeckles. *Mol. Cell* **2009**, *33*, 717–726.

95. Hube, F.; Velasco, G.; Rollin, J.; Furling, D.; Francastel, C. Steroid receptor RNA activator protein binds to and counteracts sra RNA-mediated activation of myod and muscle differentiation. *Nucleic Acids Res.* **2011**, *39*, 513–525.

96. Caretti, G.; Schiltz, R.L.; Dilworth, F.J.; Di Padova, M.; Zhao, P.; Ogryzko, V.; Fuller-Pace, F.V.; Hoffman, E.P.; Tapscott, S.J.; Sartorelli, V. The RNA helicases p68/p72 and the noncoding RNA sra are coregulators of myod and skeletal muscle differentiation. *Dev. Cell* **2006**, *11*, 547–560.

97. Caretti, G.; Lei, E.P.; Sartorelli, V. The dead-box p68/p72 proteins and the noncoding RNA steroid receptor activator sra: Eclectic regulators of disparate biological functions. *Cell Cycle* **2007**, *6*, 1172–1176.

98. Willingham, A.T.; Orth, A.P.; Batalov, S.; Peters, E.C.; Wen, B.G.; Aza-Blanc, P.; Hogenesch, J.B.; Schultz, P.G. A strategy for probing the function of noncoding RNAs finds a repressor of nfat. *Science* **2005**, *309*, 1570–1573.

99. Sharma, S.; Findlay, G.M.; Bandukwala, H.S.; Oberdoerffer, S.; Baust, B.; Li, Z.; Schmidt, V.; Hogan, P.G.; Sacks, D.B.; Rao, A. Dephosphorylation of the nuclear factor of activated t cells (nfat) transcription factor is regulated by an RNA-protein scaffold complex. *Proc. Natl. Acad. Sci. USA* **2011**, *108*, 11381–11386.

100. Niland, C.N.; Merry, C.R.; Khalil, A.M. Emerging roles for long non-coding RNAs in cancer and neurological disorders. *Front. Genet.* **2012**, *3*, doi:10.3389/fgene.2012.00025.

101. Cabianca, D.S.; Casa, V.; Bodega, B.; Xynos, A.; Ginelli, E.; Tanaka, Y.; Gabellini, D. A long ncRNA links copy number variation to a polycomb/trithorax epigenetic switch in fshd muscular dystrophy. *Cell* **2012**, *149*, 819–831.

102. Schmidt, L.H.; Spieker, T.; Koschmieder, S.; Schaffers, S.; Humberg, J.; Jungen, D.; Bulk, E.; Hascher, A.; Wittmer, D.; Marra, A.; *et al.* The long noncoding malat-1 RNA indicates a poor prognosis in non-small cell lung cancer and induces migration and tumor growth. *J. Thorac. Oncol. Off. Publ. Int. Assoc. Study Lung Cancer* **2011**, *6*, 1984–1992.

103. Tripathi, V.; Ellis, J.D.; Shen, Z.; Song, D.Y.; Pan, Q.; Watt, A.T.; Freier, S.M.; Bennett, C.F.; Sharma, A.; Bubulya, P.A.; *et al.* The nuclear-retained noncoding RNA malat1 regulates alternative splicing by modulating sr splicing factor phosphorylation. *Mol. Cell* **2010**, *39*, 925–938.

104. Wilusz, J.E.; JnBaptiste, C.K.; Lu, L.Y.; Kuhn, C.D.; Joshua-Tor, L.; Sharp, P.A. A triple helix stabilizes the 3' ends of long noncoding RNAs that lack poly(a) tails. *Genes Dev.* **2012**, *26*, 2392–2407.

105. Gutschner, T.; Hammerle, M.; Diederichs, S. Malat1—A paradigm for long noncoding RNA function in cancer. *J. Mol. Med.* **2013**, *91*, 791–801.

106. Langley, B.; Thomas, M.; Bishop, A.; Sharma, M.; Gilmour, S.; Kambadur, R. Myostatin inhibits myoblast differentiation by down-regulating myod expression. *J. Biol. Chem.* **2002**, *277*, 49831–49840.

107. Rios, R.; Carneiro, I.; Arce, V.M.; Devesa, J. Myostatin is an inhibitor of myogenic differentiation. *Am. J. Physiol. Cell Physiol.* **2002**, *282*, C993–C999.

108. Zhang, B.; Arun, G.; Mao, Y.S.; Lazar, Z.; Hung, G.; Bhattacharjee, G.; Xiao, X.; Booth, C.J.; Wu, J.; Zhang, C.; *et al.* The lncRNA malat1 is dispensable for mouse development but its transcription plays a cis-regulatory role in the adult. *Cell Rep.* **2012**, *2*, 111–123.

109. Eissmann, M.; Gutschner, T.; Hammerle, M.; Gunther, S.; Caudron-Herger, M.; Gross, M.; Schirmacher, P.; Rippe, K.; Braun, T.; Zornig, M.; *et al.* Loss of the abundant nuclear non-coding RNA malat1 is compatible with life and development. *RNA Biol.* **2012**, *9*, 1076–1087.

110. Prasanth, K.V.; Prasanth, S.G.; Xuan, Z.; Hearn, S.; Freier, S.M.; Bennett, C.F.; Zhang, M.Q.; Spector, D.L. Regulating gene expression through RNA nuclear retention. *Cell* **2005**, *123*, 249–263.

111. Ahn, A.H.; Kunkel, L.M. The structural and functional diversity of dystrophin. *Nat. Genet.* **1993**, *3*, 283–291.

112. Muntoni, F.; Torelli, S.; Ferlini, A. Dystrophin and mutations: One gene, several proteins, multiple phenotypes. *Lancet Neurol.* **2003**, *2*, 731–740.

113. Torelli, S.; Ferlini, A.; Obici, L.; Sewry, C.; Muntoni, F. Expression, regulation and localisation of dystrophin isoforms in human foetal skeletal and cardiac muscle. *Neuromuscul. Disord. NMD* **1999**, *9*, 541–551.

114. Brioschi, S.; Gualandi, F.; Scotton, C.; Armaroli, A.; Bovolenta, M.; Falzarano, M.S.; Sabatelli, P.; Selvatici, R.; D'Amico, A.; Pane, M.; *et al.* Genetic characterization in symptomatic female dmd carriers: Lack of relationship between x-inactivation, transcriptional dmd allele balancing and phenotype. *BMC Med. Genet.* **2012**, *13*, doi:10.1186/1471-2350-13-73.

115. Tran, T.H.; Zhang, Z.; Yagi, M.; Lee, T.; Awano, H.; Nishida, A.; Okinaga, T.; Takeshima, Y.; Matsuo, M. Molecular characterization of an x(p21.2;q28) chromosomal inversion in a duchenne muscular dystrophy patient with mental retardation reveals a novel long non-coding gene on xq28. *J. Hum. Genet.* **2013**, *58*, 33–39.

116. Lee, J.T. Lessons from x-chromosome inactivation: Long ncRNA as guides and tethers to the epigenome. *Genes Dev.* **2009**, *23*, 1831–1842.

117. Rinn, J.L.; Kertesz, M.; Wang, J.K.; Squazzo, S.L.; Xu, X.; Brugmann, S.A.; Goodnough, L.H.; Helms, J.A.; Farnham, P.J.; Segal, E.; *et al.* Functional demarcation of active and silent chromatin domains in human hox loci by noncoding RNAs. *Cell* **2007**, *129*, 1311–1323.

118. Rukov, J.L.; Shomron, N. MicroRNA pharmacogenomics: Post-transcriptional regulation of drug response. *Trends Mol. Med.* **2011**, *17*, 412–423.

119. Mishra, P.J. The miRNA-drug resistance connection: A new era of personalized medicine using noncoding RNA begins. *Pharmacogenomics* **2012**, *13*, 1321–1324.

120. Giovannetti, E.; van der Velde, A.; Funel, N.; Vasile, E.; Perrone, V.; Leon, L.G.; de Lio, N.;

Avan, A.; Caponi, S.; Pollina, L.E.; *et al.* High-throughput microRNA (mirRNA) arrays unravel the prognostic role of mir-211 in pancreatic cancer. *PLoS One* **2012**, *7*, e49145.

121. Gagan, J.; Dey, B.K.; Dutta, A. MicroRNAs regulate and provide robustness to the myogenic transcriptional network. *Curr. Opin. Pharmacol.* **2012**, *12*, 383–388.

122. Twayana, S.; Legnini, I.; Cesana, M.; Cacchiarelli, D.; Morlando, M.; Bozzoni, I. Biogenesis and function of non-coding RNAs in muscle differentiation and in duchenne muscular dystrophy. *Biochem. Soc. Trans.* **2013**, *41*, 844–849.

13

Cascading *cis*-Cleavage on Transcript from *trans*-Acting siRNA-Producing Locus 3

Changqing Zhang [1,†,*], **Guangping Li** [2,†], **Jin Wang** [3], **Shinong Zhu** [1] **and Hailing Li** [1]

[1] College of Horticulture, Jinling Institute of Technology, Nanjing 210038, China;
E-Mails: zsn@jit.edu.cn (S.Z.); lihailing@jit.edu.cn (H.L.)
[2] College of Forest Resources and Environment, Nanjing Forestry University, Nanjing 210037,
China; E-Mail: liguangping108@sina.com
[3] State Key Laboratory of Pharmaceutical Biotechnology, School of Life Sciences,
Nanjing University, Nanjing 210093, China; E-Mail: jwang@nju.edu.cn

[†] These authors contributed equally to this work.

[*] Author to whom correspondence should be addressed; E-Mail: zhang_chq2002@sohu.com;

Abstract: The production of small RNAs (sRNAs) from phased positions set by microRNA-directed cleavage of *trans*-acting-siRNA-producing locus (TAS) transcript has been characterized extensively; however, the production of sRNAs from non-phased positions remains unknown. We report three *cis*-cleavages that occurred in *TAS3* transcripts in *Vitis vinifera*, by combining high-throughput sRNA deep sequencing information with evolutional conservation and genome-wide RNA degradome analysis. The three *cis*-cleavages can be deciphered to generate an orderly cleavage cascade, and can also produce distinct phasing patterns. Each of the patterns, either upstream or downstream of the *cis*-cleaved position, had a set of sRNAs arranged in 21-nucleotide increments. Part of the cascading *cis*-cleavages was also conserved in *Arabidopsis thaliana*. Our results will enhance the understanding of the production of sRNAs from non-phased positions that are not set by microRNA-directed cleavage.

Keywords: *trans*-acting siRNA; *cis*-cleavage; ta-siRNA-producing locus; miRNA

1. Introduction

In plants, many endogenous small RNAs (sRNAs), including microRNAs (miRNAs), heterochromatic small interfering RNAs (siRNAs), natural antisense siRNAs, and *trans*-acting siRNAs (ta-siRNAs), play important roles in regulating gene expression networks [1,2]. The sRNAs are also valuable tools for functional genomics studies. Usually, the sRNAs silence gene expression by either degrading mRNA or repressing translation but, in a few cases, they also generate a population of secondary siRNAs. The ta-siRNAs are secondary siRNAs that are produced by a miRNA-targeted trigger that bridges the pathways of miRNA and siRNA regulation. ta-siRNAs can regulate plant development, metabolism, and responses to biotic and abiotic stresses, and thus have received more attention in the recent decade [3–5].

During the biogenesis of ta-siRNA, a single-stranded RNA is transcribed from a ta-siRNA-producing locus (TAS) and then cleaved by a phase-initiator (a miRNA or, in some cases, a ta-siRNA). Then, RNA-dependent RNA polymerase 6 (RDR6)-dependent conversion of the resulting fragments into double-stranded RNA and its subsequent cleavage by dicer-like 4 (DCL4) at every ~21 nucleotide (nt) relative to the phase-initiator cleavage site generates ~21-nt phased sRNAs. Some of the phased sRNAs become ta-siRNAs by binding argonaute (AGO) proteins to direct a *trans*-cleavage of targeted mRNAs [5–7]. Plant TASs can be classified into at least eight families, based on initiator-dependence, sequence similarity, and target gene identity. *TAS1* and *TAS2* are targets of miR173 and their ta-siRNAs can target the pentatricopeptide repeat genes [8]. *TAS3* is a target of miR390 and its ta-siRNA can target the auxin response factor gene family [8]. The initiator of *TAS4* is miR828, and the *TAS4* ta-siRNA can target the MYB transcription factor gene family [9]. *TAS5* is triggered by miR482 and its ta-siRNA can target the Bs4 resistance gene [10]. miR156 and miR529 initiate *TAS6*, which targets an mRNA that encodes a zinc finger protein [11]. miR828 initiates *TAS7*, which can target 13 genes, including genes that encode the leucine-rich receptor protein kinase-like protein and a calcium-transporting ATPase [12]. *At1g63130* is a pentatricopeptide repeat gene that was reported to be cleaved by *TAS2*-derived ta-siR2140 [13]. *TAS3* is flanked by two miR390 binding sites; one of which can be cleaved by the interaction of miR390 and AGO7, and another that is non-cleavable. Both binding sites are critical for the biogenesis of the *TAS3* ta-siRNAs. In contrast, other TASs have only a single miRNA binding site and are cleaved by the interaction of miRNA/ta-siRNA and AGO1. Recently, AGO2 has been reported to mediate *cis*-cleavage of *TAS1c*, although its slicer activity has not been demonstrated so far [14]. Taken together, it might be expected that many of the sequenced sRNAs could be mapped onto the phased positions set by the phase-initiator. Yet, while many of the sRNAs were successfully mapped, unexpectedly, many were mapped onto non-phased positions; that is, the intervals between the phased positions [12,13]. These sRNAs have been called "non-phased sRNAs" and how they are produced remains unclear.

The recent publication of the degradome library generated from cleaved mRNA fragments and sRNA libraries generated by high-throughput deep sequencing has enabled the study of all the cleavages that occur in a TAS transcript [15,16]. Here, we studied *cis*-cleavage of grapevine *TAS3* and found a cascading *cis*-cleavage, which produce sRNAs from the so-called non-phased positions and broaden the known scope of non-phased sRNA production.

2. Results and Discussion

2.1. Overview of Small RNA Distribution on TAS3 from Vitis vinifera

Previously we reported that the *TAS3* from *Vitis vinifera* (vvi*TAS3*) can be targeted by vvi-miR390 to trigger ta-siRNA production in grapevine [12]. Here, to determine the distribution of sRNAs on the vvi*TAS3* transcript, a sRNA library from grapevine leaves (GEO: GSM458927) was used. To improve mapping confidence, only the sRNAs that mapped to a single site on the whole *V. vinifera* genome were used, because they could be attributed with certainty to a particular locus.

As a result of the mapping, we detected 131 unique sRNAs, representing 3969 reads, which matched perfectly to vvi*TAS3* (Figure 1). The 5' ends of the reads occupied 79 positions on the transcript. Only 14 (18%) of the positions were found to belong to phased positions set by vvi-miR390 when a 1-nt offset from the phased positions was allowed. After filtering out sRNAs that had TPM (tags per million) values of five or less, some unique sRNAs that mapped to non-phased positions remained (Figure 1). The percentage of phased sRNA positions increased from 18% to 40%. Together, these results showed that some sRNAs are really generated from non-phased positions, and might even be functional because of the relatively high levels at which they are often expressed.

Figure 1. Abundance distribution of sRNAs mapped to vvi*TAS3*. The number of reads with a 5' end at each position is plotted. Bars above the sequence represent sense reads; those below represent antisense reads. The cleavage site of vvi-miR390 is marked by a vertical dotted line. Vertical gray lines indicate the miRNA-set 21-nt phased cleavage, allowing a 1-nt offset. Regions with TPMs greater than five are not shown.

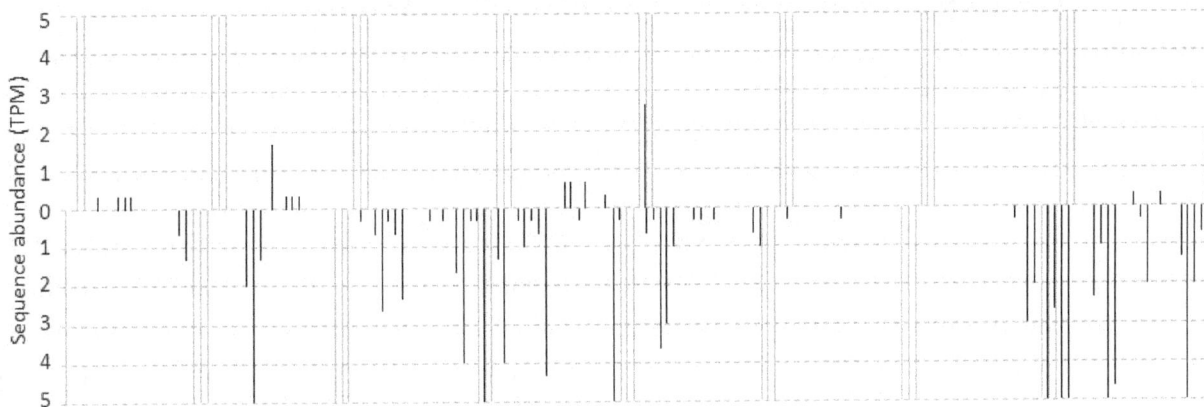

2.2. Computational Prediction and Validation of cis-Cleavages

Recent reports have shown that many functional siRNAs belong to a class of 21–22-nt 5'U/A sRNAs [14]. Therefore, we filtered out the potential siRNAs from the mapped sRNAs by limiting the length of the reads to 21 nt and the 5' end to U/A. As a result, we detected 35 sRNAs that were mapped to the antisense strand that passed the rule.

Additionally, it has been reported that many functional cleaved positions tend to be conserved through evolution, and they have been found to be highly conserved in alignments of genomic sequences from different species [8]. To identify conserved positions in the vvi*TAS3* transcript, we compiled a dataset of *TAS3* sequences from eight dicotyledonous plants; namely, *V. vinifera*, *Ricinus communis*, *Populus trichocarpa*, *Arabidopsis thaliana*, *Malus domestica*, *Fragaria vesca*, *Prunus persica*, and *Glycine max*, and aligned them using ClustalX2 [17] with the default parameters (Figure 2). The multiple sequence alignment showed that there are no insertions or deletions among the *TAS3* sequences, except for a one-nucleotide deletion in vvi*TAS3* between position 116 and 117. The conserved positions were filtered by requiring each position to be conserved in at least six of the species, and to correspond to the 10th position of the 35 candidate *cis*-acting sRNAs. We found that 19 of the 35 sRNAs passed these rules.

Figure 2. Alignment of *TAS3* DNA sequences from eight dicotyledonous plant species. The numbers in the last row indicate the number of species with the same nucleotide as vvi*TAS3* at each of the marked positions. Only numbers over six are shown. vvi, *Vitis vinifera*; rco, *Ricinus communis*; ptr, *Populus trichocarpa*; ath, *Arabidopsis thaliana*; mdo, *Malus domestica*; fve, *Fragaria vesca*; ppe, *Prunus persica*; gma, *Glycine max*.

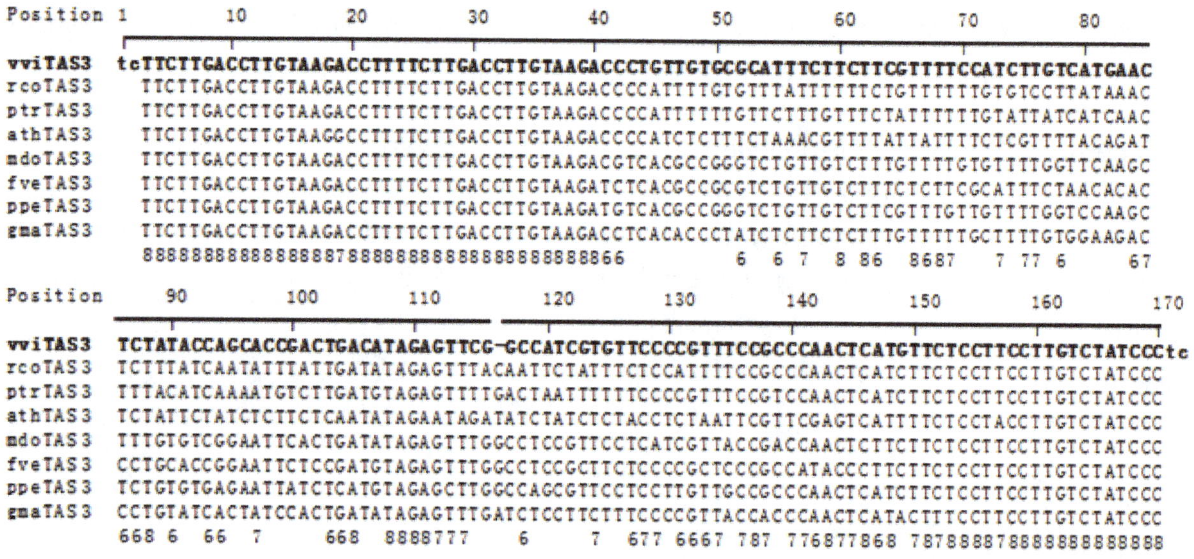

Finally, using a parallel sequencing sRNA library that contained degradome tags from grapevine leaves [18], we validated the predicted *cis*-cleaved positions on the vvi*TAS3* transcript by requiring that the *cis*-cleaved positions overlapped with the 5' end of the RNA degradation fragment mapped onto the TAS. As a result, three of the positions, 63, 85, and 138, were validated. The corresponding *cis*-acting siRNAs (ca-siRNAs) were three 21-nt 5'U ca-siRNAs and one 22-nt 5'U ca-siRNA (Table 1). It has been reported that the size of the sRNAs and the 5'-terminal nucleotide are critical for the sorting of AGO. AGO1 binds 21-nt 5'U sRNAs, but in some cases, it also binds 22-nt 5'U sRNAs [14]. In Arabidopsis, the 22-nt 5'U 3'D10(-) from *TAS1c* has been reported to mediate its *cis*-cleavage by binding to AGO1, and the 21-nt 5'U 3'D6(-) from *TAS1c* has also been shown to mediate *TAS1c* cleavage by binding to AGO1; however, in this case, the cleaved site is not its original site [14]. In this study, we found three ca-siRNAs that were 21-nt 5'U sRNAs and one that was a 22-nt 5'U sRNA. These results implied that the four ca-siRNAs were all loaded to AGO1.

Table 1. Validated *cis*-cleaved positions on vvi*TAS3* transcript and their ca-siRNAs.

cis-cleaved position	ca-siRNA	
	Name	Sequence
63	ca-siRM72	UGGAAAACGAAGAAGAAAUGC
85	ca-siRM94	UGGUAUAGAGUUCAUGACAAG
138	ca-siRM147	UGAGUUGGGCGGAAACGGGGA
		UGAGUUGGGCGGAAACGGGGAA

It has been demonstrated that two miR390 binding sites located on each side of *TAS3* are critical for *TAS3* ta-siRNAs biogenesis [8]. Therefore, we looked for ca-siRNA targeting sites on vvi*TAS3* and the flanking 300 bp upstream and downstream of the gene where the two miR390 binding sites were located. We found one targeted site on vvi*TAS3* for each of the ca-siRNAs. This finding suggested that *cis*-cleavage might use a different mechanism from the mechanism used by miR390 to initiate the cleavage of vvi*TAS3*.

We used the same method and criteria to examine the antisense strand that was not targeted by miR390. Surprisingly, no ca-siRNA targeting sites were detected on the antisense strand of vvi*TAS3*. The asymmetrical distribution between the targeted and non-targeted strand might imply that the non-targeted strand is readily in a double-stranded RNA form and is constantly processed by DCL4 and, therefore, protected from *cis*-cleavage. In addition, it might support the hypothetically biological function of *cis*-cleavage, *i.e.*, inactivating TAS transcription to feedback control ta-siRNA's production, as the sense strand was a template strand synthesizing antisense strand, so it would be more effective when the *cis*-cleavages preferred to occur in sense strand rather than in antisense strand.

2.3. Cascading cis-Cleavages

After identifying the *cis*-cleaved positions and their ca-siRNAs, we investigated how ca-siRNA production is triggered. It has been shown that ca-siRM147 can be triggered by miR390 [12], but for other ca-siRNAs the triggers remain unclear because they are out of the register set by miR390. To identify possible initiators, we investigated the distribution of *cis*-cleaved position and the locations of ca-siRNA 5' ends on the vvi*TAS3* transcript. We found that the 5' end of ca-siRM72 occurred precisely at the register set by ca-siRM147 in which the cleaved site of ca-siRM147 located on sense strand was 65 nt away from the 5' end of ca-siRM72 on antisense strand. The 5' end of ca-siRM94 was offset by 1 nt from the register set by ca-siRM147. Because the cleavage of phasing sRNAs often occurs within 1–2 nt of the phased position [8,13], we propose that the production of ca-siRM147, ca-siRM72, and ca-siRM94 is triggered by miR390, ca-siRM147, and ca-siRM147, respectively. The cascading *cis*-cleavage that we have proposed is shown schematically in Figure 3.

Figure 3. Cleavage cascades generated by *cis*-cleavages of siRNA on the sense strand of vvi*TAS3*. The four numbers in the brackets indicate the raw abundance of ca-siRNA in grapevine berries, leaves, inflorescences, and tendrils, respectively.

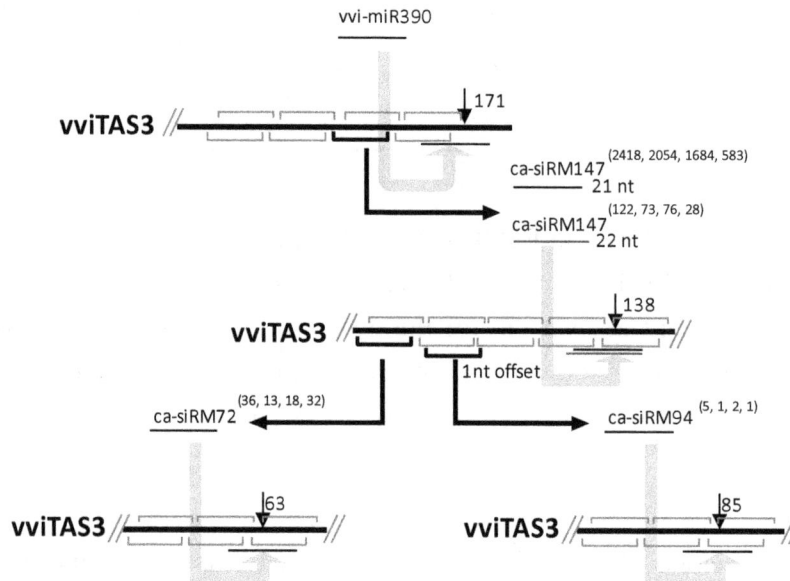

2.4. The Accumulated Levels of ca-siRNAs

When we analyzed the accumulated levels of the ca-siRNAs in the grapevine leaf, berry (GEO: GSM458930), inflorescence (GEO: GSM458929), and tendril (GEO: GSM458928) libraries, we found that the levels were in agreement with the cleavage cascades. The abundance of the ca-siRNAs that were located upstream of the cascade was always higher than the abundance of the ca-siRNAs that were downstream. For example, in the grapevine leaf library, the 21-nt ca-siRM147 located upstream had 2054 sequenced reads, while ca-siRM72, located downstream of ca-siRM147 cleavage, had only 13 sequenced reads. Moreover, the ca-siRNAs that occurred precisely at the register were always more abundant than those that occurred out of the register. For example, in the grapevine leaf library, ca-siRM72 had 13 sequenced reads, while ca-siRM94, which was shifted by 1 nt from the phased positions set by ca-siRM147, had only one sequenced read. Similar results were obtained for the other three tissues (Figure 3).

2.5. cis-*Cleavages Produced sRNAs in Increments of Approximately 21 nt*

To test whether or not *cis*-cleavages can also produce phased sRNAs in increments of approximately 21 nt, we searched for sRNAs with 5' ends that overlapped the predicted phased and non-phased positions by allowing an offset of 1 nt. As expected, each *cis*-cleavage had a set of corresponding phased sRNAs arranged in ~21-nt increments upstream and downstream of the cleavage position (Table 2).

Table 2. Phased patterns set by different *cis*-cleavages.

cis-cleaved position	Upstream of *cis*-cleaved position			Downstream of *cis*-cleaved position		
	Number of phased sRNAs	Number of non-phased sRNAs	*p* value *	Number of phased sRNAs	Number of non-phased sRNAs	*p*-value
63	2	12	1.11×10^{-1}	3	47	4.92×10^{-1}
85	5	27	**7.54×10^{-3}**	4	30	1.00×10^{-1}
138	8	41	**4.71×10^{-4}**	3	16	7.43×10^{-2}

* *p*-values less than 0.01 are in bold font.

To test whether or not the phased patterns produced from *cis*-cleavages were statistically significant, we developed an improved equation (see Experimental Section) by modifying previous algorithms [12,13,19] to evaluate the phasing pattern set by *cis*-cleavage. First, the new equation is not constrained by the previous 231-bp length requirement [12,13,19], but requires only a multiple of 21 nt, which provides a more accurate TAS evaluation, especially for TASs that are longer or shorter than 231 bp. Second, our equation uses a variable *s* to reflect the maximum offset from a phase position [12,19], making the evaluation more flexible. This equation could be applied to TAS identification in the future. Using our improved algorithm [12,13,19], we determined that two *cis*-cleavages had significant phasing patterns (*p*-value < 0.01). When the number of sRNAs located in the phased positions set by ca-siRNAs was counted, we found that 54 unique sRNA located in non-phased positions set by miRNA390 were included in the phasing patterns. These results suggested that some common processes might be used for both miRNA-mediated ta-siRNA production and the *cis*-cleavage of siRNA.

2.6. The Conservation of ca-siRNAs and Cascading cis-Cleavages

To examine the conservation of *cis*-cleavages further, we first looked for the presence of ca-siRNAs in the sRNA datasets of *V. vinifera*, *A. thaliana*, *M. domestica*, and *P. persica* downloaded from the Gene Expression Omnibus (GEO) or the plant MPSS databases. We found that although the accumulation levels of the four ca-siRNAs varied in the different species, they were expressed in all four species, except for 22-nt ca-siRNA147, which was not detected in *P. persica* (peach) (Table 3).

Table 3. Conservation of ca-siRNAs in the sRNA datasets of four species.

ca-siRNAs		Average of normalized abundance (TPM)			
Name	Length	Grapevine	Apple	Peach	Arabidopsis
ca-siRM72	21 nt	6	2	1	1
ca-siRM94	21 nt	1	3	14	1
ca-siRM147	21 nt	419	32	9	2
	22 nt	18	1	0	2

We then evaluated the corresponding *cis*-cleavages based on the Col7d samples (GEO: GSE20197) and the degradome library (GEO: GSM280227) from Arabidopsis, and found that, two *cis*-cleavages occurred in positions 85 and 139 (equivalent to position 138 in vvi*TAS3* because of the nucleotide deletion in the vvi*TAS3* sequence) were also validated on ath*TAS3* (Figure 4). The accumulated levels of ca-siRNAs in Col7d sample were also in agreement with the cleavage cascades. In which, the 21-nt and

22-nt ca-siRM147 located upstream had, respectively, three and two sequenced reads, while ca-siRM94, located downstream of ca-siRM147 cleavage, had one sequenced reads. In a previous study [5], it was suggested that *cis*-cleavage occurred on position 139 in ath*TAS3*, although the ca-siRNA and functional sRNAs were not found. Here, we detected the ca-siRNA and a secondary ca-siRNA product (ca-siRNA94), which we believe provides enough evidence to establish the *cis*-cleavage on *TAS3*. The findings reported here for ath*TAS3* indicate that cascading *cis*-cleavage is conserved.

Figure 4. Cleavage cascades generated by *cis*-cleavages of siRNA on the sense strand of ath*TAS3*. The triangle indicates the position of the nucleotide deletion in vvi*TAS3*. The number in the bracket indicates the raw abundance of ca-siRNA in Col7d (GEO: GSE20197).

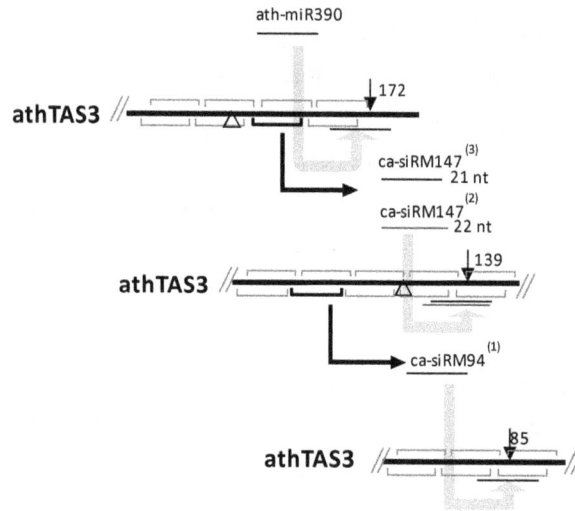

3. Experimental Section

3.1. Sources of sRNA Libraries

In this study, we used four deep sequencing sRNA datasets; namely, two degradome library and two sRNA libraries. All the datasets were downloaded from the Gene Expression Omnibus (GEO). The GEO accession numbers for these libraries are given in the Results section.

3.2. Evaluation of Phasing Patterns Set by cis-Cleavage

Once a *cis*-cleaved position was determined, the numbers of phased and non-phased positions were counted upstream and downstream of the cleavage sites respectively. Phased positions refer to positions arranged in 21-nt increments relative to the cleavage position as well as to positions shifted by *s* nt relative to the positions of 21-nt increments. Non-phased positions are all the other positions. The *p*-value of each detected phasing pattern was calculated based on a random hyper-geometric distribution using an improved equation based on previously used algorithms [12,13,19,20]:

$$P(k_1) = \sum_{x=k_1}^{\min\left(k_1+k_2,\frac{2L}{21}-1\right)} \frac{\binom{(2L-1)-(\frac{2L}{21}-1)\times(2s-1)}{k_2}\binom{\frac{2L}{21}-1}{k_1}}{\binom{(2L-1)-(\frac{2L}{21}-1)\times 2s}{k_1+k_2}} \tag{1}$$

Where L is the length of the detected pattern and is a multiple of 21, K_1 is the number of phased positions having sRNA hits, K_2 is the number of non-phased positions having sRNA hits, and s is the maximum allowed offset from the phase position.

3.3. Expressional Conservation of ca-siRNAs

The expressional conservation of the grapevine ca-siRNAs was investigated by performing a search against 84 sRNA libraries from grapevine, apple, Arabidopsis, and peach. The sRNA libraries of grapevine (GEO: GSE18405) and apple (GEO: GSE36065) were downloaded from the GEO and the sRNA libraries of Arabidopsis and peach were used from the MPSS databases [21]. The normalized abundance is the raw expression value divided by the total number of signatures and multiplied by 1,000,000.

4. Conclusions

In this work, we reexamined the distribution of sRNAs on vvi*TAS3* using a stringent threshold that used only the sRNAs that mapped to a single site on the whole *V. vinifera* genome and that had normalized abundant values of one or more TPM. Our results showed that the non-phased positions were indeed located by some of the uniquely mapped sRNAs. We identified three *cis*-cleavages that directed by four ca-siRNAs at positions 63, 85, and 138 on vvi*TAS3* by combining computational predictions and validation. We found that three *cis*-cleavages, together with their ca-siRNAs, formed a cascading *cis*-cleavage. The accumulated levels of four ca-siRNAs in the berry, leaf, inflorescence, and tendril libraries of *V. vinifera* also agreed with the cascade. A comparative analysis showed that the expression levels of the four ca-siRNAs were conserved among grapevine, apple, peach, and Arabidopsis, and part of the *cis*-cascade was also identified in Arabidopsis. We also found that sRNAs were located at the phased positions set by ca-siRNA. These results broaden the known scope of non-phased sRNA production. We also developed an improved equation by modifying previous algorithms to evaluate the phasing pattern set by *cis*-cleavage. It could be applied to TAS identification in the future.

Acknowledgments

This work was supported by the National Science Foundation of China (Grant No. 31171273) and the Qing Lan Project of Jiangsu Province, China.

References

1. Axtell, M.J. Classification and comparison of small RNAs from plants. *Annu. Rev. Plant Biol.* **2013**, *64*, 137–159.
2. Chen, X. Small RNAs and their roles in plant development. *Annu. Rev. Cell Dev. Biol.* **2009**, *25*, 21–44.
3. Peragine, A.; Yoshikawa, M.; Wu, G.; Albrecht, H.L.; Poethig, R.S. SGS3 and SGS2/SDE1/RDR6 are required for juvenile development and the production of trans-acting siRNAs in *Arabidopsis*. *Genes Dev.* **2004**, *18*, 2368–2379.

4. Vazquez, F.; Vaucheret, H.; Rajagopalan, R.; Lepers, C.; Gasciolli, V.; Mallory, A.C.; Hilbert, J.L.; Bartel, D.P.; Crete, P. Endogenous trans-acting siRNAs regulate the accumulation of *Arabidopsis* mRNAs. *Mol. Cell* **2004**, *16*, 69–79.

5. Allen, E.; Xie, Z.; Gustafson, A.M.; Carrington, J.C. microRNA-directed phasing during trans-acting siRNA biogenesis in plants. *Cell* **2005**, *121*, 207–221.

6. Gasciolli, V.; Mallory, A.C.; Bartel, D.P.; Vaucheret, H. Partially redundant functions of *Arabidopsis* DICER-like enzymes and a role for DCL4 in producing trans-acting siRNAs. *Curr. Biol.* **2005**, *15*, 1494–1500.

7. Yoshikawa, M.; Peragine, A.; Park, M.Y.; Poethig, R.S. A pathway for the biogenesis of trans-acting siRNAs in *Arabidopsis*. *Genes Dev.* **2005**, *19*, 2164–2175.

8. Axtell, M.J.; Jan, C.; Rajagopalan, R.; Bartel, D.P. A two-hit trigger for siRNA biogenesis in plants. *Cell* **2006**, *127*, 565–577.

9. Rajagopalan, R.; Vaucheret, H.; Trejo, J.; Bartel, D.P. A diverse and evolutionarily fluid set of microRNAs in *Arabidopsis thaliana*. *Genes Dev.* **2006**, *20*, 3407–3425.

10. Li, F.; Orban, R.; Baker, B. SoMART: A web server for plant miRNA, tasiRNA and target gene analysis. *Plant J.* **2012**, *70*, 891–901.

11. Arif, M.A.; Fattash, I.; Ma, Z.; Cho, S.H.; Beike, A.K.; Reski, R.; Axtell, M.J.; Frank, W. DICER-LIKE3 activity in *Physcomitrella patens* DICER-LIKE4 mutants causes severe developmental dysfunction and sterility. *Mol. Plant* **2012**, *5*, 1281–1294.

12. Zhang, C.; Li, G.; Wang, J.; Fang, J. Identification of trans-acting siRNAs and their regulatory cascades in grapevine. *Bioinformatics* **2012**, *28*, 2561–2568.

13. Chen, H.M.; Li, Y.H.; Wu, S.H. Bioinformatic prediction and experimental validation of a microRNA-directed tandem trans-acting siRNA cascade in *Arabidopsis*. *Proc. Natl. Acad. Sci. USA* **2007**, *104*, 3318–3323.

14. Rajeswaran, R.; Aregger, M.; Zvereva, A.S.; Borah, B.K.; Gubaeva, E.G.; Pooggin, M.M. Sequencing of RDR6-dependent double-stranded RNAs reveals novel features of plant siRNA biogenesis. *Nucleic Acids Res.* **2012**, *40*, 6241–6254.

15. Howell, M.D.; Fahlgren, N.; Chapman, E.J.; Cumbie, J.S.; Sullivan, C.M.; Givan, S.A.; Kasschau, K.D.; Carrington, J.C. Genome-wide analysis of the RNA-DEPENDENT RNA POLYMERASE6/DICER-LIKE4 pathway in Arabidopsis reveals dependency on miRNA- and tasiRNA-directed targeting. *Plant Cell* **2007**, *19*, 926–942.

16. German, M.A.; Pillay, M.; Jeong, D.H.; Hetawal, A.; Luo, S.; Janardhanan, P.; Kannan, V.; Rymarquis, L.A.; Nobuta, K.; German, R.; *et al.* Global identification of microRNA-target RNA pairs by parallel analysis of RNA ends. *Nat. Biotechnol.* **2008**, *26*, 941–946.

17. Larkin, M.A.; Blackshields, G.; Brown, N.P.; Chenna, R.; McGettigan, P.A.; McWilliam, H.; Valentin, F.; Wallace, I.M.; Wilm, A.; Lopez, R.; *et al.* Clustal W and Clustal X version 2.0. *Bioinformatics* **2007**, *23*, 2947–2948.

18. Pantaleo, V.; Szittya, G.; Moxon, S.; Miozzi, L.; Moulton, V.; Dalmay, T.; Burgyan, J. Identification of grapevine microRNAs and their targets using high-throughput sequencing and degradome analysis. *Plant J. Cell Mol. Biol.* **2010**, *62*, 960–976.

19. Dai, X.; Zhao, P.X. pssRNAMiner: A plant short small RNA regulatory cascade analysis server. *Nucleic Acids Res.* **2008**, *36*, W114–W118.

20. Zhang, C.; Wang, J.; Hua, X.; Fang, J.; Zhu, H.; Gao, X. A mutation degree model for the identification of transcriptional regulatory elements. *BMC Bioinforma.* **2011**, *12*, doi:10.1186/1471-2105-12-262.

21. Nakano, M.; Nobuta, K.; Vemaraju, K.; Tej, S.S.; Skogen, J.W.; Meyers, B.C. Plant MPSS databases: Signature-based transcriptional resources for analyses of mRNA and small RNA. *Nucleic Acids Res.* **2006**, 34, D731–D735.

14

MicroRNA Transcriptomes Relate Intermuscular Adipose Tissue to Metabolic Risk

Jideng Ma [1,†], Shuzhen Yu [1,†], Fengjiao Wang [1], Lin Bai [1], Jian Xiao [1], Yanzhi Jiang [2], Lei Chen [3], Jinyong Wang [3], Anan Jiang [1], Mingzhou Li [1,*] and Xuewei Li [1,*]

[1] Institute of Animal Genetics & Breeding, College of Animal Science & Technology, Sichuan Agricultural University, Ya'an 625014, China; E-Mails: jideng_ma@sina.com (J.M.); yushuzhen1988@126.com (S.Y.); wangfengjiaosicau@gmail.com (F.W.); blin16@126.com (L.B.); jianxiao112@163.com (J.X.); lingdang317@163.com (A.J.)

[2] College of Life and Basic Sciences, Sichuan Agricultural University, Ya'an 625014, China; E-Mail: jiangyz04@163.com

[3] Chongqing Academy of Animal Science, Chongqing 402460, China; E-Mails: sicau.chen@gmail.com (L.C.); kingyou@vip.sina.com (J.W.)

[†] These authors contributed equally to this work.

[*] Authors to whom correspondence should be addressed; E-Mails: mingzhou.li@163.com (M.L.); xuewei.li@sicau.edu.cn (X.L.);

Abstract: Intermuscular adipose tissue is located between the muscle fiber bundles in skeletal muscles, and has similar metabolic features to visceral adipose tissue, which has been found to be related to a number of obesity-related diseases. Although various miRNAs are known to play crucial roles in adipose deposition and adipogenesis, the microRNA transcriptome of intermuscular adipose tissue has not, until now, been studied. Here, we sequenced the miRNA transcriptomes of porcine intermuscular adipose tissue by small RNA-sequencing and compared it to a representative subcutaneous adipose tissue. We found that the inflammation- and diabetes-related miRNAs were significantly enriched in the intermuscular rather than in the subcutaneous adipose tissue. A functional enrichment analysis of the genes predicted to be targeted by the enriched miRNAs also indicated that intermuscular adipose tissue was associated mainly with immune and inflammation responses. Our results suggest that the intermuscular adipose tissue should be recognized as a potential metabolic risk factor of obesity.

Keywords: intermuscular adipose tissue (IMAT); metabolic risk; miRNA; pig; immune response; inflammation response; obesity; transcriptome

1. Introduction

Adipose tissues (ATs) play a vital role in energy homeostasis and process the largest energy reserve in the body of animals. The rapidly expanding adipokine family is secreted by ATs [1], and, as a result, AT has been identified as an endocrine organ that influences a variety of physiological and pathological processes (such as immunity and inflammation) [2,3] that are involved in the development of metabolic diseases such as cardiovascular disease and type 2 diabetes mellitus [4–6]. Functional and metabolic differences between the visceral and subcutaneous ATs have been well documented. Subcutaneous AT mainly affects metabolic processes, while visceral AT has been identified as a metabolic risk factor for obesity. Recent studies have revealed that the intermuscular adipose tissue (IMAT), which is located between the muscle fiber bundles in skeletal muscles, has similar functional and metabolic features as the visceral ATs [7,8]. Indeed, IMAT was found in greater amounts than visceral AT in acromegaly patients despite their increased muscle mass, suggesting that increased amounts of AT in muscles might be associated with growth hormone-induced insulin resistance [9].

MicroRNAs (miRNAs) are endogenous small non-coding RNAs that modulate gene expression at a post-transcriptional level by binding to the 3' untranslated region (3'-UTR) of the target mRNAs [10]. During the past decade, various miRNAs that play crucial roles in adipose deposition and adipogenesis have been identified. Typically, miR-143 was identified as a pro-adipogenic modulator during pre-adipocyte differentiation [11,12]. MiR-103 [13] and the miR-17-92 cluster [14] were reported to accelerate adipocyte differentiation. MiR-27a [15], miR-27b [16], miR-448 [17] and miR-15a [18] were demonstrated to suppress adipogenic differentiation. MiR-519d [19], miR-335 [20] and miR-377 [21] were associated with lipid metabolism disorders. However, features of the miRNA transcriptome of IMAT have yet to be investigated.

Sus scrofa (pig) is emerging as an ideal biomedical model for obesity and metabolic disorders in human because of the similarity in metabolic features and proportional organ sizes in these two species [22]. To decipher the unique metabolic and functional features of IMAT, we sequenced the miRNA transcriptomes of porcine IMAT by small RNA-sequencing and compared it with a representative subcutaneous adipose tissue, superficial abdominal subcutaneous adipose tissue (sASAT). We identified various known, conserved, and putative novel porcine miRNAs in these two tissues. Notably, the sASAT-enriched miRNAs were related mainly to lipid metabolic homeostasis, while the IMAT-enriched miRNAs were related mainly to inflammation and diabetes, and the target genes of the IMAT-enriched miRNAs were primarily associated with inflammatory and diabetes processes. Together, these findings indicated the metabolic risk of IMAT. Our results will contribute to studies into the role of IMAT in obesity-related metabolic disease.

2. Results and Discussion

2.1. Transcriptome Sequencing Data

We used a small RNA-sequencing approach to sequence the miRNA transcriptomes of porcine IMAT and sASAT and obtained 17.76 million (M) and 18.50 M raw reads, respectively. More than 80% of the raw reads passed the quality filters (see Methods) and were termed the high-quality reads (IMAT: 14.93 M, 84.09%; sASAT: 14.92 M, 81.59%) (Supplementary Table S1). The high-quality reads in both transcriptomes exhibited the canonical size range distribution that is common to mammalian miRNAs (Figure 1a). The vast majority of the reads were 21–23 nucleotides (nt) in length. The 22-nt reads accounted for 61.50% of all the high-quality reads, followed by the 21-nt (14.00%) and 23-nt (13.99%) reads. This result indicated the reliability of using the small RNA-sequencing approach to generate miRNA reads as candidates for further analysis.

Figure 1. Description of miRNAs in two adipose tissues. (**a**) Length distribution of the high-quality reads; (**b**) Distribution of read counts of the identified miRNAs; (**c**) Distribution of read counts in the three defined miRNA groups; (**d**) Copy numbers of the top 10 miRNAs with highest read counts. IMAT, intermuscular adipose tissue; sASAT, superficial abdominal subcutaneous adipose tissue.

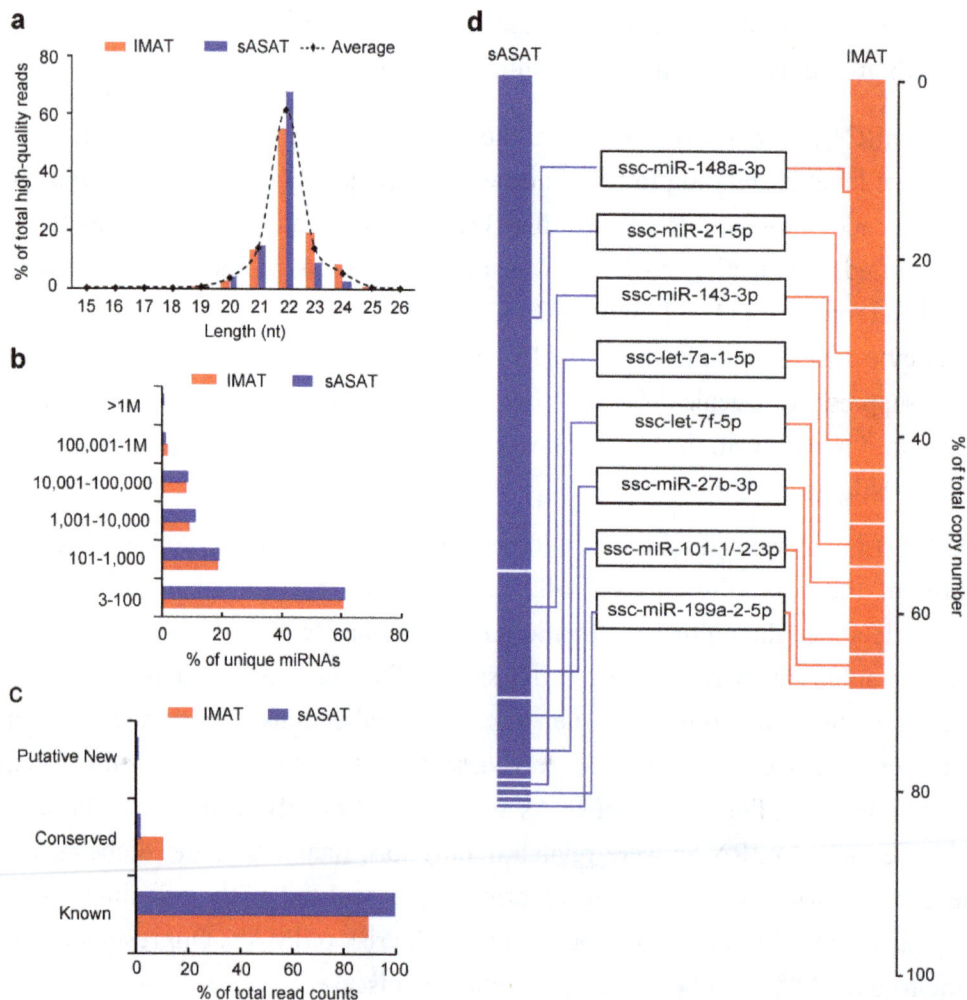

2.2. MiRNA Profiling of sASAT and IMAT

A total of 597 mature miRNAs corresponding to 453 miRNA precursors (pre-miRNAs) were identified in the two libraries by mapping them to the pig genome. In agreement with previous reports [23,24], we found that all the miRNA classes consisted of multiple mature variants (the isomiRs). The most abundant isomiR in each class was picked as the reference sequence for that class [23] based on the evidence that there was a significant positive correlation between the counts of most abundant isomiR and the total counts of all isomiRs in the same class (IMAT: Spearman's $r = 0.98, p < 10^{-5}$; sASAT: Spearman's $r = 0.97, p < 10^{-4}$).

The identified mature miRNAs and their precursors were divided into three subgroups according to alignment criteria (Supplementary Table S2) as: (1) Porcine known miRNAs: 297 miRNAs mapped to 176 known porcine pre-miRNAs; specifically, 210 were in miRBase 18.0 [25] and 87 were novel miRNA*s; (2) Porcine conserved miRNAs: 107 miRNAs mapped to 71 other known mammalian pre-miRNAs in miRBase 18.0 and these pre-miRNAs mapped to the pig genome. These miRNAs were labeled with the names of the corresponding conserved miRNAs; (3) Porcine putative new miRNAs: 230 miRNAs (longer than 18 nt and unmapped to any known mammalian pre-miRNAs in miRBase 18.0) encompassing 206 candidate pre-miRNAs that were predicted RNA hairpins derived from the pig genome, and were labeled PPN (Porcine putative new). Notably, there are the distinct pre-miRNAs coding the identical mature miRNAs, which resulting in 617 miRNAs (i.e., reference sequence) corresponding to 597 unique miRNA sequences (Supplementary Table S3).

The identified miRNAs exhibited a large dynamic range of read counts ranging from 3 to millions. The vast majority of miRNAs (IMAT: 61.11%; sASAT: 61.03%) were in low abundance (3 to 100 read counts) and belonged mainly to the porcine conserved and putative new miRNA groups. Only a few miRNAs (IMAT: 2.11%; sASAT: 1.17%) were in high abundance (>100,000 read counts) and they belonged mainly to the porcine known miRNA group (Figure 1b,c). This result suggests that the low-abundance conserved and putative new miRNAs may have escaped from previous detection efforts.

We found that the top ten miRNAs with the highest abundance contributed 67.46% and 82.94% of the total counts in the IMAT and sASAT libraries, respectively, and eight miRNAs were shared by two libraries in the top 10 positions (Figure 1d). The high abundance of these miRNAs implies that they may have housekeeping cellular roles and may be the main regulatory miRNAs in adipogenesis [11,26,27] and cellular basal metabolism [28,29]. For example, let-7a-5p [12], miR-148a-3p [26], miR-21-5p [27], miR-143-3p [30] and miR-101-3p [13] have been reported to be up-regulated during 3T3-L1 pre-adipocyte differentiation, whereas miR-27b-3p was found to be down-regulated during adipogenesis of human multipotent adipose-derived stem cells [31] and miR-199a-5p was up-regulated in subcutaneous AT in obese versus non-obese individuals [13].

2.3. Inflammation- and Diabetes-Related miRNAs Enriched in IMAT

More than half of the unique miRNAs (351 of 597, 58.79%) were co-expressed in IMAT and sASAT. Only 171 (28.64%) and 75 (12.56%) of the unique miRNAs were expressed specifically in IMAT and sASAT, respectively (Figure 2a and Supplementary Table S3). It was well-known that miRNAs function in a dose-dependent manner [32], thus the less abundant miRNAs (<1000 read counts in both libraries) were considered to be less important and were filtered out. Of the 110 more abundant unique miRNAs (>1000 read counts in either library), 53 (48.18%) were determined to be differentially expressed (DE) between IMAT and sASAT using the IDEG6 program [33] (Figure 2a and Supplementary Table S4). The changes in expression patterns of the top 14 DE miRNAs with the highest read counts showed significant positive correlations between the q-PCR results and the small RNA-sequencing data (Person's $r = 0.894$, $p < 10^{-4}$), again highlighting the reliability of the small RNA-sequencing approach (Figure 2b). Moreover, in the process of q-PCR validation, we also found that all expression levels of selected miRNAs obtained by q-PCR within the biological replicates were highly correlated and with very low deviation, which not only indicated the high repeatability and reliability of the q-PCR approach but also reflected the high purity of our experimental samples (Supplementary Table S5).

Figure 2. Characteristics of the differentially expressed (DE) miRNAs between porcine sASAT and IMAT. (**a**) Distribution of 597 unique miRNAs between sASAT (blue) and IMAT (yellow). The red circle represents the 110 miRNAs with read counts >1000 in either of the two libraries. The dashed circles indicate the 45 IMAT-enriched (**left**) and eight sASAT-enriched (**right**) miRNAs ($p < 0.001$); (**b**) q-PCR validation for the top 14 DE miRNAs with highest read counts between IMAT and sASAT. Pearson's correlation was used to determine the relationship between the q-PCR and small RNA-seq results for miRNA expression levels. IMAT-NE and sASAT-NE represent normalized expression levels for the miRNAs in the IMAT and sASAT libraries, respectively; (**c**) The differential expression of 19 inflammation- and diabetes-related miRNAs between IMAT and sASAT.

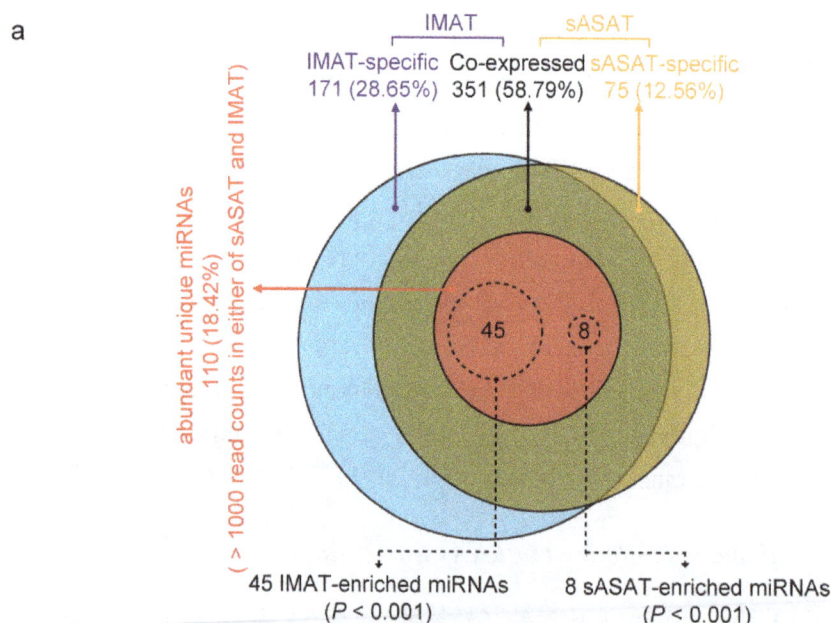

Figure 2. *Cont.*

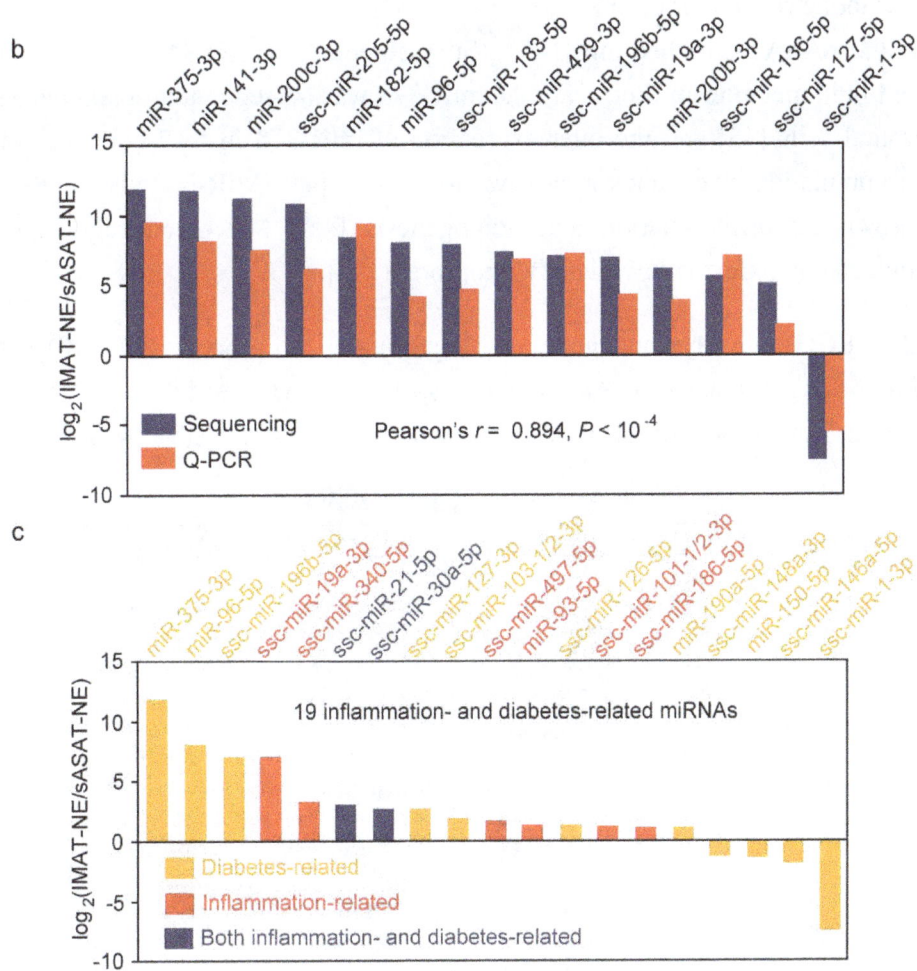

Notably, many of the DE miRNAs (19 out of 53, 35.84%) were associated with inflammation and diabetes based on the annotations assigned using the Pathway Central database (SA Biosciences, Frederick, MD, USA) (Figure 2c). Eight inflammation-related and 9 diabetes-related miRNAs were present in higher abundance in the IMAT transcriptome compared with the sASAT transcriptome (Figure 2c). MiR-21 was found to be over-expressed at the inflammation site [34] and it has been suggested that miR-21 could act as a biomarker for inflammation in the aging process and cardiovascular disease [35]. MiR-101 was reported to be related to inflammation and chondrocyte extracellular matrix degradation [36]. Circulating miR-30a was up-regulated in diabetes patients and has been associated with insulin resistance [20]. The ectopic high expression of miR-103 could induce impaired glucose homeostasis or, conversely, the silencing of miR-103 could improve glucose homeostasis and insulin sensitivity [37].

In addition, various IMAT-enriched miRNAs related to pathological responses were found (Supplementary Table S4). For example, miR-182 and miR-183 (members of the miR-183-96-182 cluster) are well-characterized oncomiRs that can promote the clonal expansion of activated helper T lymphocytes [38,39]. MiR-200b, miR-200c and miR-141 (members of the miR-200 family) were reported to be significantly altered in bladder [40] and breast cancers [41]. MiR-191 was suggested as a biomarker for the diagnosis and prognosis of acute myeloid leukemia [42]. These results suggest that

IMAT is associated mainly with inflammation- and diabetes-related responses, and should be deemed as a potential metabolic risk factor of obesity.

In contrast, the sASAT-enriched miRNAs (Supplementary Table S4) were mainly related to adipogenesis and lipid metabolism. For example, miR-378 was up-regulated in adipogenesis of human AT-derived stromal cells [43] and the over-expression of miR-378 in ST2 cell line was reported to promote lipid accumulation by enhancing *de novo* lipogenesis [44]. MiR-365 was revealed as a central regulator of brown fat differentiation and adipogenesis [45]. MiR-146a regulated mainly lipid accumulation induced by oxidized low-density lipoprotein [46].

Figure 3. KEGG pathways and gene ontology biological process (GO-BP) categories enriched in the target genes of the top eight sASAT- and IMAT-enriched miRNAs. GO-BP is the GO terms under the biological process ontology.

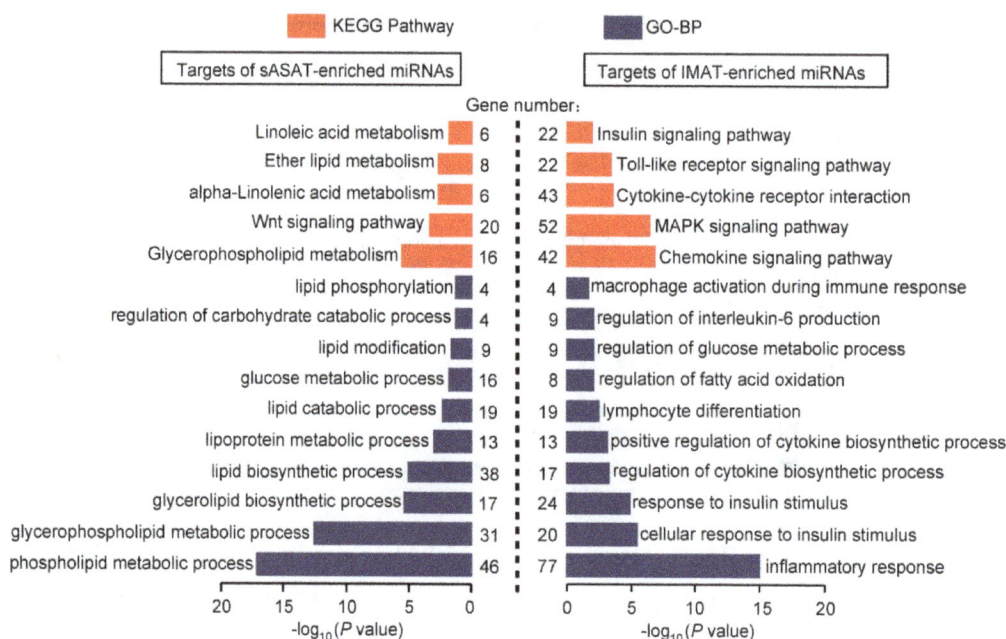

2.4. Functional Enrichment Analyses of miRNA Target Genes

To further highlight the distinct functional features of IMAT and sASAT, the target genes of the top eight DE miRNAs enriched in sASAT (967 mRNA genes) and IMAT (1707 mRNA genes) were predicted using PicTar [47], TargetScan human 6.2 [48] and MicroCosm Targets (version 5.0) [49] (Supplementary Table S6), and analyzed using DAVID [50] to determine whether or not they were enriched for specific functional categories and pathways. Similar to the finding for the DE miRNAs, the target genes of the IMAT-enriched miRNAs were primarily associated with inflammatory and diabetes-related processes, such as "inflammatory response" (77 genes, $p = 1.21 \times 10^{-15}$), "cellular response to insulin stimulus" (20 genes, $p = 4.14 \times 10^{-6}$), "lymphocyte differentiation" (19 genes, $p = 3.75 \times 10^{-3}$), "regulation of interleukin-6 production" (9 genes, $p = 1.17 \times 10^{-2}$), "macrophage activation during immune response" (4 genes, $p = 2.74 \times 10^{-2}$), "chemokine and toll-like signaling pathways" (42 genes, $p = 1.61 \times 10^{-7}$) and "insulin signaling pathway" (22 genes, $p = 1.51 \times 10^{-2}$). In contrast, the target genes of the sASAT-enriched miRNAs were mainly associated with lipid and energy metabolism, such as "glycerophospholipid metabolic process" (46 genes, $p = 5.10 \times 10^{-18}$), "lipid biosynthetic process" (38 genes, $p = 7.36 \times 10^{-6}$), "glucose metabolic process" (16 genes, $p = 1.53 \times 10^{-2}$) and "Wnt signaling pathway" (20 genes, $p = 5.27 \times 10^{-4}$) (Figure 3). These results further suggested that while sASAT is mainly involved in metabolic homeostasis, IMAT is susceptive to inflammation and should be regarded as a potential metabolic risk factor.

3. Experimental Section

3.1. Animals and Sample Collection

Three 210-day-old female Landrace pigs with normal weight (111.67 ± 1.15 kg) were used. The piglets were weaned simultaneously at 28 ± 1 day of age. A starter diet provided 3.40 Mcal·kg^{-1} metabolisable energy (ME), 20.00% crude protein and 1.15% lysine from the thirtieth to sixtieth day after weaning. From the 61st to the 120th day, the diet contained 3.40 Mcal·kg^{-1} ME, 17.90% crude protein and 0.83% lysine. From the 121st to 210th day, the diet contained 3.40 Mcal kg^{-1} ME, 15.00% crude protein and 1.15% lysine. The animals were allowed access to feed and water *ad libitum* and lived under the same normal conditions.

The macroscopic IMAT were directly separated from the regions that were beneath the biceps femoris muscle fascia of porcine hind leg. Since IMAT preparation could be easily contaminated by its surrounding tissues, we paid maximum attention to eliminate the others especially, such as connective tissue and muscle tissue, and all samples were resected from central part of tissue block. The sASAT were from the subcutaneous tissue of central abdomen near the last rib. All samples were immediately frozen in liquid nitrogen and stored at −80 °C before total RNA extraction.

3.2. Small RNA Libraries Construction and High-throughput Sequencing

Total RNA was extracted using the *mir*Vana™ miRNA isolation kit (Ambion, Austin, USA) following the manufacturer's protocol. The integrity of total RNA was also tested via analysis by Bioanalyzer 2100 and RNA 6000 Nano LabChip Kit (Agilent, Palo Alto, CA, USA) with RIN number >6.0.

For a certain adipose tissue, equal amounts (5 μg) of total RNA isolated from three pigs were mixed. Approximately 15 μg of small RNA-enriched total RNA was prepared for Illumina sequencing. In general, the processing by Illumina consisted of the following successive steps: the small RNA ranged from 14 to 40 nt were purified by polyacrylamide gel electrophoresis (PAGE) and ligated specific adapters followed by polyacrylamide gel purification. Then the modified small RNA was reverse transcribed and amplified by RT-PCR. Finally, the enriched cDNA was sequenced on Genome Analyzer Instrument (GAI, Illumina, San Diego, CA, USA). The small RNA-sequencing data discussed in this publication have been deposited in NCBI's Gene Expression Omnibus and are accessible through GEO Series accession number GSE30334.

3.3. Analysis of Small RNA-Sequencing Data

The raw reads were processed using Illumina's Genome Analyzer Pipeline software and subsequently handled as described by Li *et al.* with some improvement [23]. After trimming off the sequencing adapters, the resulting reads was successively filtered by read length (only the read with the of 14 to 27 nt were retained), sequence component (containing <80% A, C, G or T; containing no more than two N (undetermined bases)) and copy numbers (the low-abundance reads (only the read with >3 counts were retained). Then the retained reads were searched against the NCBI [51], Rfam [52] and Repbase database [53] to remove porcine known classes of RNAs (*i.e.*, mRNA, rRNA, tRNA, snRNA, snoRNA and repeats). The sequencing reads survived from above strict filter rules were deemed as "high-quality reads".

The high-quality reads were mapped to the pig genome (Sscrofa9) using NCBI Local BLAST following five steps in order: (1) map the high-quality reads to the 228 known porcine pre-miRNAs (encoding 257 miRNAs) and then to 6716 known pre-miRNAs (encoding 7952 miRNAs) from 24 other mammals in miRBase 18.0; (2) map the mapped high-quality reads to pig genome to obtain their genomic locations and annotations in Ensembl release 59 (Sscrofa 9, April 2009); (3) cluster the unmapped sequences in step 1 that mapped to the pig genome as putative novel miRNAs; and (4) predict hairpin RNA structures of the high-quality reads in step 3 from the adjacent 60 nt sequences in either direction from the pig genome using UNAFold [54]. To avoid ambiguous reads that have been assigned to multiple positions in pig genome, only reads longer than 18 nt in length were included in step 4.

3.4. miRNA Differential Expression Analysis

Program IDEG6 [33] was employed for detecting the DE miRNAs between two libraries. A unique miRNA is considered to be differentially expressed when it simultaneously obtains $p < 0.001$ under three statistical tests (a Audic-Claverie test, a Fisher exact test and a Chi-squared 2×2 test) with the Bonferroni correction.

3.5. Prediction and Functional Annotation of miRNA Target Genes

The potential targets of a certain miRNA were predicted by PicTar [47], TargetScan human 6.2 [48] and, MicroCosm Targets Version 5.0 [49], and the pairwise overlaps of results from three programs composed the final predicted targets. The predictions were according to the interactions of human mRNA-miRNA due to the absence of porcine miRNAs in current version of above-mentioned algorithm. The gene ontology biological process (GO-BP) terms and KEGG pathway terms enriched in predicted target genes were determined using a DAVID bioinformatics resources [50].

3.6. Q-PCR Validation

The expression changes of 14 selected miRNAs were validated by an EvaGreen-based High-Specificity miRNA qRT-PCR Detection Kit (Stratagene, La Jolla, CA, USA) on the CFX96™ Real-Time PCR Detection System (Bio-Rad, Hercules, CA, USA). The q-PCR validation were carried out on three biological replicates.The primer pairs were available in Supplementary Table S7. Three endogenous control genes (U6 snRNA, 18S rRNA and Met-tRNA) [23] were used in this assay. The $\Delta\Delta$Ct method was used to determine the expression level differences between surveyed samples. Normalized factors (NF) of three endogenous control genes and relative quantities of objective miRNAs were analyzed using the qBase software [55].

4. Conclusions

We have generated reliable miRNA transcriptomes of porcine sASAT and IMAT, and identified many known and novel miRNAs using small RNA-sequencing approach. We found that inflammation- and diabetes-related miRNAs were enriched in IMAT compared with sASAT, which indicated the metabolic risk of the IMAT. A functional enrichment analysis of genes targeted by the enriched miRNAs also indicated that IMAT was mainly associated with the immune and inflammation response and may be a potential metabolic risk factor of obesity. The current study provides data that can be used in future studies to investigate the metabolic role of IMAT in obesity-related metabolic dysfunction. Our findings will also help promote the further development of the pig model for human metabolic research. It is also worth noting that further detailed comparision of IMAT between obese and non-obese individuals will be necessary and beneficial to decipher the role of miRNAs in adipogenesis and IMAT-related metabolic diseases.

Acknowledgments

This work was supported by grants from the National High Technology Research and Development Program of China (863 Program) (2013AA102502), the Specialized Research Fund of Ministry of Agriculture of China (NYCYTX-009), the Project of Provincial Twelfth Five Years' Animal Breeding of Sichuan Province (2011YZGG15), and the National Special Foundation for Transgenic Species of China (2011ZX08006-003) to X.L. and M.L., the Chongqing Fund for Distinguished Young Scientists (CSTC2010BA1007).

References

1. Zhang, Y.; Proenca, R.; Maffei, M.; Barone, M.; Leopold, L.; Friedman, J.M. Positional cloning of the mouse obese gene and its human homologue. *Nature* **1994**, *372*, 425–432.

2. Kadowaki, T.; Yamauchi, T.; Kubota, N.; Hara, K.; Ueki, K.; Tobe, K. Adiponectin and adiponectin receptors in insulin resistance, diabetes, and the metabolic syndrome. *J. Clin. Invest.* **2006**, *116*, 1784–1792.

3. Lago, F.; Dieguez, C.; Gómez-Reino, J.; Gualillo, O. The emerging role of adipokines as mediators of inflammation and immune responses. *Cytokine Growth Factor Rev.* **2007**, *18*, 313–325.

4. Arner, P. Insulin resistance in type 2 diabetes-role of the adipokines. *Curr. Mol. Med.* **2005**, *5*, 333–339.

5. Dogru, T.; Sonmez, A.; Tasci, I.; Bozoglu, E.; Yilmaz, M.I.; Genc, H.; Erdem, G.; Gok, M.; Bingol, N.; Kilic, S. Plasma visfatin levels in patients with newly diagnosed and untreated type 2 diabetes mellitus and impaired glucose tolerance. *Diabetes Res. Clin. Pract. Suppl.* **2007**, *76*, 24–29.

6. Berg, A.H.; Scherer, P.E. Adipose tissue, inflammation, and cardiovascular disease. *Circ. Res.* **2005**, *96*, 939–949.

7. Gallagher, D.; Kuznia, P.; Heshka, S.; Albu, J.; Heymsfield, S.B.; Goodpaster, B.; Visser, M.; Harris, T.B. Adipose tissue in muscle: A novel depot similar in size to visceral adipose tissue. *Am. J. Clin. Nutr.* **2005**, *81*, 903–910.

8. Boettcher, M.; Machann, J.; Stefan, N.; Thamer, C.; Häring, H.U.; Claussen, C.D.; Fritsche, A.; Schick, F. Intermuscular adipose tissue (IMAT): Association with other adipose tissue compartments and insulin sensitivity. *J. Magn. Reson. Imaging* **2009**, *29*, 1340–1345.

9. Freda, P.U.; Shen, W.; Heymsfield, S.B.; Reyes-Vidal, C.M.; Geer, E.B.; Bruce, J.N.; Gallagher, D. Lower visceral and subcutaneous but higher intermuscular adipose tissue depots in patients with growth hormone and insulin-like growth factor I excess due to acromegaly. *J. Clin. Endocrinol. Metab.* **2008**, *93*, 2334–2343.

10. Nelson, P.; Kiriakidou, M.; Sharma, A.; Maniataki, E.; Mourelatos, Z. The microRNA world: Small is mighty. *Trends Biochem. Sci.* **2003**, *28*, 534–540.

11. Xie, H.; Lim, B.; Lodish, H.F. MicroRNAs induced during adipogenesis that accelerate fat cell development are downregulated in obesity. *Diabetes* **2009**, *58*, 1050–1057.

12. Kajimoto, K.; Naraba, H.; Iwai, N. MicroRNA and 3T3-L1 pre-adipocyte differentiation. *RNA* **2006**, *12*, 1626–1632.

13. Ortega, F.J.; Moreno-Navarrete, J.M.; Pardo, G.; Sabater, M.; Hummel, M.; Ferrer, A.; Rodriguez-Hermosa, J.I.; Ruiz, B.; Ricart, W.; Peral, B. MiRNA expression profile of human subcutaneous adipose and during adipocyte differentiation. *PLoS One* **2010**, *5*, e9022.

14. Wang, Q.; Li, Y.C.; Wang, J.; Kong, J.; Qi, Y.; Quigg, R.J.; Li, X. miR-17-92 cluster accelerates adipocyte differentiation by negatively regulating tumor-suppressor Rb2/p130. *Proc. Natl. Acad. Sci. USA* **2008**, *105*, 2889–2894.

15. Lin, Q.; Gao, Z.; Alarcon, R.M.; Ye, J.; Yun, Z. A role of miR-27 in the regulation of adipogenesis. *FEBS J.* **2009**, *276*, 2348–2358.

16. Kim, S.Y.; Kim, A.Y.; Lee, H.W.; Son, Y.H.; Lee, G.Y.; Lee, J.-W.; Lee, Y.S.; Kim, J.B. miR-27a is a negative regulator of adipocyte differentiation via suppressing *PPARγ* expression. *Biochem. Biophys. Res. Commun.* **2010**, *392*, 323–328.

17. Kinoshita, M.; Ono, K.; Horie, T.; Nagao, K.; Nishi, H.; Kuwabara, Y.; Takanabe-Mori, R.; Hasegawa, K.; Kita, T.; Kimura, T. Regulation of adipocyte differentiation by activation of serotonin (5-HT) receptors *5-HT2AR* and *5-HT2CR* and involvement of microRNA-448-mediated repression of *KLF5*. *Mol. Endocrinol.* **2010**, *24*, 1978–1987.

18. Andersen, D.C.; Jensen, C.H.; Schneider, M.; Nossent, A.Y.; Eskildsen, T.; Hansen, J.L.; Teisner, B.; Sheikh, S.P. MicroRNA-15a fine-tunes the level of Delta-like 1 homolog (*DLK1*) in proliferating 3T3-L1 preadipocytes. *Exp. Cell Res.* **2010**, *316*, 1681–1691.

19. Martinelli, R.; Nardelli, C.; Pilone, V.; Buonomo, T.; Liguori, R.; Castanò, I.; Buono, P.; Masone, S.; Persico, G.; Forestieri, P. miR-519d overexpression is associated with human obesity. *Obesity* **2012**, *18*, 2170–2176.

20. Ferland-McCollough, D.; Ozanne, S.; Siddle, K.; Willis, A.; Bushell, M. The involvement of microRNAs in Type 2 diabetes. *Biochem. Soc. Trans.* **2010**, *38*, 1565.

21. Wang, Q.; Wang, Y.; Minto, A.W.; Wang, J.; Shi, Q.; Li, X.; Quigg, R.J. MicroRNA-377 is up-regulated and can lead to increased fibronectin production in diabetic nephropathy. *FASEB J.* **2008**, *22*, 4126–4135.

22. Spurlock, M.E.; Gabler, N.K. The development of porcine models of obesity and the metabolic syndrome. *J. Nutr.* **2008**, *138*, 397–402.

23. Li, M.; Xia, Y.; Gu, Y.; Zhang, K.; Lang, Q.; Chen, L.; Guan, J.; Luo, Z.; Chen, H.; Li, Y. MicroRNAome of porcine pre-and postnatal development. *PLoS One* **2010**, *5*, e11541.

24. Xie, S.-S.; Li, X.-Y.; Liu, T.; Cao, J.-H.; Zhong, Q.; Zhao, S.-H. Discovery of porcine microRNAs in multiple tissues by a Solexa deep sequencing approach. *PLoS One* **2011**, *6*, e16235.

25. Kozomara, A.; Griffiths-Jones, S. miRBase: Integrating microRNA annotation and deep-sequencing data. *Nucleic Acids Res.* **2011**, *39*, D152–D157.

26. Qin, L.; Chen, Y.; Niu, Y.; Chen, W.; Wang, Q.; Xiao, S.; Li, A.; Xie, Y.; Li, J.; Zhao, X. A deep investigation into the adipogenesis mechanism: Profile of microRNAs regulating adipogenesis by modulating the canonical Wnt/β-catenin signaling pathway. *BMC Genomics* **2010**, *11*, 320.

27. Kim, Y.J.; Hwang, S.J.; Bae, Y.C.; Jung, J.S. MiR-21 regulates adipogenic differentiation through the modulation of *TGF-β* signaling in mesenchymal stem cells derived from human adipose tissue. *Stem Cells* **2009**, *27*, 3093–3102.

28. Hoekstra, M.; van der Lans, C.A.; Halvorsen, B.; Gullestad, L.; Kuiper, J.; Aukrust, P.; van Berkel, T.J.; Biessen, E.A. The peripheral blood mononuclear cell microRNA signature of coronary artery disease. *Biochem. Biophys. Res. Commun.* **2010**, *394*, 792–797.

29. Ro, S.; Park, C.; Young, D.; Sanders, K.M.; Yan, W. Tissue-dependent paired expression of miRNAs. *Nucleic Acids Res.* **2007**, *35*, 5944–5953.

30. Esau, C.; Kang, X.; Peralta, E.; Hanson, E.; Marcusson, E.G.; Ravichandran, L.V.; Sun, Y.; Koo, S.; Perera, R.J.; Jain, R. MicroRNA-143 regulates adipocyte differentiation. *J. Biol. Chem.* **2004**, *279*, 52361–52365.

31. Karbiener, M.; Fischer, C.; Nowitsch, S.; Opriessnig, P.; Papak, C.; Ailhaud, G.; Dani, C.; Amri, E.-Z.; Scheideler, M. microRNA miR-27b impairs human adipocyte differentiation and targets *PPARγ*. *Biochem. Biophys. Res. Commun.* **2009**, *390*, 247–251.

32. Carlsbecker, A.; Lee, J.-Y.; Roberts, C.J.; Dettmer, J.; Lehesranta, S.; Zhou, J.; Lindgren, O.; Moreno-Risueno, M.A.; Vatén, A.; Thitamadee, S. Cell signalling by microRNA165/6 directs gene dose-dependent root cell fate. *Nature* **2010**, *465*, 316–321.

33. Romualdi, C.; Bortoluzzi, S.; d'Alessi, F.; Danieli, G.A. IDEG6: A web tool for detection of differentially expressed genes in multiple tag sampling experiments. *Physiol. Genomics* **2003**, *12*, 159–162.

34. Okayama, H.; Schetter, A.; Harris, C. MicroRNAs and inflammation in the pathogenesis and progression of colon cancer. *Dig. Dis.* **2012**, *30*, 9–15.

35. Olivieri, F.; Spazzafumo, L.; Santini, G.; Lazzarini, R.; Albertini, M.C.; Rippo, M.R.; Galeazzi, R.; Abbatecola, A.M.; Marcheselli, F.; Monti, D. Age-related differences in the expression of circulating microRNAs: miR-21 as a new circulating marker of inflammaging. *Mech. Ageing Dev.* **2012**, *133*, 675–685.

36. Dai, L.; Zhang, X.; Hu, X.; Zhou, C.; Ao, Y. Silencing of microRNA-101 prevents IL-1b-induced extracellular matrix degradation in chondrocytes. *Arthritis Res. Ther.* **2012**, *14*, R268.

37. Trajkovski, M.; Hausser, J.; Soutschek, J.; Bhat, B.; Akin, A.; Zavolan, M.; Heim, M.H.; Stoffel, M. MicroRNAs 103 and 107 regulate insulin sensitivity. *Nature* **2011**, *474*, 649–653.

38. Stittrich, A.-B.; Haftmann, C.; Sgouroudis, E.; Kühl, A.A.; Hegazy, A.N.; Panse, I.; Riedel, R.; Flossdorf, M.; Dong, J.; Fuhrmann, F. The microRNA miR-182 is induced by IL-2 and promotes clonal expansion of activated helper T lymphocytes. *Nat. Immunol.* **2010**, *11*, 1057–1062.

39. Sarver, A.L.; Li, L.; Subramanian, S. MicroRNA miR-183 functions as an oncogene by targeting the transcription factor *EGR1* and promoting tumor cell migration. *Cancer Res.* **2010**, *70*, 9570–9580.

40. Wiklund, E.D.; Bramsen, J.B.; Hulf, T.; Dyrskjøt, L.; Ramanathan, R.; Hansen, T.B.; Villadsen, S.B.; Gao, S.; Ostenfeld, M.S.; Borre, M. Coordinated epigenetic repression of the miR-200 family and miR-205 in invasive bladder cancer. *Int. J. Cancer* **2011**, *128*, 1327–1334.

41. Isacke, C. MicroRNA-200 family modulation in distinct breast cancer phenotypes. *PLoS One* **2012**, *7*, e47709.

42. Garzon, R.; Volinia, S.; Liu, C.-G.; Fernandez-Cymering, C.; Palumbo, T.; Pichiorri, F.; Fabbri, M.; Coombes, K.; Alder, H.; Nakamura, T. MicroRNA signatures associated with cytogenetics and prognosis in acute myeloid leukemia. *Blood* **2008**, *111*, 3183–3189.

43. Zaragosi, L.-E.; Wdziekonski, B.; Brigand, K.L.; Villageois, P.; Mari, B.; Waldmann, R.; Dani, C.; Barbry, P. Small RNA sequencing reveals miR-642a-3p as a novel adipocyte-specific microRNA and miR-30 as a key regulator of human adipogenesis. *Genome Biol.* **2011**, *12*, R64.

44. Gerin, I.; Bommer, G.T.; McCoin, C.S.; Sousa, K.M.; Krishnan, V.; MacDougald, O.A. Roles for miRNA-378/378* in adipocyte gene expression and lipogenesis. *Am. J. Physiol. Endocrinol. Metab.* **2010**, *299*, E198–E206.

45. Sun, L.; Xie, H.; Mori, M.A.; Alexander, R.; Yuan, B.; Hattangadi, S.M.; Liu, Q.; Kahn, C.R.; Lodish, H.F. Mir193b-365 is essential for brown fat differentiation. *Nat. Cell Biol.* **2011**, *13*, 958–965.

46. Yang, K.; He, Y.S.; Wang, X.Q.; Lu, L.; Chen, Q.J.; Liu, J.; Sun, Z.; Shen, W.F. MiR-146a inhibits oxidized low-density lipoprotein-induced lipid accumulation and inflammatory response via targeting toll-like receptor 4. *FEBS Lett.* **2011**, *585*, 854–860.

47. Krek, A.; Grün, D.; Poy, M.N.; Wolf, R.; Rosenberg, L.; Epstein, E.J.; MacMenamin, P.; da Piedade, I.; Gunsalus, K.C.; Stoffel, M. Combinatorial microRNA target predictions. *Nat. Genet.* **2005**, *37*, 495–500.

48. Lewis, B.P.; Burge, C.B.; Bartel, D.P. Conserved seed pairing, often flanked by adenosines, indicates that thousands of human genes are microRNA targets. *Cell* **2005**, *120*, 15–20.

49. Griffiths-Jones, S.; Saini, H.K.; van Dongen, S.; Enright, A.J. miRBase: Tools for microRNA genomics. *Nucleic Acids Res.* **2008**, *36*, D154–D158.

50. Huang, D.W.; Sherman, B.T.; Lempicki, R.A. Systematic and integrative analysis of large gene lists using DAVID bioinformatics resources. *Nat. Protoc.* **2008**, *4*, 44–57.

51. Pruitt, K.D.; Tatusova, T.; Klimke, W.; Maglott, D.R. NCBI Reference Sequences: Current status, policy and new initiatives. *Nucleic Acids Res.* **2009**, *37*, D32–D36.

52. Gardner, P.P.; Daub, J.; Tate, J.G.; Nawrocki, E.P.; Kolbe, D.L.; Lindgreen, S.; Wilkinson, A.C.; Finn, R.D.; Griffiths-Jones, S.; Eddy, S.R.; *et al.* Rfam: Updates to the RNA families database. *Nucleic Acids Res.* **2009**, *37*, D136–D140.

53. Kohany, O.; Gentles, A.J.; Hankus, L.; Jurka, J. Annotation, submission and screening of repetitive elements in Repbase: RepbaseSubmitter and Censor. *BMC Bioinforma.* **2006**, *7*, 474.

54. Markham, N.R.; Zuker, M. UNAFold: Software for nucleic acid folding and hybridization. *Methods Mol. Biol.* **2008**, *453*, 3–31.

55. Hellemans, J.; Mortier, G.; de Paepe, A.; Speleman, F.; Vandesompele, J. qBase relative quantification framework and software for management and automated analysis of real-time quantitative PCR data. *Genome Biol.* **2007**, *8*, R19.

Permissions

List of Contributors

Yong Zhuang, Xiao-Hui Zhou and Jun Liu
Institute of Vegetable Crops, Jiangsu Academy of Agricultural Sciences, Nanjing 210014, China

Toshihiro Kushibiki, Takeshi Hirasawa, Shinpei Okawa and Miya Ishihara
Department of Medical Engineering, National Defense Medical College 3-2 Namiki, Tokorozawa, Saitama 359-8513, Japan

Li-Ling Lin
Department of Life Science, National Taiwan University, Taipei 106, Taiwan

Chia-Chi Wu
Institute of Molecular and Cellular Biology, National Taiwan University, Taipei 106, Taiwan

Hsuan-Cheng Huang
Institute of Biomedical Informatics, Center for Systems and Synthetic Biology, National Yang-Ming University, Taipei 112, Taiwan

Huai-Ju Chen and Hsu-Liang Hsieh
Institute of Plant Biology, National Taiwan University, Taipei 106, Taiwan

Hsueh-Fen Juan
Department of Life Science, National Taiwan University, Taipei 106, Taiwan
Institute of Molecular and Cellular Biology, National Taiwan University, Taipei 106, Taiwan
Graduate Institute of Biomedical Electronic and Bioinformatics, National Taiwan University, Taipei 106, Taiwan

Paolo Martini, Gabriele Sales, Enrica Calura and Chiara Romualdi
Department of Biology, University of Padova, Via G. Colombo 3, Padova 35121, Italy

Mattia Brugiolo
C.R.I.B.I. Biotechnology Centre, University of Padova, Via U. Bassi 58/B, Padova 35121, Italy

Gerolamo Lanfranchi and Stefano Cagnin
Department of Biology, University of Padova, Via G. Colombo 3, Padova 35121, Italy
C.R.I.B.I. Biotechnology Centre, University of Padova, Via U. Bassi 58/B, Padova 35121, Italy

Nina Hauptman and Damjan Glavač
Department of Molecular Genetics, Institute of Pathology, University of Ljubljana, SI-1000 Ljubljana, Slovenia

Annalisa Pacilli, Claudio Ceccarelli, Davide Treré and Lorenzo Montanaro
Department of Experimental, Diagnostic and Specialty Medicine, University of Bologna, Sant'Orsola-Malpighi University Hospital, via Massarenti, 9, Bologna 40138, Italy

Lars Maegdefessel
Department of Medicine, Karolinska Institute, Stockholm SE-17176, Sweden

Joshua M. Spin, Matti Adam, Uwe Raaz, Ryuji Toh, Futoshi Nakagami and Philip S. Tsao
Division of Cardiovascular Medicine, Stanford University, Stanford, CA 94305-5406, USA

Epaminondas Doxakis
Basic Neurosciences Division, Biomedical Research Foundation of the Academy of Athens, Soranou Efesiou 4, Athens 11527, Greece

Sara Tomaselli, Barbara Bonamassa and Angela Gallo
Laboratory of RNA Editing, Onco-haematology Department, Bambino Gesù Children's Hospital, IRCCS, Piazza S. Onofrio 4, Rome 00165, Italy

Anna Alisi and Valerio Nobili
Hepato-Metabolic Disease Unit and Liver Research Unit, Bambino Gesù Children's Hospital, IRCCS, Piazza S. Onofrio 4, Rome 00165, Italy

Franco Locatelli
Laboratory of RNA Editing, Onco-haematology Department, Bambino Gesù Children's Hospital, IRCCS, Piazza S. Onofrio 4, Rome 00165, Italy
Department of Pediatric Science, Università di Pavia, Strada Nuova 65, Pavia 27100, Italy

Daniela Schwarzenbacher, Marija Balic and Martin Pichler
Division of Clinical Oncology, Department of Medicine, Medical University of Graz, Auenbruggerplatz 15, 8036 Graz, Austria

Emilia Kozlowska, Wlodzimierz J. Krzyzosiak and Edyta Koscianska
Department of Molecular Biomedicine, Institute of Bioorganic Chemistry, Polish Academy of Sciences, Noskowskiego 12/14 Str., 61-704 Poznan, Poland

Daniela Erriquez
Department of Pharmacy and Biotechnology, University of Bologna, Bologna 40126, Italy

Giovanni Perini
Department of Pharmacy and Biotechnology, University of Bologna, Bologna 40126, Italy
Health Sciences and Technologies–Interdepartmental Center for Industrial Research, University of Bologna, Bologna 40064, Italy

Alessandra Ferlini
Section of Microbiology and Medical Genetics, Department of Medical Sciences, University of Ferrara, Ferrara 44100, Italy

Changqing Zhang, Shinong Zhu and Hailing Li
College of Horticulture, Jinling Institute of Technology, Nanjing 210038, China

Guangping Li
College of Forest Resources and Environment, Nanjing Forestry University, Nanjing 210037, China

Jin Wang
State Key Laboratory of Pharmaceutical Biotechnology, School of Life Sciences, Nanjing University, Nanjing 210093, China

Jideng Ma, Shuzhen Yu, Fengjiao Wang, Lin Bai, Jian Xiao, Anan Jiang, Mingzhou Li and Xuewei Li
Institute of Animal Genetics & Breeding, College of Animal Science & Technology, Sichuan Agricultural University, Ya'an 625014, China

Yanzhi Jiang
College of Life and Basic Sciences, Sichuan Agricultural University, Ya'an 625014, China

Lei Chen and Jinyong Wang
Chongqing Academy of Animal Science, Chongqing 402460, China

Index